OXFORD STUDIES IN BYZANTIUM

Oxford Studies in Byzantium consists of scholarly monographs and editions on the history, literature, thought, and material culture of the Byzantine world.

Basil II and the Governance of Empire (976–1025)
Catherine Holmes

Holy Fools in Byzantium and Beyond
Sergey A. Ivanov

A Byzantine Encyclopaedia of Horse Medicine
The Sources, Compilation, and Transmission of the *Hippiatrica*
Anne McCabe

George Akropolites: *The History*
Translated with an Introduction and Commentary by
Ruth Macrides

The Trophies of the Martyrs
An Art Historical Study of Early Christian Silver Reliquaries
Galit Noga-Banai

The Chronicle of Morea
Historiography in Crusader Greece
Teresa Shawcross

Tilling the Hateful Earth

Agricultural Production and Trade in the Late Antique East

By

MICHAEL DECKER

OXFORD
UNIVERSITY PRESS

OXFORD
UNIVERSITY PRESS

Great Clarendon Street, Oxford OX2 6DP

Oxford University Press is a department of the University of Oxford.
It furthers the University's objective of excellence in research, scholarship,
and education by publishing worldwide in

Oxford New York

Auckland Cape Town Dar es Salaam Hong Kong Karachi
Kuala Lumpur Madrid Melbourne Mexico City Nairobi
New Delhi Shanghai Taipei Toronto

With offices in

Argentina Austria Brazil Chile Czech Republic France Greece
Guatemala Hungary Italy Japan South Korea Poland Portugal
Singapore Switzerland Thailand Turkey Ukraine Vietnam

Published in the United States
by Oxford University Press Inc., New York

Reprinted 2011

ISBN 978-0-19-956528-3

Printed in the United Kingdom by
Lightning Source UK Ltd., Milton Keynes

Contents

Acknowledgements

Many colleagues, friends, and organizations assisted with the preparation of this book. I cannot hope to thank all those people who helped me to complete this project, and I beg their pardon for any omission. I gratefully acknowledge the guidance and assistance of Dr Marlia Mango, my academic supervisor at Oxford. Through her influence I found a home in studying the Byzantine landscape. During my time at Oxford I benefited from conversations and discussions of Byzantine history and archaeology with Cyril Mango, James Howard Johnston, and Elizabeth Jeffreys. Raffi Frankel was warm and giving of advice and knowledge about all things having to do with mills and presses, and Andrew Wilson offered great friendship and insight into ancient water systems. During my time at Rice University, Michael Maas was an affable and insightful colleague, and David Cook kindly and patiently offered sessions of Arabic lessons.

I offer special thanks to my colleagues Sean Kingsley, whose companionship and intellectual input have always been welcomed, to David Milson, who was always ready to discuss and work through all things Late Antique, and to Eric Cooper, who always offered steadfast friendship, coffee, and computing expertise as I grappled with this project.

The Fulbright Scholar Program supported me during work in Egypt, Jordan, and Syria in 2008 while I further developed this work. The University of South Florida Department of History and University of South Florida Publications Council helped to fund my research and creation of the maps. My colleagues in the history department of the University of South Florida offered me an environment in which I could flourish and develop my ideas.

I owe an enormous debt to Bryan Ward-Perkins, who with great fortitude, patience, and care helped me to develop my manuscript. The text was greatly improved under his guidance, and for this I am particularly grateful. All imperfections that remain are solely my own.

Finally, I would like to thank my parents and parents-in-law, who wholeheartedly supported my education materially and spiritually. My greatest appreciation goes to my wife, Katy, who lived with me and my late antique farmers for more years than either of us care to count.

Figures

Tables

Maps

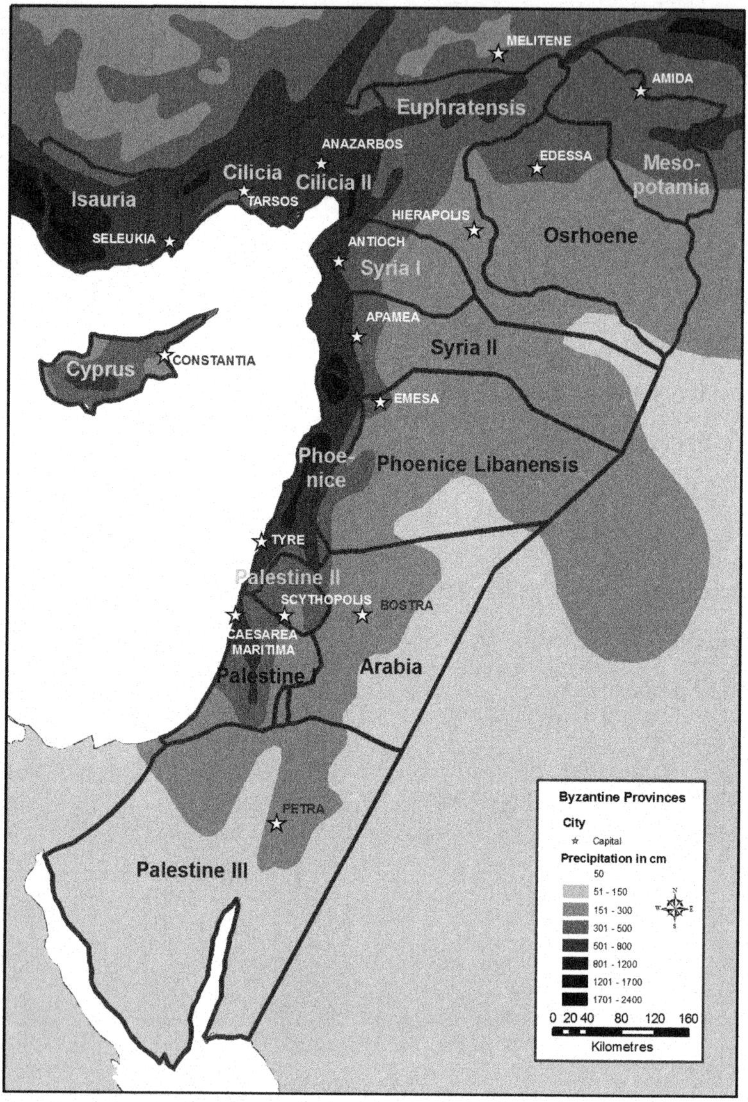

Map 1 Oriens: approximate provincial boundaries and precipitation

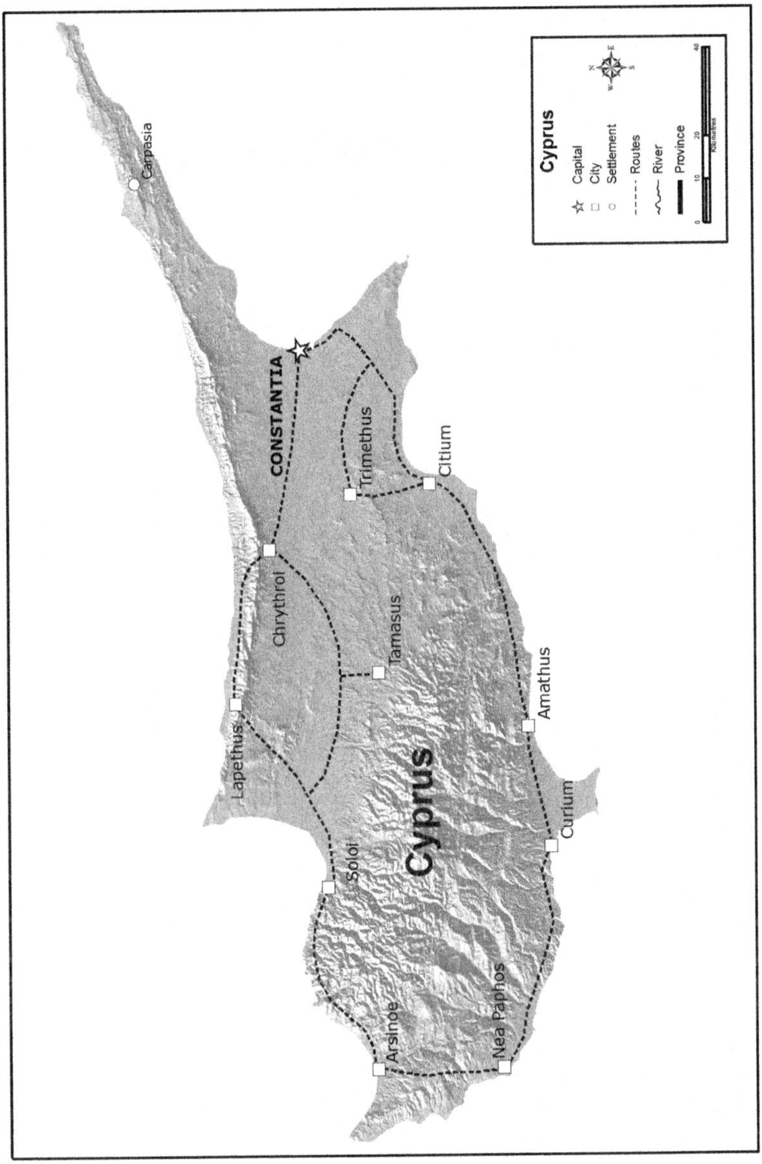
Carpasia
CONSTANTIA
Trimethus
Citium
Chrythroi
Tamasus
Amathus
Lapethus
Cyprus
Soloi
Curium
Arsinoe
Nea Paphos
Cyprus
Capital
City
Settlement
Routes
River
Province
0 10 20 40
Kilometres

Map 2 Cyprus

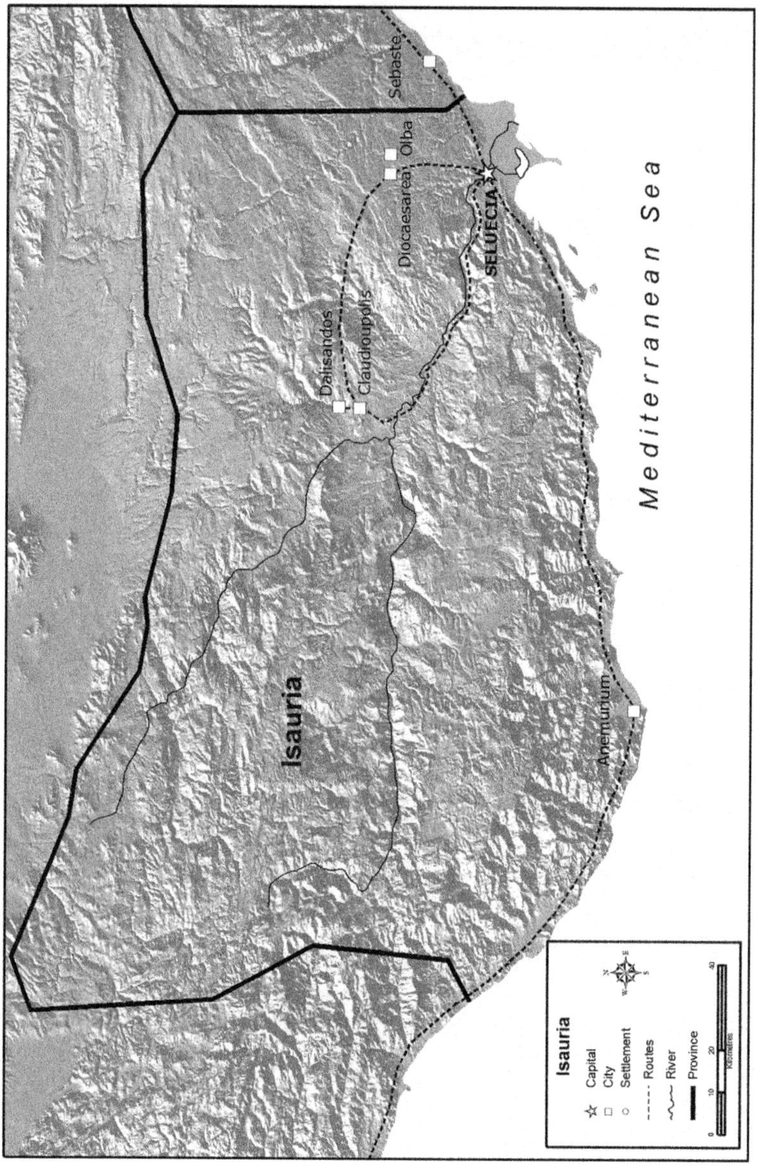

Map 3 Isauria

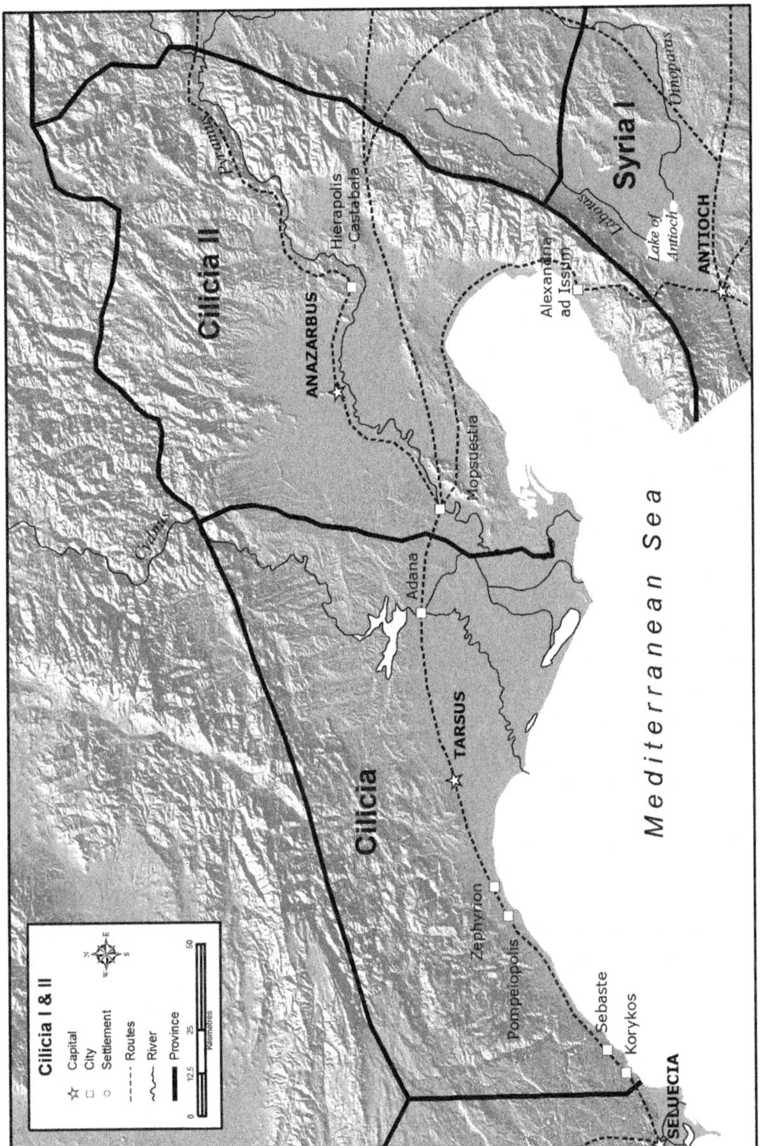

Map 4 Cilicia I and II

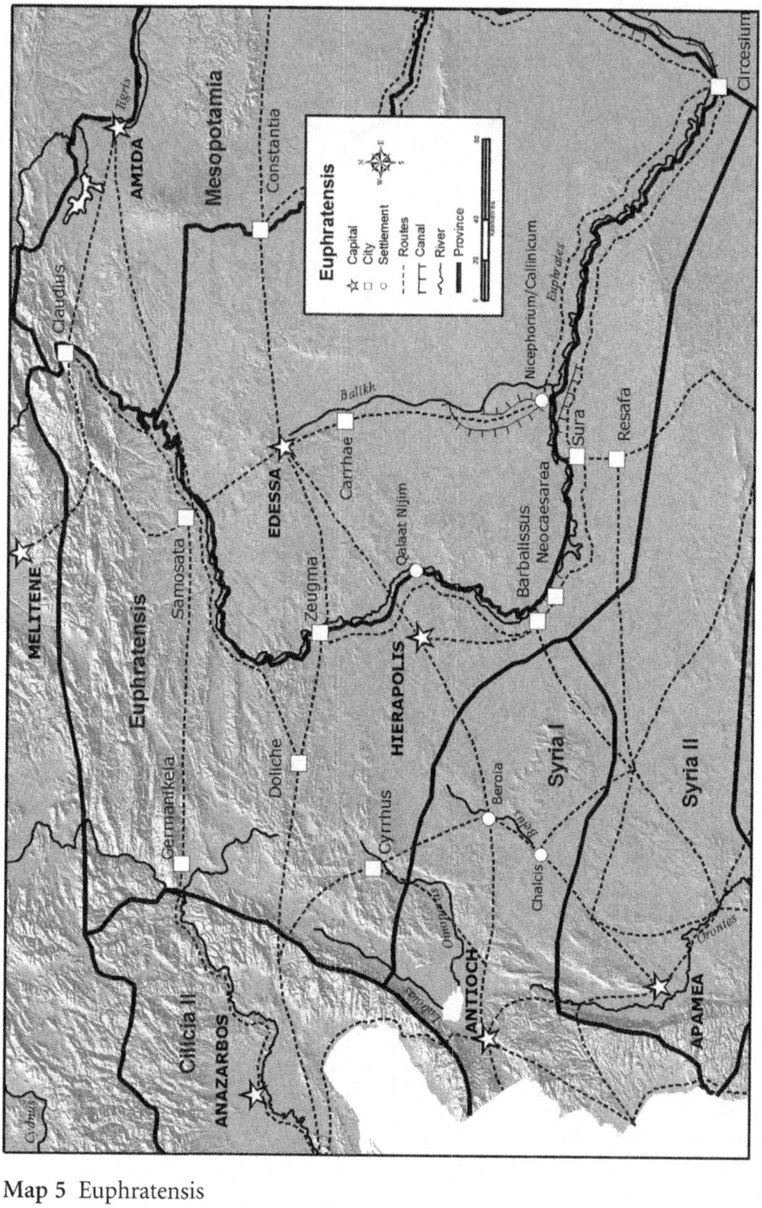

Map 5 Euphratensis

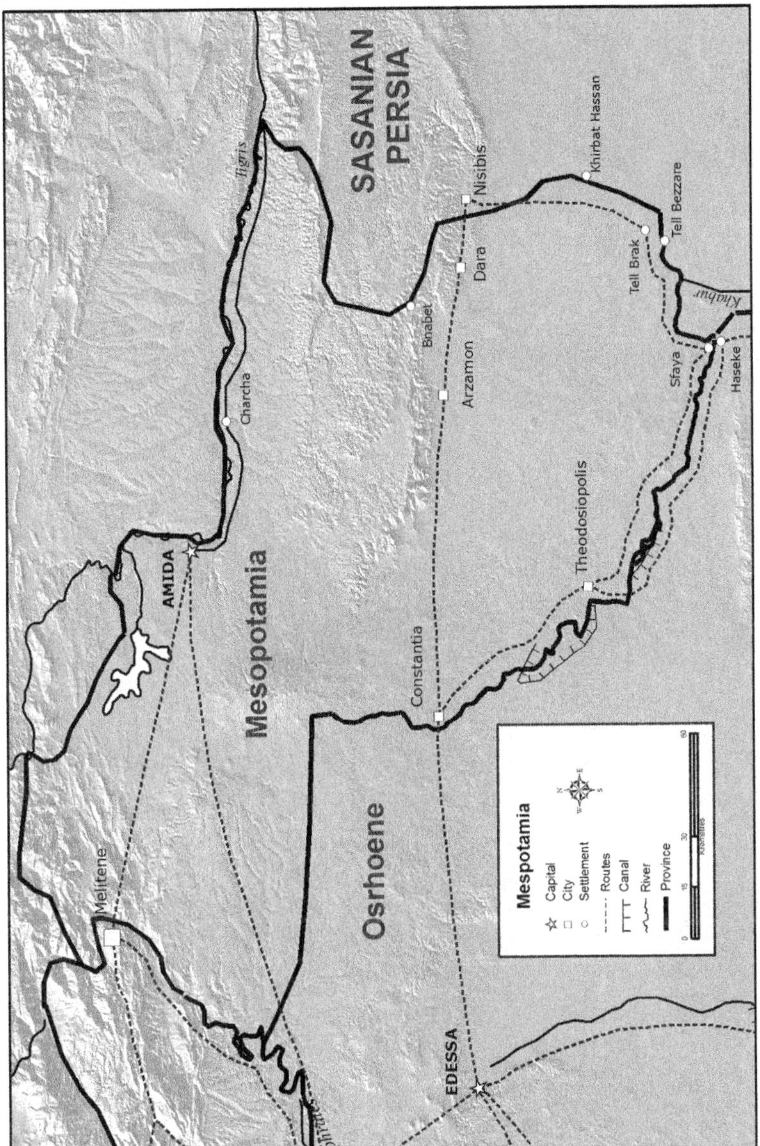

Map 6 Mesopotamia

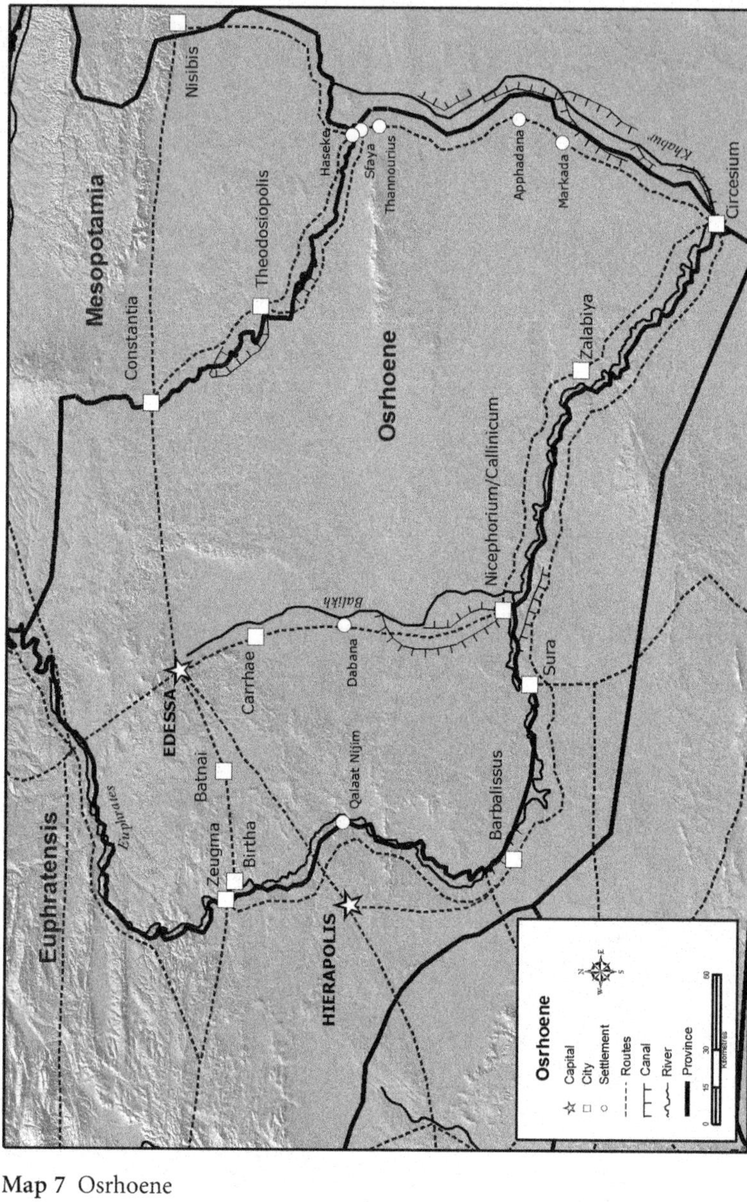

Map 7 Osrhoene

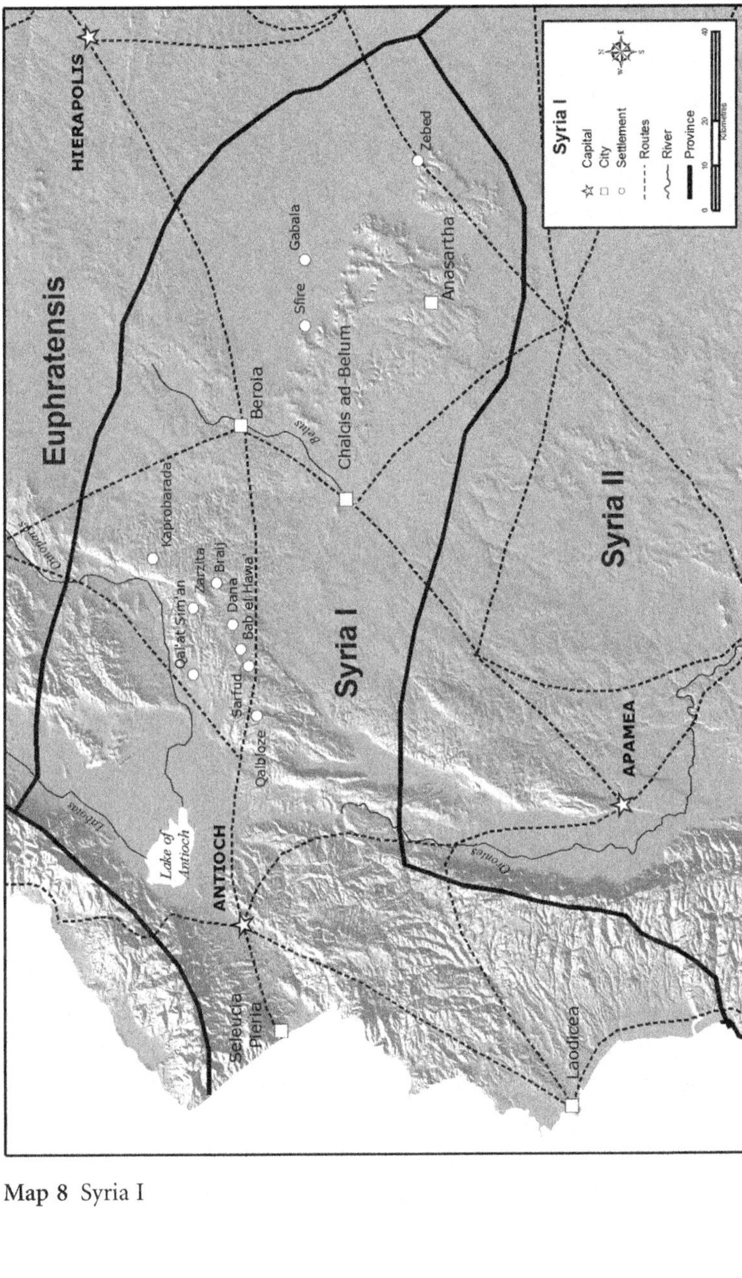
Euphratensis
HIERAPOLIS
Syria I
Syria II
ANTIOCH
APAMEA
Lake of Antioch
Seleucia Pieria
Laodicea
Beroia
Chalcis ad-Belum
Anasartha
Zebed
Gabala
Sfire
Kaprobarada
Qal'at Sim'an
Zarzita
Brad
Dana
Bab el-Hawa'
Sarfud
Qalbloze
Orontes
Belus
Labotas
Oinoparas
Syria I
Capital
City
Settlement
Routes
River
Province
Kilometres

Map 8 Syria I

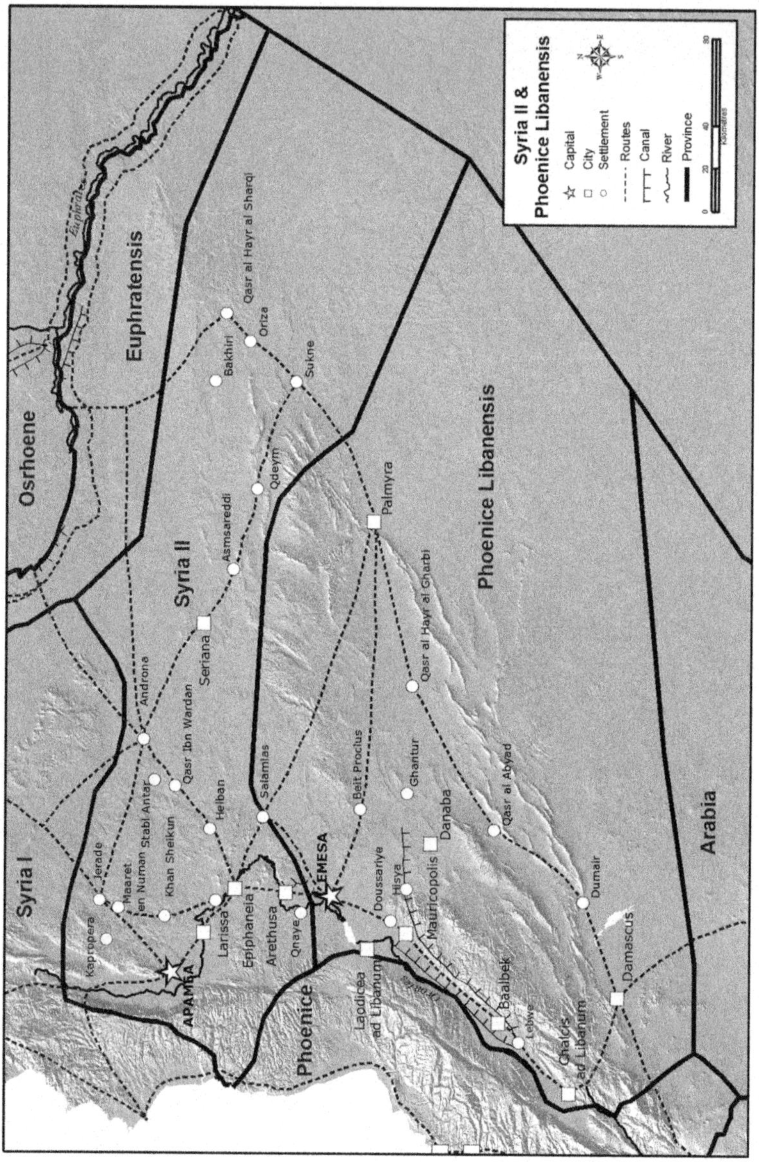

Map 9 Syria II and Phoenice Libanensis

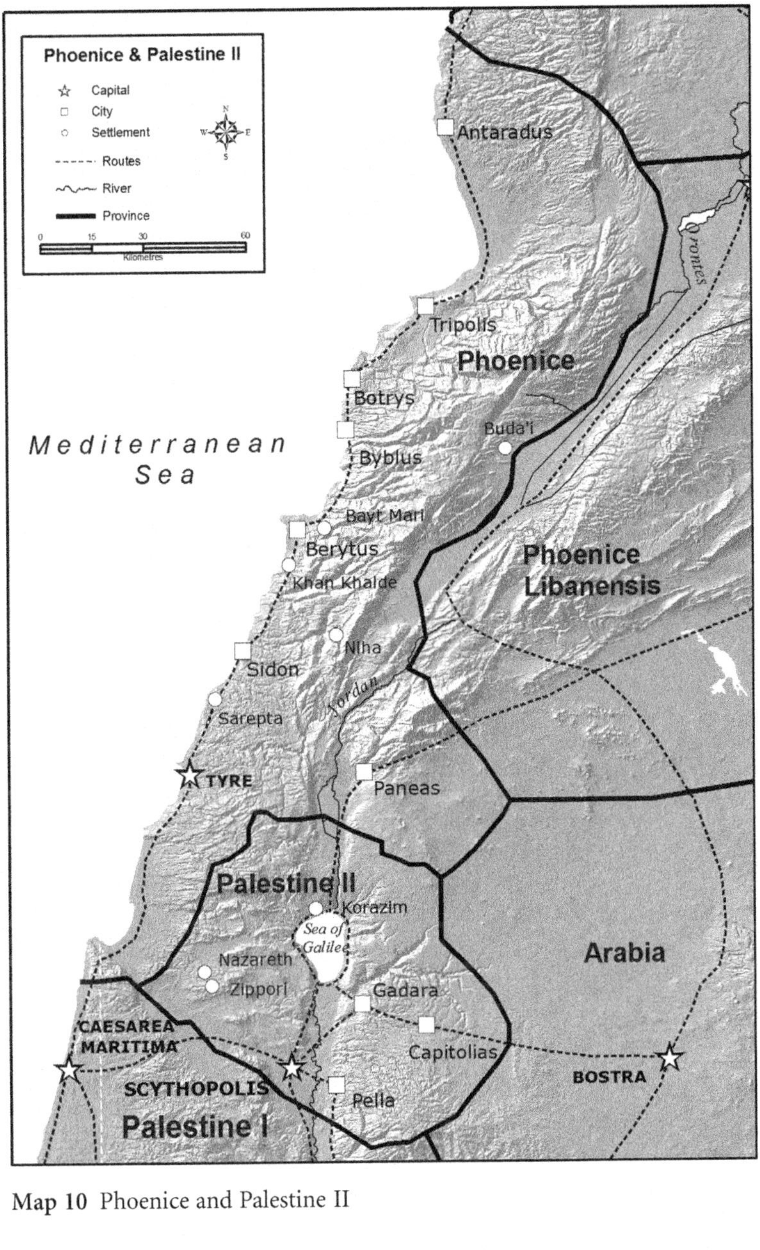

Map 10 Phoenice and Palestine II

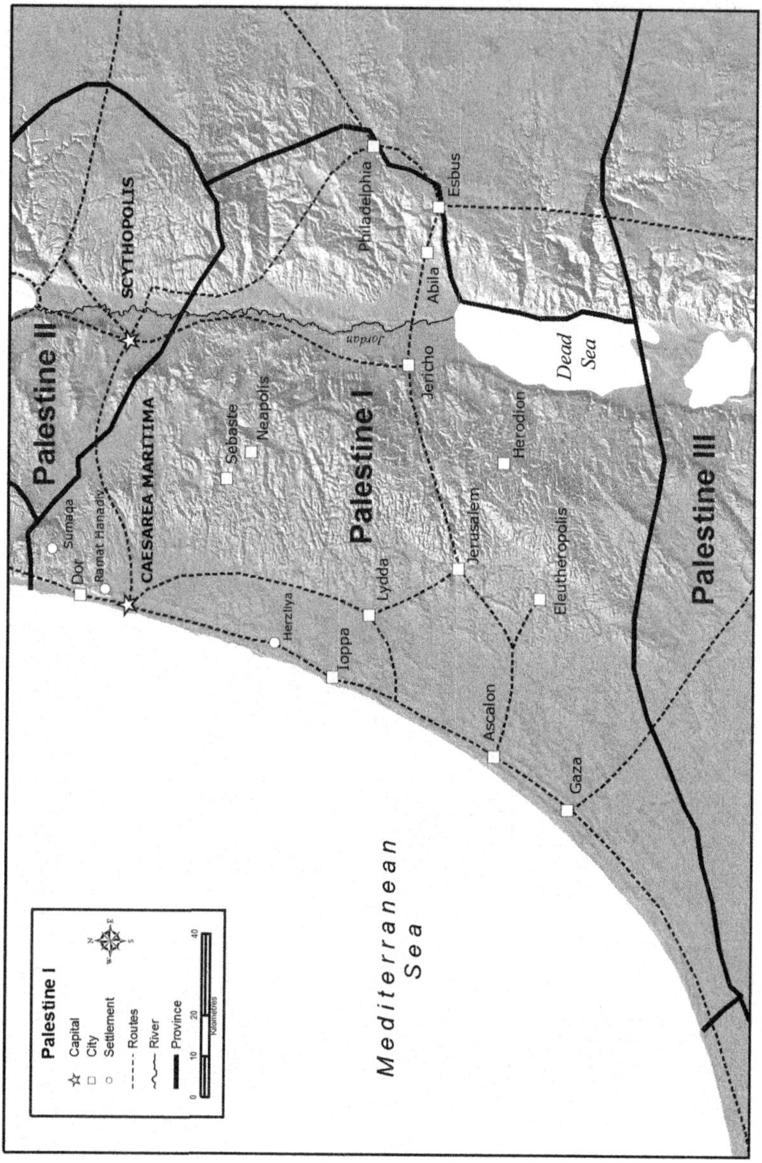

Map 11 Palestine I

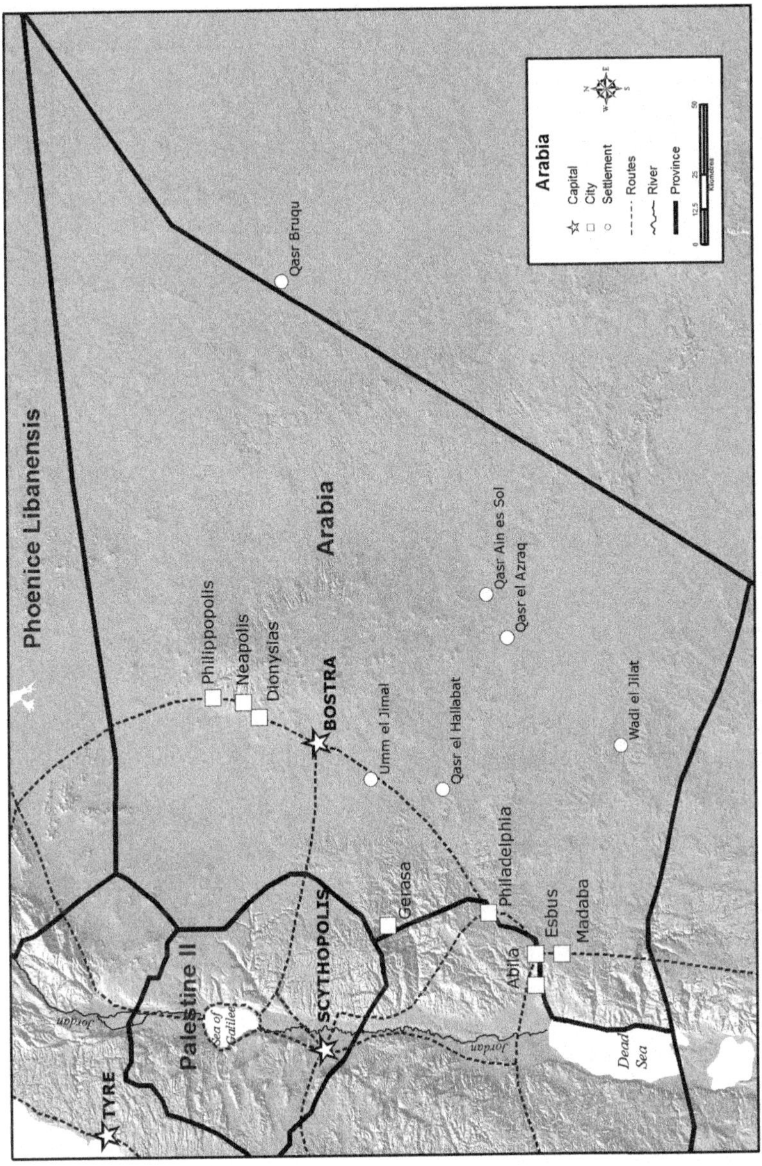

Map 12 Arabia

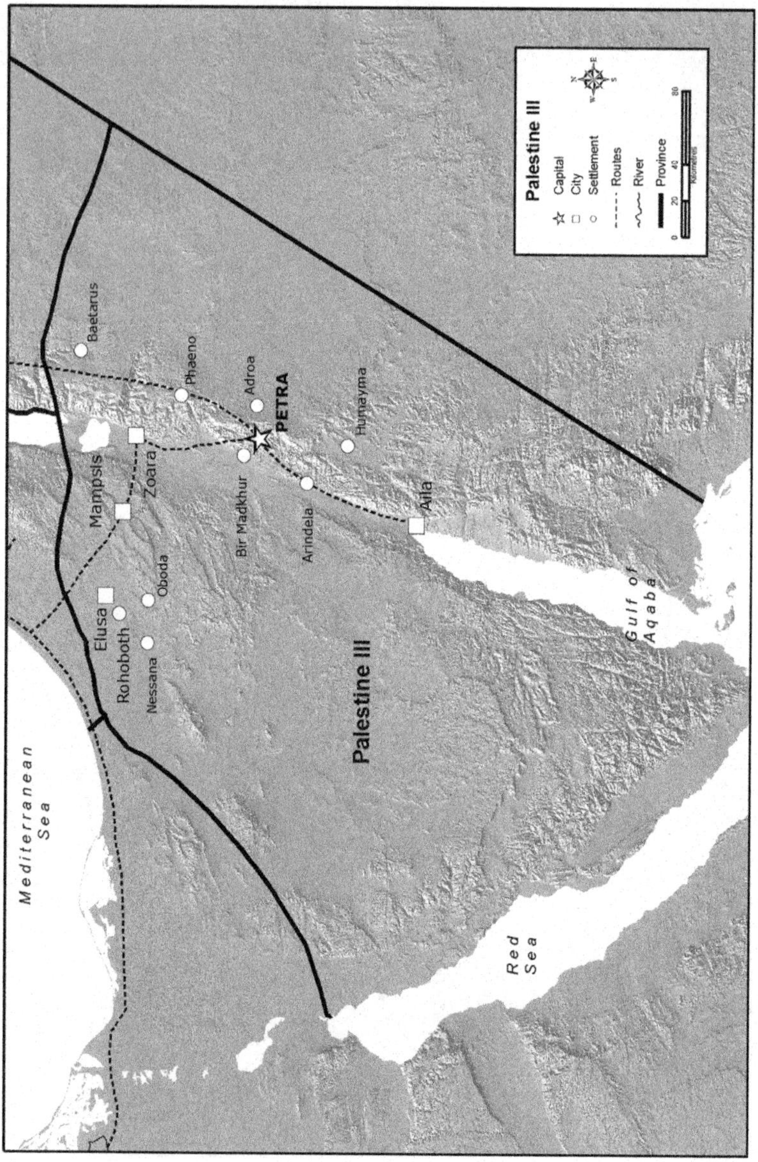

Map 13 Palestine III

Abbreviations

AAAS	*Annales archéologiques arabes syriennes*
ADAJ	*Annual of the Department of Antiquities of Jordan*
AJA	*American Journal of Archaeology*
AHR	*American Historical Review*
AnatSt	*Anatolian Studies*
ANRW	H. Temporini (ed.), *Aufstieg und Niedergang der römischen Welt* (Berlin 1972–)
AntAfr	*Antiquités africaines*
APAW	*Abhandlungen der Preussischen Akademie der Wissenschaften* (Berlin)
'Αρχ Πόντ.	*'Αρχεῖον Πόντου*
ASciPhi	*Arabic Science and Philosophy*
BARev	*Biblical Archaeology Review*
BASOR	*Bulletin of the American Schools of Oriental Research*
BCH	*Bulletin de Correspondence Hellénique*
BF	*Byzantinische Forschungen*
BiblArch	*Biblical Archaeologist*
BMGS	*Byzantine and Modern Greek Studies*
BSOAS	*Bulletin of the School of Oriental and African Studies*
BZ	*Byzantinische Zeitschrift*
CJ	*Codex Justinianus*
CQ	*Classical Quarterly*
CR	*Classical Review*
CSEL	*Corpus scriptorum ecclesiasticorum Latinorum*
CTh	*Codex Theodosianus*
CurrAnthr	*Current Anthropology*
Digest	*Digesta Iustiniani Augusti*
DM	*Damaszener Mitteilungen*
DOP	*Dumbarton Oaks Papers*
EchCl	*Echos du monde classique. Classical Views*
EHR	*English Historical Review*

EI[2]	*Encyclopaedia of Islam*, 2nd edn (Leiden)
ESAR	T. Frank and T. R. S. Broughton, *Economic Survey of Ancient Rome* (Baltimore, 1933–40)
ESI	*Excavations and Surveys in Israel*
GRBS	*Greek, Roman and Byzantine Studies*
HSCP	*Harvard Studies in Classical Philology*
IEJ	*Israel Exploration Journal*
IGLS	*Inscriptions grecques et latines de Syrie*
IGRR	*Inscriptiones graecae ad res romanas pertinentes*
IJNA	*International Journal of Nautical Archaeology*
ILS	*Inscriptiones Latinae Selectae*
JAridEnviron	*Journal of Arid Environments*
JEH	*Journal of Economic History*
JESHO	*Journal of the Economic and Social History of the Orient*
JFA	*Journal of Field Archaeology*
JHistGeogr	*Journal of Historical Geography*
JInterdiscipHist	*Journal of Interdisciplinary History*
JNES	*Journal of Near Eastern Studies*
JÖB	*Jarhbuch der Österreichischen Byzantinistik*
JRA	*Journal of Roman Archaeology*
JRS	*Journal of Roman Studies*
LCL	Loeb Classical Library
LibSt	*Libyan Studies*
M	H. Danby, *The Mishnah* (Oxford, 1933)
MAA	*Mediterranean Archaeology and Archeometry*
MGH	*Monumenta Germaniae historica*
NatGeogRes	*National Geographic Research*
NEAEHL	*The New Encyclopedia of Archaeological Excavations in the Holy Land*
OCD[3]	S. Hornblower and A. Spawforth, *The Oxford Classical Dictionary*, 3rd edn (Oxford, 1995)
ODB	*The Oxford Dictionary of Byzantium*
OJA	*Oxford Journal of Archaeology*
OKS	*Ostkirchliche Studien*
P.Ness.	Nessana Papyri
P.Petra	Petra Papyri

P.Ryl.	Rylands Papyri
PastPres	*Past and Present*
PBSR	*Papers of the British School at Rome*
PEQ	*Palestine Exploration Quarterly*
PG	J.-P. Migne, *Patrologiae cursus completus, Series graeca* (Paris, 1857–66)
PL	J.-P. Migne, *Patrologiae cursus completus, Series latina* (Paris, 1844–64)
ProcBritAc	*Proceedings of the British Academy*
RB	*Revue Biblique*
RBN	*Revue belge de numismatique et de sigillographie*
RDAC	*Report of the Department of Antiquities Cyprus*
RE	*Paulys Real-Encyclopädie der classischen Altertumswissenschaft*
RÉByz	*Revue des études byzantines*
SBHeid	*Sitzungsberichte der Heidelberger Akademie der Wissenschaften, Philosophisch-historische Klasse*
SC	*Sources Chrétiennes*
SEG	*Supplementum epigraphicum graecum*
SHAJ	*Studies in the History and Archaeology of Jordan*
Syr.Rom.Law.	*The Syro-Roman Lawbook*, ed. and trans. A. Vööbus (Stockholm, 1982)
TAPS	*Transactions of the American Philosophical Society*
TB	I. Epstein (ed.), *Babylonian Talmud* (London, 1935–52)
TIB	TIB Cilicia = F. Hild and H. Hellenkemper, *Kilikien und Isaurien* (Tabula Imperii Byzantini, 5; Vienna, 1990)
TJ	J. Neusner (trans.), *The Talmud of the Land of Israel: A Preliminary Translation and Explanation* (Chicago, 1982–94)
Waddington	W. H. Waddington, *Inscriptions grecques et latines de Syrie* (Paris, 1870, reprint Rome, 1968)
ZDMG	*Zeitschrift der deutschen morgenländischen Gesellschaft*
ZDPV	*Zeitschrift des deutschen Palästina-Vereins*

Introduction

Agriculture was the foundation of the ancient Mediterranean economy. As in the states and empires that preceded it, the vast majority of people in the Roman empire engaged in farming, and the agrarian character of much of the Mediterranean and Middle Eastern life persisted into the twentieth century. Despite this, the study of the countryside of this ancient region remains intellectually open and an exciting ground for further exploration.

The present work aims to bring to life the late antique landscape. In order to do so, I explore the physical environment of the whole, or parts, of a number of modern countries (Cyprus, Israel and the Palestinian Territories, Jordan, Lebanon, Syria, and Turkey) along the eastern Mediterranean coastlands located from Sinai to south-eastern Turkey. The inland territory ranges from the Tigris and Euphrates rivers in the north to the coast of the Red Sea in the south. To the east and west, the limits of my interests extend from the southern scarp of the Taurus to the Syrian Desert. Natural and longstanding boundaries enclose this area and create a certain unity: mountains and an imperial enemy in Persia flanked the north and north-west. The desert lay to the east; while to the south and south-east the desert, the sea, and Egypt stood as physical and notional borders. During the Tetrarchy, the Romans expressed their conception of this unity in their grouping of a 'baker's dozen' of provinces into an administrative superprovince—the diocese of Oriens.

Within the lines drawn by these imperial authorities, there were great variations amongst people and landscape, but also there was commonality. The coastlands and deserts were never far from any settlement, nor far from one another. People living in Palmyra were more aware of the deserts, but they knew something of the sea as well, since their caravans off-loaded in southern Mesopotamia on the Persian Gulf. One could leave the blissful seaside of Seleucia, pass by the suburban villas of Daphne and the plains of Antioch, and before taking the evening meal, traipse into a chain of arid hills. It pays to invoke such contrasting images as they were constants in the warp and weft of the landscape: river and desert, mountain and plain, hill and valley. Over most of the length of the upper Euphrates valley, one could stand on any eminence and see a band of greenery a few bowshots in width and the steppe land beyond. This variety in landscape and the precarious nature of farming such a challenging environment shaped much of the experience of farmers and rural dwellers in all periods, but particularly those of Late Antiquity.

In Chapter 1, I examine the physical and developed environment in which the majority of rural dwellers lived, the constructed milieu of the village, and the general character and scale of landholding. I shall then study the most common agrarian pursuits, namely the growing and processing of the Mediterranean staples of grain, wine, and olive oil. Beyond understanding the basics of their production, I seek to explain whether or not ancient agriculture was static, or if one can reasonably argue that there occurred changes which may have significantly affected the economy. For this reason, this study focuses on the place of the basic staples of grain, wine, and olive oil in subsistence and in the wider economy of the Late Roman–Early Byzantine world.[1] In addition, the presumed nature of the ancient (and late antique and early medieval) economies as primitive, characterized by a lack of commoditization and a weak market, has generated questions addressed in this work regarding the role of the market in the exchange of basic goods. The possibility of specialization and open transfer of staples, the capacities of producers to meet fundamental demand, and the farmers' adaptation to changing conditions all require investigation.

The period of interest extends from the fourth through the seventh centuries. The starting-point roughly parallels the foundation of Constantinople and the Tetrarchic reforms, both of which arguably mark the beginning of 'Late Antiquity', 'Late Rome', or 'Early Byzantium', depending on one's perspective. I do not end abruptly with the Muslim conquests of the 630s, but consider the state of the landscape through the end of the seventh century. For this purpose, I have been able to make limited use of Arabic material, supplemented by archaeological data. This relatively long-term approach allows one to weigh-in on questions prominent in the wider, secondary literature of the late antique world. Among issues of interest are whether economic decline

[1] Throughout this work, 'Late Roman', 'Early Byzantine', and 'Late Antiquity' are used more or less interchangeably to denote the fourth to seventh centuries AD. In those instances when a different span of time is implied, as in the case of some of the periodization described in various archaeological reports from Israel, when e.g. 'Late Roman' may mean third century, I have attempted to draw the reader's attention to the difference implied. All dates, unless specified, are AD. Since transliteration of Greek, Syriac, Arabic, and Hebrew vary considerably, I have tended to cite places and people as I encountered them in the secondary literature and attempt to remain internally consistent. For abbreviations of redactions of the Talmud, I follow B.-Z. Rosenfeld and J. Menirav, 'Methods of Pricing and Price Regulation in Roman Palestine in the Third and Fourth centuries', *JAOS* 121/3 (2001). Abbreviations of tractates of the Mishnah and Talmud follow those set out by I. Epstein, *The Babylonian Talmud* (London, 1935).

afflicted the eastern empire at the time of the Muslim conquests and whether wider economic changes accompanied the Muslim takeover of the Levant.[2]

The Written Sources

When compared to modern archives, the sources used in this study are varied, but not copious. Nothing akin to the substantial body of papyri from Egypt survives in the diocese of Oriens, but one can build up a coherent picture by appealing to a fair number of texts and the considerable corpus of material evidence.[3] The former include a limited number of papyri, namely those discovered in the excavations of the 1930s from Nessana in the Negev, Israel. In addition, valuable insight is afforded by the recently discovered carbonized archive from the church in Petra, Jordan. These documents, dating to the sixth century, will, upon their full publication, shed additional light on rural life. Other textual evidence can be found particularly in the lives of saints.

In northern Syria and Mesopotamia, saints' lives, notably those chronicled by *Lives of the Eastern Saints*, a sixth-century Syriac work of John of Ephesus, contain precious, though limited, information. The *Life of Simeon Stylite the Younger* of the sixth century also offers some details on rural conditions. Among the more important seventh-century sources are *vitae* from neighbouring regions. The stories detailing the exploits of Theodore of Sykeon and John the Almsgiver Patriarch of Alexandria, and George of Choziba in Palestine, touch on economic matters. While this short list is by no means an exhaustive account of the hagiographic works, it shows that they cover a substantial chronological and geographical range. Other Christian works, such as the epistolary and homiletic output of fourth- and fifth-century churchmen like Basil of Caesarea, Gregory of Nyssa, Jerome, Theodoret of Cyrrhus, and John Chrysostom, contain further occasional references to economic activity and life in rural areas.

[2] See M. McCormick, *Origins of the European Economy: Communications and Commerce, A.D. 300–900* (Cambridge, 2001), 27 ff. who sees the east as sharing in the generally poor diet, sanitation and health that the western lands of the Roman empire endured, but not until the sixth and seventh centuries.

[3] Egypt itself, which lies beyond the scope of this study, has been the subject of recent studies by J. Banaji, *Agrarian Change in Late Antiquity: Gold, Labour, and Aristocratic Dominance* (Oxford, 2001) and P. Sarris, 'The Origins of the Manorial Economy: New Insights from Late Antiquity', *EHR* 129/2 (2004); P. Sarris, 'Rehabilitating the Great Estate: Aristocratic Property and Economic Growth in the Late Antique East', in W. Bowden, L. Lavan, and C. Machado (eds.), *Recent Research on the Late Antique Countryside* (Leiden and Boston, 2004); P. Sarris, *Economy and Society in the Age of Justinian* (Cambridge, 2006).

By no means has this study exhausted the possibilities this literature contains; the nature of the texts makes it difficult to find specific pieces of data. From the standpoint of a historian interested in the countryside, secular ancient historians such as Ammianus Marcellinus in the fourth century, and Procopius of Caesarea and Agathias in the sixth century, generally suffer from their concern with political narrative. They do sometimes provide rare gems of information regarding life outside city and court. Religious histories and chronicles, such as those of Socrates and Sozomen, offer similar anecdotal evidence. These are joined by universal chronicle writings such as the sixth-century work of John Malalas, and those from the Syriac tradition, like the sixth-century work attributed to Joshua the Stylite and the larger, eighth-century *Chronicle* of Pseudo-Dionysius of Tell Maḥrē. The latter provides exceptionally good detail for agrarian conditions in late antique Mesopotamia.

Alongside these more prominent genres, I have consulted other literary and legal works that range fairly widely, such as the *Saturnalia* of the fourth- to fifth-century Latin writer Macrobius, the panegyrical and rhetorical works of the fourth-century Themistius, and the sixth-century Choricius of Gaza and Corippus. The fourth-century letters and orations of Libanius are of considerable value for information about the state of the land in northern Syria. The Theodosian and Justinianic Codes, as well as the later informal 'Farmer's Law', describe conditions in rural areas of the empire, particularly from a fiscal standpoint.

Epigraphic material provides a rich body of evidence. The collection of the *Inscriptions Greques et Latines de Syrie* (*IGLS*) is a fundamental source, but it has yet to be systematically explored. Other epigraphic surveys from Cilicia and various sites in Palestine also provide information on the countryside. Minor use has been made of lead seals. Lead seals can shed light on administrative oversight of trade and taxation, but they are not as informational about the early period of Byzantine history in the East as they are for Anatolia in the Middle Byzantine period.

Each of the written sources, of course, brings difficulties. The major problem is their lack of interest in rural affairs, but this also offsets some of the usual problems of omission and bias standard in historiography. In some instances, rural conditions are platforms for criticism of elites, such as the homilies of Chrysostom and in the *Anecdota* of Procopius but these are relatively infrequent. Most of the information provided—and explored here—is often a matter of supplementary details to the narrative, and not its focus. Potential problems of exaggeration or bias are of secondary concern in a study like this. Since elites paid little attention to life outside of cities, it is this dearth, rather than the actual reports of these sources, that make them difficult to use. In those instances when they do offer comment on rural conditions, their perspective is that of an estate owner. Such is the case with Libanius. But with the rise of Christianity and the championing of the poor by bishops like Basil and Chrysostom, some recognition of the life of modest farmers may be seen. Of course, in such instances of moralizing, exaggeration is possible.

The problem of elite dominance of the sources is somewhat mitigated through the use of the agricultural handbooks (discussed in the Appendix) which, although produced by estate owners, comment on universal agrarian experiences. Similarly, Rabbinic sources provide considerable information of conditions that prevailed among all strata of Jewish society within the East. These sources describe a fine cross-section of rural life: there are references to the crops, equipment, and trading activities of universal importance across the eastern landscape. The Jerusalem Talmud, edited in Palestine around the year 400, and the Babylonian Talmud, assembled in Mesopotamia and edited in several phases (predominantly around 500), are invaluable sources for daily life generally beyond the Jewish communities that produced them.[4]

The Archaeological Evidence

Unlike the textual items, the amount of physical evidence is steadily increasing, and future exploration promises to advance our knowledge considerably. Most of the relevant archaeological material comes from the southern part of the former Oriens—modern Israel and Jordan. The major challenge in writing an agricultural history based largely on physical data is that our coverage to date is incomplete. Although substantial portions of the former Oriens have been the subject of some sort of archaeological work, the region is vast and thus there remain large gaps in our knowledge.

Another limitation is the variable quality of the corpus itself. Publications of well-stratified excavations are common from these regions, but there are numerous instances in which work conditions have been less-than-ideal. Stratigraphic data and reliable ceramic series are still missing from many areas and scientific surveys that utilize the latest strategies of remote sensing, off-site survey, and spatial analysis, are sparse.

Chronological controls and micro-stratigraphic methods are frequently lacking from excavations. Surveys are infrequently paired with excavations

[4] J. Neusner, *Introduction to Rabbinic Literature* (New York and London, 1994); H. L. Strack, G. Stemberger, and M. N. A. Bockmuehl, *Introduction to the Talmud and Midrash* (Edinburgh, 1991).

that could provide adequate ceramic chronologies and detailed geomorphological studies. Both with survey and excavation, quantified studies of ceramic assemblages, and complete publications, are rare. In order to offset some of the difficulties that these deficiencies present, I have adopted a catholic approach to the evidence: rather than exclude the results from certain work (especially older, less-advanced archaeological work), my approach rather seeks to cross-check data on a regional and inter-regional scale. I have attempted to avoid relying too heavily on emergency/salvage-excavation or survey work, though in some instances such work does provide valuable supporting information.

In cases where abundant archaeological material is not available, other data help to offset some of the inadequacies of the material record. For example, in the Syrian surveys undertaken by the Princeton Expedition, the results were, by today's standards, rather random and unscientific. Sites with considerable surviving architectural remains were sought out by Howard Crosby Butler's team. Incomplete though his approach was, Butler's work provides a valuable catalogue of material and complements other datasets, including near-contemporary architectural surveys by Jean Lassus, modern surveys, the epigraphic surveys of the *IGLS*, and the findings from Butler's own extensive epigraphic research. Such checks are not always available, but I have attempted to build up a picture of the late antique countryside that consults with and weighs as many different types of evidence as possible. Generally, I have favoured case studies of those regions where the work has been relatively abundant and where I can craft a composite picture from both written and material evidence. This approach, which tests the strengths of the data and spreads the risk of error by interrogating multiple sources, provides an effective foundation for this present work.

Finally, the quantitative approach that I favour in this work plays a substantial part in my analysis and conclusions. In the past, I have been struck by the way that this method helps to frame issues that are often difficult to grasp, and how the conclusions frequently spark reaction. The latter result is partly due to the discomfort caused by a scholarly discourse that seems to present the past in absolute terms, the manner in which numbers are usually perceived. I am fully aware of the true limitations and the rudimentary nature of my calculations, and urge the reader to take these constraints under consideration. My attempt to quantify data is a useful illustration of the possible, even the probable, but ultimately only a portion of the hypothesis that I offer.

1

The Land
Climate and Geography

CLIMATE AND GEOGRAPHY

It is impossible to be certain of the nature of the climate of late antique Oriens. Though there is a general consensus that conditions over the whole of the period were not greatly different from those today, we still lack clarity about short-term weather patterns and local climatic variance. Yet some assessment of the Roman–Byzantine environment is possible. The intent of this separate and brief discussion is not to treat climate and geography merely as a backdrop against which the human actors played their parts. Such reductionism would ignore the important place of the environment in the daily lives of the inhabitants of Oriens. It would also negate the critical impact of climate and geographical circumstances in the *longue durée*. To some degree, the choice of crops, settlement patterns, tillage methods, animal selection, and a host of other farming choices were all responses to environmental concerns. These cares are therefore explicit and often implicit within discussion throughout this book and underpin many basic assumptions.

The present Mediterranean climate that dominates Oriens has been described as 'capricious', and is characterized by violence and rapid development.[1] Most of Oriens is arid or semi-arid with extreme summer heat, with many places prone to droughts and floods.[2] The Mediterranean climate today is characterized by hot, dry summers and comparatively cooler, moist winters with rainfall three times that of summer. Within the broader framework of this climate, there is significant diversity, accentuated by the seasonal variables. As Grove and Rackham note, the start of the hydrologist's year in the

[1] R. King, L. J. Proudfoot, and B. J. Smith, *The Mediterranean: Environment and Society* (London, 1997), 30. The Mediterranean as a climatic and economic unit is discussed fully in P. Horden and N. Purcell, *The Corrupting Sea: A Study of Mediterranean History* (Oxford, 2000).

[2] The definition of an arid landscape employed here is of one which receives on average of less than 300 mm of precipitation annually, while a semi-arid landscape receives 300–600 mm of precipitation annually. See FAO (ed.), *Improving Productivity of Dryland Areas* (Rome, 1987).

Mediterranean corresponds to the Byzantine New Year.[3] Rains begin shortly after 1 September, but are often sparse, rarely excessive, and sometimes they do not occur at all. Across the eastern Mediterranean, annual fluctuations in rainfall are common and often dramatic. In the Negev, for example, approximately one out of three years will be one of severe drought.[4] The coastal regions of Syria–Phoenicia–Palestine today typically receive 600–1000 mm of rain annually, whilst precipitation weakens dramatically inland; within a hundred kilometres of the coast, Damascus is already a desert city.

The Climate of Oriens in the Roman–Early Islamic Period: From Optimum to Disaster?

The climate of Mediterranean Late Antiquity was probably not dissimilar to that of today.[5] As today, there was long-term and short-term weather variability which significantly affected, and in some cases shaped, human activity on the land. In the earliest part of our period, the climate of the eastern Mediterranean world was particularly clement. The trend in northern Europe was toward warmer conditions; the extension of vine cultivation into Roman Britain is a well-known indicator of this climatic peak. In the Levant, the Roman climatic optimum from *c.* 300 BC to AD 300 indicates cooler temperatures and more rainfall vital to the agricultural success of marginal, typically dry, regions. Wood samples recovered from the early Roman assault ramp at Masada display depletion on both ^{13}C and ^{18}O isotopes, signs of cooler, wetter conditions than those prevalent today.[6] During this time, the advance of olive cultivation and the retreat of oak forests in the eastern Mediterranean region indicate an expansion of cultivation into former woodland, due in part to more rainfall.[7] Based on the finds from eight sites spread throughout Israel, Hirschfeld has argued that wetter weather dominated the region from about AD 300 to 500.[8] While Hirschfeld's notions of a wetter Oriens in the Early Byzantine period would seem to correlate well with what

[3] A. T. Grove and O. Rackham, *The Nature of Mediterranean Europe: An Ecological History* (New Haven, 2001), 52 n. 2.

[4] M. Evenari, L. Shanan, and N. Tadmor, *The Negev: The Challenge of a Desert*, 2nd edn (Cambridge, Mass., and London, 1982), 300–35.

[5] D. D. Gilbertson, 'Explanations: Environment as Agency', in G. Barker (ed.), *Farming the Desert: The UNESCO Libyan Valleys Archaeological Survey* (Paris, 1996), 315. Mediterranean climate patterns similar to that of today have perhaps persisted since the third or second millennium BC. See Grove and Rackham, *Nature of Mediterranean Europe*, 166.

[6] A. Issar and D. Yakir, 'Isotopes from Wood Buried in the Roman Siege Ramp of Masada: The Roman Period's Colder Climate', *BiblArch* 60/2 (1997).

[7] According to the pollen samples noted by A. Issar, *Water Shall Flow from the Rock: Hydrogeology and Climate in the Lands of the Bible* (Berlin, 1990), 178.

[8] Y. Hirschfeld, 'A Climatic Change in the Early Byzantine Period? Some Archaeological Evidence', *PEQ* 136/2 (2004).

we know of broad settlement expansion into the previously dry margins, as we see below, other data are contradictory.

Mediterranean weather apparently became more unpredictable and hostile after the fourth century. Some climatic and material evidence supports Hirschfeld's notions of marked climate and environmental change, which have yet to be fully investigated and understood.[9] Vita-Finzi noted that the extensive post-Roman (Younger Fill) alluvial deposition in Tripolitania, Cyrenaica, Morocco, Spain, Italy, Greece, and Jordan apparently coincided with a period of cold, wet weather in Europe following the end of the Roman period.[10] Whereas much of the Mediterranean faced cold and episodes of flooding following the Roman optimum, the climate of Greater Syria seems to have become hotter and drier, a telltale sign being decreasing water levels of the Dead Sea.[11] The Madaba Map, produced in AD 560–5, depicts a Dead Sea seriously depleted by a series of droughts or high human consumption of the waters of the Jordan.[12] The image of a droughty Syria-Palestine is buttressed by geological work around the Dead Sea that points to increasing dryness, especially from AD 600 to 700.[13] It thus seems that the farmers who built many of the impressive wadi-farming systems still visible in the Negev, and the hydraulic features of the Syrian steppe, did so under unfavourable, dry climatic conditions. Even in the best of days, the Negev and the eastern portions of the Syrian provinces were marginal landscapes, more friendly to flocks than to farm. Extensive dry-farming was never viable in the Negev,[14] and yet it was in the sixth–seventh centuries that hydraulic farming and

[9] Y. Hirschfeld, 'The Crisis of the Sixth Century: Climatic Change, Natural Disasters, and the Plague', *MAA* 6/1 (2006).

[10] C. Vita-Finzi, *The Mediterranean Valleys: Geological Changes in Historical Times* (London, 1969).

[11] H. J. Bruins, 'Comparative Chronology of Climate and Human History in the Southern Levant from the Late Chalcolithic to the Early Arab Period', in O. Bar-Yosef and R. S. Kra (eds.), *Late Quaternary Chronology and Paleoclimates of the Eastern Mediterranean* (Tucson, Ariz., 1994), 308.

[12] D. Amiram, 'The Madaba Mosaic Map as a Climate Indicator for the Sixth Century', *IEJ* 47/2 (1997). Due to the settlement peak described below, the uptake of water for domestic, agricultural, and industrial purposes from the Jordan was probably also at a high point, contributing to declining Dead Sea levels.

[13] A. Frumkin et al., 'Middle Holocene Environmental Change Determined from the Salt Caves of Mount Sedom, Israel', in O. Bar-Yosef and R. S. Kra (eds.), *Late Quaternary Chronology and Paleoclimates of the Eastern Mediterranean* (Tucson, Ariz., 1994), 323, fig. 6. In the hill country near the Dead Sea pine and oak replaced olives immediately after the Byzantine period and forest regenerated, evidence indicative of a retrenchment of settlement: C. Heim, N. Nowaczyk, and J. F. W. Negendank, 'Near East Desertification: Evidence from the Dead Sea', *Naturwissenschaften* 84 (1997).

[14] Even if present-day rainfall over much of the Negev were to double, the entire region would still remain an agriculturally marginal landscape, limited by interannual rainfall variability, little good arable, and high evapirotranspiration rates that hinder farming endeavours.

settlement prosperity reached their peak. The Negev villages were built with obvious self-confidence echoed in the large-scale building along the fringes of the Syrian Desert further north. Only around 670 AD did the climate of Greater Syria perhaps enter another phase of cooler and wetter weather.[15]

Some short-term variation in the climatic record of the Roman–Byzantine East is also discernible, and recently Koder has noted that unstable atmospheric conditions at certain moments during Late Antiquity led to agricultural upheaval.[16] Another study links the *aurora borealis* and traumatic weather events, and focuses on sunspot activity, today considered a major climatic influence.[17] The appearance in the climatic record of the so-called '536 event' has sparked new debate within scholarly circles of climate history. Literary sources, such as Cassiodorus, noted the weakness of the summer sun and the poor harvest of the following year. The dry fog that blanketed Europe around AD 536 occurred during a period of decline in solar activity that resulted in fifteen years of reduced tree growth.[18] During the decade or so immediately following the '536 event', it appears that weather patterns became volatile resulting in a brief period of cooler weather.[19] That there should be several cooler, wetter periods (as well as violent, one-time episodes) during this time of generally sustained warmth and desiccation is not surprising. Such interannual weather changes are familiar to us and their causes better

On the Negev and its settlement development, see D. Bar, 'Frontier and Periphery in Late Antique Palestine', *GRBS* 44/4 (2004); D. Bar, 'Population, Settlement and Economy in Late Roman and Byzantine Palestine', *BSOAS* 67/3 (2004).

[15] N. Shehadeh, 'The Climate of Jordan in the Past and Present', in A. Hadidi (ed.), *SHAJ* 2 (Amman, 1985), 28. See also the archaeoclimatic modelling for the Middle East undertaken by R. U. Bryson and R. A. Bryson, 'Application of a Global Volcanicity Time-Series on High-Resolution Paleoclimatic Modeling of the Eastern Mediterranean', in A. Issar and N. Brown (eds.), *Water, Environment and Society in Times of Climatic Change: Contributions from an International Workshop Within the Framework of International Hydrological Program (IHP) UNESCO, held at Ben-Gurion University, Sede Boker, Israel from 7–12 July 1996* (Dordrecht, 1998). E. Netzer, 'Population Growth and Decline in the Northern Part of Eretz-Israel during the Historical Period as Related to Climatic Changes', in A. Issar and N. Brown (eds.), *Water, Environment and Society in Times of Climatic Change: Contributions from an International Workshop Within the Framework of International Hydrological Program (IHP) UNESCO, Held at Ben-Gurion University, Sede Boker, Israel from 7–12 July 1996* (Dordrecht, 1998), notes the apparent synchronization in higher levels of the Dead Sea, indicative of higher rainfall conditions, and settlement based on climatological and archaeological data across northern Israel.

[16] J. Koder, 'Climatic Change in the Fifth and Sixth Centuries?', in P. Allen and E. Jeffreys (eds.), *The Sixth Century End or Beginning?* (Brisbane, 1996), 275–6.

[17] P. Farquharson, 'Byzantium, Planet Earth and the Solar System', in P. Allen and E. Jeffreys (eds.), *The Sixth Century: End or Beginning?* (Brisbane, 1996), 266–7.

[18] J. D. Gunn, *The Years Without Summer: Tracing A.D. 536 and its Aftermath* (Oxford, 2000).

[19] See M. G. L. Baillie, 'Dendrochronology Raises Questions About the Nature of the AD 536 Dust-Veil Event', *The Holocene* 4/2 (1994); M. G. L. Baillie, 'Patrick, Comets and Christianity', *Emania* 13 (1995) for the dendrochronological data.

known; El Niño is a particularly famous modern example. Whilst it remains difficult to determine both the causes and the long-term effects of such events in the distant past, we need little reminder what devastation could be wrought by nature on the livelihood of farmers great and small in antiquity. The severe drought of 611 that caused a famine in Mesopotamia is one of many instances in which chroniclers have noted the calamitous effects of poor weather.[20]

Geography

The north-western limits of Oriens were the Taurus and Anti-Taurus barriers that abut the Cilician plain. These mountainous areas dominate the landscape as one travels east to the Tur 'Abdin, which overlooks Roman Mesopotamia (defined by the northern portions of the Euphrates and Tigris valleys—this area was referred to as al-Jazira ('the island') by the Arabs).[21] As a rule, the upland country was more sparsely populated than the well-watered coastal plains, and the interior steppes were less densely settled than the coastlands. In Late Antiquity, the north-western portions of the Jazira steppe region formed a parched buffer zone, the interface of Sasanian Persia and Rome. In this area, cereal culture was possible, but, like today, often depended on irrigation centred in the city and village. On the whole, the northern Mesopotamian steppes remained agriculturally underdeveloped and home mainly to pastoralists. In areas like the Massyas valley (the Biqā^{c}), and the western slopes of the Lebanon and Anti-Lebanon mountains, drylands extended nearly to the coast, and agricultural development lagged behind other more geographically favoured areas. As in the desert lands further to the east and north, the Biqā^{c} shared in remarkable agricultural transformation during the fourth through seventh centuries, and seems to have been relatively prosperous up to the period of the Muslim conquests.

The Syrian Desert begins roughly where precipitation falls below 250 mm. Along this line lies a broad band of steppe land from the Euphrates and Tigris valleys to Palestine. This area, particularly the landscape around Epiphaneia (Hama), shows considerable evidence of intensive and extensive settlement and exploitation. Along its north–south border, the steppe met a number of minor ranges of hills. South of Aleppo (ancient Beroia), the volcanic Basalt Massif circumscribes the Jebels Hass and Šbeit, whilst to the west the Limestone Massif (Massif of Belus) comprises a long chain of calcareous hills. South of the expansive plain of Chalcis lie the hills of the Jebel il 'Ala and

20 M. Morony, 'Michael the Syrian as a Source for Economic History', *Hugoye* 3/2 (2000), par. 6.

21 Also referred to as Upper Mesopotamia, distinct from Persian–Early Islamic Lower Mesopotamia (Babylonia).

the Palmyrene hills around Epiphaneia and Palmyra. Still further south lies the Hauran, a fertile plain dominated by the jebel of the same name. West of the Hauran, the basaltic Golan rises, and the Biqā^{c} opens between the Anti-Lebanon and Lebanon Mountains that run parallel to the coast. To the south are the Moabite and Kerak plateau, the Arava rift, and the Red Sea. Palestine's geography differs little from that of Syria-Mesopotamia. The limestone ridges of the Judaean Hills form the backbone of Palestine, stretching from the Negev to the foothills of the Hermon and thence to the Lebanon and Anti-Lebanon ranges. The latter two mountain chains rise suddenly from the steppe, block rainfall, and communication from the seaboard, and create sharply contrasting climates and landscapes between the coast and interior.

This highly varied, often harsh landscape and climate dictated settlement patterns and land management. Human effort was needed to enhance the physical surroundings and offset short- and long-term climatic severity. How farmers achieved this is the subject of much of this book. However, in their struggle with the environment, the arsenal available to ancient and medieval cultivators was limited. Wheat and barley, for example, were fundamental to the Roman–Byzantine diet in the settled lands; without grain, life was not possible. But for a variety of reasons, cereals were especially vulnerable to negative factors in climate and geography. Farmers could attempt to overcome limited drought conditions or poor soil by growing barley instead of wheat, but some cereal crop had to be grown. When barley proved unreliable, abandonment or irrigation were amongst the limited set of choices available to late antique farmers. With the other crops of the Mediterranean triad of grain, wine, and olive oil, the effects of weather were less important: one did not need wine or olive oil to survive, but the limits of cereal cultivation determined the extent of settlement.

Water

Most of Oriens supported dry-farming. The 200 mm annual rainfall threshold (see Map 1) is a critical one: below this limit, dry-farming of cereals is not possible. Some regions, such as the Syrian steppe, the southern and eastern regions of Roman Mesopotamia, southern and eastern Arabia, and the southern limits of Palestine III around Aila, received insufficient rainfall for the dry-farming of grain. In these places, subsistence agriculture could only succeed with irrigation. Bands of higher rainfall sweep to the north and north-east of the heart of the diocese, north of the great Syrian steppe and south of the Euphrates. Although portions of the Jazira (the Upper Mesopotamian steppe) receive more than 200 mm annual rainfall, the further south and east one travels, the lower the precipitation. Rainfall is, even in areas theoretically sufficient for dry-farming, irregular and unreliable, thus making irrigated agriculture a necessity throughout much of eastern Syria.[22] The mountain region of the Tur 'Abdin, with its 300 mm annual precipitation, is unique in Roman Mesopotamia. On the Euphrates, the 200 mm rainfall threshold is crossed already by the time one reaches Barbalissus (Meskene). Several cities of importance lay beyond this line: Neocaesarea (Dibsi Faraj), Sura, Callinicum (Raqqa), Zenobia (Halabiye), and Circesium (Qarqisiye) amongst them. Similarly levels of precipitation plummet east of the Palestinian coast; the northern Negev settlement of Elusa (Halutza) receives annual average precipitation of 200 mm, but nearby Oboda and Sobata (Shivta) receive 83 and 86 mm, respectively. In the winter, snowfall is common at higher elevations, whilst summer temperatures may reach 45° in Mesopotamia; such temperatures create high evapirotranspiration rates which cause plants and animals rapidly to lose moisture.[23]

In the extreme north of Oriens, Mesopotamia possesses water resources from the Tigris, as well as the Euphrates and its tributaries. Around the Tur 'Abdin, rainwater percolates through the limestone hills and emerges in many springs at the foot of the escarpment. These sources provide water for human inhabitants and feed numerous watercourses that merge into the Balikh (ancient Balissus), Jaghjagh (ancient Mygdonius), and Khabur (ancient Aboras). All of these watercourses eventually join the Euphrates. Amongst the tributaries, the Balikh, Khabur, and Jaghjagh offered opportunities for Roman agricultural exploitation. They flow slowly through the wide expanse of alluvial soils of the Jazira and have low banks, making water distribution by canal, or with machinery, to nearby fields relatively easy. Of these rivers, the Khabur, being at the Persian frontier, was the most important to the Romans. Its flow of 40 m^3 per second and its relative regularity from season to season made it especially useful. The Khabur's low alluvial terrace could be watered easily, and although the upper terrace would have proved more challenging, it too was exploitable.[24] The hydraulic resources of this region made it one of the more fertile and developed of the East.

Other northern rivers were important in the agricultural and commercial life of the empire. The Batman Su, a tributary of the Tigris on the Persian

22 C. Akhruf, 'Syria', in I. A. O. A. Economists (ed.), *World Atlas of Agriculture* (Novara, 1973), ii.477.

23 Several regions in Oriens experience relatively high levels of dewfall. The Negev in one year had dewfall (37 mm) in excess of measured rainfall. Whilst some believe that precipitation in the form of dew has no bearing in agricultural production, others cite studies show that plants benefit from the condensation of water vapour. Evenari, Shanan, and Tadmor, *The Negev*, 258; G. Le Strange, *Palestine Under the Moslems, A Description of Syria and the Holy Land from A.D. 650 to 1500* (London, 1890); P. Mayerson, *The Ancient Agricultural Regime of Nessana and the Central Negeb* [*sic*] (London, 1960), 10–11, notes that Muqaddasi reports that 'In Palestine, during the summer-time, every night, when the south wind is blowing, dew falls, and in such quantities that the gutters of the Aksâ Mosque are set to run.'

24 J. Lauffray, *Halabiyya-Zenobia I: Place forte du limes oriental et la haute-Mésopotamie au VIe siècle* (Paris, 1983), 59.

border, remained in Roman hands throughout the period and irrigated the narrow valley in which it was confined.[25] The Tigris watered the region from Melitene (Malatya) to Amida (Diyarbakir). The ancient Pyramus, Sarus, and Cydnus water the vast crescent of the Cilician Plain. They probably were the most productive of rivers for irrigation purposes, as they flowed through the hinterlands of major late antique population centres noted for their fertility: Mopsuestia, Anazarbus, and Tarsus.[26]

The Euphrates has an average annual flow of 710 m^3 per second and thus has the capacity to irrigate vast areas of land, but irrigation and urban use of the river presented severe challenges to ancient peoples.[27] The season of the highest rate of flow is in the springtime, when the river frequently peaks around 2,100 m^3 per second. When the harvest approached there often was little need for additional water. During sowing and the early stages of plant growth, the fields required applications of water. During this time the Euphrates was typically at its seasonal low and therefore difficult to exploit. There were additional difficulties facing ancient hydraulic engineers. The waters of the Upper and Middle Euphrates have high banks that dominate the riverside over many kilometres before the river strikes the vast alluvium of Iraq and becomes more manageable. When the water was low, these high banks presented an even greater impediment to lifting water. Another serious problem was the wall of rock along much of the right bank, which prevented canal construction along the south side of the river. In the region of Zenobia, there are basalt and gypsoid cliffs, known as 'The Strangler', that constrict the river between narrow bluffs. These features meant that Roman cities could not tap the huge potential of the river over much of its length, but had to be content with limited, more local canal systems, which consequently inhibited the size of settlement.

The Orontes flows north from the Lebanon range to the Mediterranean near Antioch. Springs supply the river, and their steady flow results in lower

[25] G. Algaze et al., 'The Tigris–Euphrates Archaeological Reconnaissance Project: A Preliminary Report of the 1989–1990 Seasons', *Anatolica* 17 (1991), 178.

[26] Alongside the above-mentioned should be included innumerable springs, small watercourses, seasonal wadis, and perennial streams throughout the region. Lake Tiberias, Lake Homs near ancient Emesa, and Lake Antioch probably all played a role (though impossible to determine at present) in the agricultural life of the empire. It may be only a slight exaggeration to say that nearly all of these resources were tapped for agricultural purposes at the peak of ancient settlement in the Byzantine period. In some areas, notably the basaltic Golan and Hauran, geological factors had a major influence on water exploitation and thus human settlement. The general impermeability of the upper strata of recent volcanic rock made aquifer formation irregular. In these areas, surface runoff and storage became an important element in water management. Groundwater, stored in aquifers, was accessible everywhere in hand-dug wells and was vital to development in marginal regions of low rainfall.

[27] M. G. Ionides, *The Régime of the Rivers Euphrates and Tigris* (London, 1937), 3.

seasonal variations and less chance of flooding than is common with snow-fed rivers. Before more recent hydraulic development turned the stream into a narrow, turbid ditch, the flow of the Orontes was strong. In 1940, maximum flow at Antioch exceeded 200 m³ per second.[28] In some stretches, the Orontes poses the same problems for use as the Euphrates, namely high banks, rapids, and marshes. For Roman and Byzantine era farmers, the Orontes, with its high rate of flow and proximity to the cities of Apamea and Antioch, was second in importance to the Euphrates. Not far from Antioch, and north of the flow of the Orontes, the Afrin drained the region of the Jebel Sim'an and the territory of Cyrrhus. Further east and south, the now dry river of Chalcis flowed by the city of the same name.

In the south of Oriens a number of watercourses drain into the Mediterranean from the Phoenician coast, including the Nahr al-Kalb (ancient Lycus),[29] the Nahr al-Kabir, and the Litani. All of these originate in heavily irrigated southern Phoenicia, a region noted for its fertility from the time of Strabo (16.2.16) to that of the Arab geographers.[30] The latter rivers generally irrigate bands of arable land confined within narrow valleys. Northern Palestine possessed the Crocodilion River that waters the region of Caesarea. The major river of Palestine is the Jordan, which drains much of the centre of the former provinces of Palaestina I and II. The Jordan is joined by the Yarmuk, which flows from the uplands south and east of the Sea of Galilee.

This survey illustrates two points: that there were a number of rivers that could be exploited for irrigation in Oriens and that doing so was never easy. Considerable investment in laborious projects of canal construction and maintenance, or in expensive irrigation-lifting equipment, meant that rain-fed farming was always preferred to artificial watering. As we will see below, the aggregate of textual and archaeological evidence suggests that Oriens was increasingly exploited, and that a major part of this intensification was the development of irrigation works. Nearly every major city in the eastern Mediterranean lay on, or in close proximity to, a perennial river. Typically, they were utilized not only for washing and drinking, but for farming as well.

Soil

Ancient farmers recognized that soils could improve or degrade over time and that some simply were more fertile than others. Human action is often pivotal

[28] J. Weulersse, *L'Oronte: étude de fleuve* (Tours, 1940), 39, fig. 15.

[29] J. Rougé, 'La navigation intérieure dans le Proche Orient antique', in P. Louis (ed.), *L'Homme et l'eau en Méditerranée et au Proche Orient* (Lyon, 1986), 39.

[30] Le Strange, *Palestine Under the Moslems*, 413.

in long-term sustainability of soil; erosion and deposition are often heavily affected by farming practice. Whilst it is possible that the soils of the Levant have changed substantially since Late Antiquity, we have no precise data to be certain of this. On balance, it is assumed that the overall picture is similar to today. Today, most of the soils along the coastal zone of the eastern Mediterranean are productive and easily managed. Pockets of terra rossa soils (good for vines and olives) dot the Mediterranean coastal regions, especially the uplands of Israel, Lebanon, and North Syria. Mesopotamia, like the coastal zone, is dominated by soils composed primarily of degraded limestone. These are most productive when irrigated, but also provide a good return in most dry-farming conditions. The alluvial soils of river valleys are managed easily enough and have exceptional agricultural potential when well-watered. Young, volcanic soils in the Hauran and parts of central Syria formed the backbone of the most fecund corn-lands in Oriens. In a few cases, soils offered no viable medium for cultivated plants. This was especially true of sizeable stretches of the sandy steppe lands of Syria and the deserts of *Palaestina Tertia*. More commonly, fertility was not in the first instance limited by the quality of the soils, but rather by the availability of water. Thus, as discussed in Chapter 6, access to water preoccupied the farmers in Late Antiquity as they pushed further into the margins of the empire.

When considering how the landscape was managed, it is imperative to consider how farmers assessed and understood the qualities of the earth. Their approach was not scientific, but nor was it without reason: They judged soil empirically, and gauged its production when matched with various methods of cultivation and different crops across the years. They observed characteristics of colour, texture, and other physical aspects, such as water retention and depth. *Geoponica* 2.9[31] states that the best soils were those which were black, pointing to a preference for the silty, alluvial soils common along the coast and in areas of Mesopotamia. Easily worked, deep soil was favoured for the ease of its tillage and was apparently considered best suited for trees, probably with intercultivation of cereals envisioned. Soils that were 'warm', or sweet to the taste and alluvium were considered best for vines and trees. Earth that tasted salty was to be avoided: only the date palm could survive saline soils, according to Anatolius (*Geop.* 2.10), who must have been thinking of the plantations in his native Phoenicia. In addition, the presence of certain kinds of foliage told the good cultivator the nature of the soil in a given spot and therefore what seeds and trees should be matched to the ground (2.10–2.13). Not only did the soils have to be properly paired with

[31] For a discussion of this agricultural handbook, see the Appendix.

the crops, their management and continued fertility required constant effort (see below Chapters 6–7).

Public and Marginal Zones

In addition to the cultivated zones, there were sizeable tracts of land that were non-arable, and, therefore marginal in terms of conventional agricultural activities. This is not to say that such areas were not exploited for the resources they offered; on the contrary, such zones provided an important resource within the economic mesh of late antique Oriens. Villagers amongst the numerous settlements that fronted the Syrian-Arabian desert from the Red Sea to Mesopotamia made use of these steppe lands for pasturage of flocks in a rhythm attuned to the rainfall pattern of winter and early spring. The forests in the high mountains of Lebanon were claimed by the emperor and, as imperial lands, were protected and managed for the timber they could provide.[32] Elsewhere, Mediterranean oak forests survived until Late Antiquity and provided pannage, timber, and wild game. Wilderness areas also contributed to other economic activities, such as quarrying. The concept that wilderness areas—within and adjacent to the cultivated zones—were imperial property seems to have been the rule: the agglomerations of imperial estates (the *salti* noted in the written sources) were generally located in thinly populated areas like the Negev and Isauria. In these regions, pastoralists were the most successful in exploiting the landscape. According to law quoted in the Theodosian Code (*CTh* 7.7), it seems that attempts were made to regulate and charge pastoralists for access to pastures. Given the vast extent of the frontier that bordered the desert in eastern Oriens, it is very unlikely that the Romans were successful in imposing pasture taxes on pastoralists outside the imperial *salti.*

The uncultivated margins of settlements thus provided land for grazing and forest products (such as pannage, game, timber for building, wood for fuel, leaves for animal fodder, honey, and wax). Waterways, both natural and artificial, provided drinking and irrigation water, and sometimes offered fish, reeds, ossier beds for basketry and tool-making, and waterfowl. In other 'waste' places, there were herbs and other wild edible plants collected both for use in the daily diet and as medicines.[33] The exploitation of these landscapes by agents involved in economic activities outside of conventional arable farming deserves special treatment that falls beyond the scope of this

[32] J. F. Breton, *Les inscriptions forestières d'Hadrien dans le Mont Liban* (Paris, 1980).
[33] J. M. Frayn, *Subsistence Farming in Italy* (Fontwell, 1979), 57–72.

study. It is important to stress, however, that foresters and pastoralists formed an important element within the economic framework of late antique Oriens.

The Population of Oriens

Before discussing details of the human landscape, namely how rural inhabitants were distributed across the area that formed Oriens, it is vital to consider how many people inhabited the diocese. Given the little work that exists in this arena, it is important to provide some analysis of this problem. While there have been attempts to estimate the numbers of people living in a given city, or even a group of cities, most estimates are not based on detailed demographic or statistical modelling, nor are they based upon close analysis of early modern census records from which we might hope to project figures for antiquity.[34] It is therefore difficult to go beyond the area of conjecture when we discuss the problems of the total population in the countryside of Oriens. My own estimates are based on comparative work and are not exhaustive: they represent a crude judgement of a static population and cannot take into account changes over time. It is therefore important to bear in mind that the figures discussed here are rudimentary and impressionistic. It is hoped that, at least to a degree, this discussion will underscore the fact that a serious, complex demographic study is absent in scholarship of Late Antiquity. Such a study could go far beyond the present in discussing how many people lived in the cities and countryside of Oriens, how their distribution affected economic connectivity and exchange, and how their numbers influenced land utilization and agrarian productivity.

The demography of ancient Egypt provides a comparative case useful in approximating the population of Oriens. Bagnall and Frier estimate that Roman Egypt had a population of 4–5 million, while Scheidel suggests a population peak of about 5 million in the second century.[35] In discussing population, it should be noted that Egypt was certainly the most agriculturally rich region within the empire, given its large size and the fecundity of the Nile Valley. Oriens, however, contained a larger arable land area, some of whose regions were famed for their fertility, notably the Cilician plain, Upper Mesopotamia, the Hauran, and the Jordan valley.[36] In their agricultural

[34] See for example, M. Broshi, 'The Population of Western Palestine in the Roman–Byzantine Period', *BASOR* 236 (1976); K. C. Gutwein, *Third Palestine: A Regional Study in Byzantine Urbanization* (Washington, DC, 1981); J. C. Russel, 'Ancient and Medieval Population', *TAPS* 48 (1958), 63–70.

[35] R. S. Bagnall, *Egypt in Late Antiquity* (Princeton, 1993), 56; W. Scheidel, *Death on the Nile: Disease and the Demography of Roman Egypt* (Leiden and Boston, 2001), 212.

[36] Egypt today, for example, which corresponds very roughly to its Roman counterpart in size, is currently 1,001,450 km^2, but only about 28,000 km^2 of this was habitable: see Bagnall,

potential, since the Bronze Age, these regions had proved capable of supporting large, wealthy urban communities.

From the lists of Hierocles and George of Cyprus, we possess the names of 181 settlements that may have been cities, and others that certainly were not.[37] Several cities in Oriens were quite large by ancient standards, above all Antioch and Apamea, the *metropoleis* of Syria Prima and Secunda, respectively. Other cities of note were the metropoleis of lesser provinces, such as Tarsus (Cilicia Prima), Anazarbus (Cilicia Secunda), Edessa (Osrhoene), Amida (Mesopotamia), Hierapolis (Euphratensis), Tyre (Phoenice), Emesa (Phoenice Libanensis), Bostra (Arabia), Caesarea (Palaestina Prima), Scythopolis/Bet She'an (Palaestina Secunda), and Petra (Palaestina Tertia). Economic and social importance was indubitably factored into the selection of these centres as capitals. Even Petra, whose inclusion in a list of the prosperous cities of Syria-Palestine would have raised eyebrows amongst scholars a decade ago, is now known to have maintained a vibrant local economy well into the sixth century.[38]

But how big were the large cities of the Byzantine East? Will's plausible analysis of the area of Antioch places the population *intra muros* of the city at 150,000.[39] John Chrysostom states that the city had 200,000 inhabitants, of

Egypt in Late Antiquity, 56 n. 15. Despite similar agrarian development efforts in all these countries, the combined areas of Jordan, Israel, Syria, and Lebanon today possess more than double the arable land in Egypt. In antiquity, Syria was itself able to match the arable land area of the Nile Valley, though it could not rival its fertility. When one considers the extent of the Cilician plain and Upper Mesopotamia, Oriens would have greatly eclipsed Egypt in arable area, but overall population density apparently remained lower.

37 On these lists, see A. H. M. Jones, *The Cities of the Eastern Roman Provinces* (Oxford, 1971), Appendix 3. The lists are to be treated with caution, and we know little of many sites that they mention. However, archaeological evidence offers information on many of the sites noted in the city lists. Clearly many settlements not designated village or *saltus* were, in fact, no better than small settlement enclaves. A good example is Arindela, a fort in the Arava Valley in present-day Jordan which archaeological investigation has shown to be a military post rather than a city. See D. L. Kennedy, *The Roman Army in Jordan* (London, 2000), 197–9. Nevertheless, there are many known cities in these lists, and the danger of overestimating population by assuming a certain order of magnitude of the lesser-known sites on the list is to a certain degree offset by the probable omission of important settlement sites known from late antique Oriens, such as Umm el-Jimal, Androna, Taroutia, al-Bara, and Medina el-Far. Inevitably, until the appropriate level of data collection and synthesis is developed in this milieu, our numbers will be little better than guess-work, and the present investigation is tempered by this fact. Late Roman cities and city life are treated at length by B. Ward-Perkins, 'The Cities', in A. Cameron and P. Garnsey (eds.), *The Cambridge Ancient History*, vol. 3: *The Late Empire, A.D. 337–425* (Cambridge, 1998), 371–410.

38 Z. T. Fiema, 'Economics, Administration and Demography of Late Roman and Byzantine Southern Transjordan', Ph.D. diss. (Salt Lake City, 1991); Z. T. Fiema, 'Byzantine Petra: A Reassessment', in T. S. Burns and J. W. Eadie (eds.), *Urban Centers and Rural Contexts in Late Antiquity* (East Lansing, Mich., 2001).

39 E. Will, 'Antioch sur l'Oronte: métropole de l'Asie', *Syria* 74 (1997), 99–114.

whom 100,000 were Christians.[40] Apamea probably contained a similar number within its much larger walled area (enclosing 250 ha).[41] In the early Roman period, census-takers recorded 117,000 citizens at Apamea, and this probably excluded the urban poor and slaves.[42] As a result of their pillage of the city in 573, the Persians reportedly deported 292,000 prisoners.[43] Both figures indicate the existence of a major urban centre with a population in excess of 100,000. If we accept an urbanization rate of 25 per cent—as proposed by Scheidel for Alexandria and Egypt—broadly applicable to the largest cities of Syria, then the hinterlands of Antioch and Apamea each contained approximately 350,000 people.[44] The two cities, with their respective regions, would have thus supported 1.2 million inhabitants, amounting to roughly 120 people per square kilometre, quite a high population density. Subtracting the two great urban populations in this estimate would place the rural population density at approximately 70 per square kilometre, still quite high but well below the dense settlements on the Egyptian landscape (140–180 persons per km^2).[45]

The population of other cities in Oriens may be similarly sketched. Study of Caesarea, the capital of Palaestina Prima, has elicited estimates from 35,000 to 120,000 people.[46] Caesarea covers 111.5 ha, less than half the area of Apamea. Caesarea's population was thus probably in the region of 35,000–45,000; this would place about 175,000–225,000 in the city's hinterland. Provincial cities could be of a similar size: Cyrrhus, another North Syrian city, was probably much smaller than Antioch or Apamea, but not much smaller than Caesarea. The fifth-century bishop Theodoret oversaw 800 parishes in a 1,600-square-(Roman)-mile (3,500 km^2) city territory. Cyrrhus itself probably contained 30,000–40,000 people,[47] and Cumont's estimate of 200,000 people in the whole of the Cyrrhus region therefore appears a possible maximum.[48] Abila, a provincial city in Palaestina Secunda, had an estimated

[40] F. Cumont, 'The Population of Syria', *JRS* 24 (1934), 188.

[41] Based on the calculation that an area of 1 hectare = 300 people, see M. J. Fuller, 'Abila of the Decapolis: A Roman–Byzantine City in Transjordan', Ph.D. diss. (St Louis, 1989), 240–1; Apamea would have possessed 75,000 people.

[42] Cumont, 'Population', 188. The ratio of people to city area would therefore be 468 per hectare.

[43] John of Ephesus, *The Third Part of the Ecclesiastical History of John, Bishop of Ephesus*, ed. and trans. R. Payne Smith (Oxford, 1860), 6.6.

[44] Scheidel, *Death on the Nile*, 247–8.

[45] Bagnall, *Egypt in Late Antiquity*, 56 n. 15.

[46] Based on 300 people per ha.

[47] Theodoret, ed. Azéma, *Epistulae* 113.88–90. Seventy inhabitants per sq km would put the population of urban and rural Cyrrhus at 245,000. For a provincial city like Cyrrhus, we can posit an urban : rural ratio of approximately 1:5 or less.

[48] Cumont, 'Population', 189. At 70 persons per sq km, Cyrrhus' total population would have been about 245,000, but it seems possible that outside the major cities like Constantinople,

population of 12,000–16,000 at the peak of the Roman–Byzantine period.[49] The Negev administrative cities of Mampsis and Elusa are thought to have had populations of 500 and 10,500 people, respectively, during the Byzantine period.[50] De Vries, the excavator at Umm el-Jimal, estimated that this large village in the Hauran had a population of around 5,000.[51]

From these figures, we can discern four broad tiers of population among the cities of Oriens. Behind first-rank cities like Antioch were large urban entities like Anazarbus, Mopsuestia, Tarsus, Caesarea, Edessa, Hierapolis, Seleucia-in-Isauria, Scythopolis, Amida, Emesa, Jerusalem, and Bostra. Many of these were provincial capitals and all probably had populations around 30,000. These twelve cities would thus account altogether for an urban population of 360,000 people. If these cities represented around 20 per cent of their territorial population, then the number of people living on the landscape, controlled by these dozen cities, approaches 2 million. These twelve centres do not include other urban settlements known to have been prosperous and important during Late Antiquity. We must also consider that there were numerous, substantial cities apart from the provincial capitals, such as Berytus, Damascus, Samosata, and Syrian Epiphaneia, to name a few.

The remainder of the population in Oriens was distributed throughout the third-tier settlements, similar in size to Abila, with Mampsis representing the fourth category, the smallest settlement to have civic status. Based on these population gradations and urbanization figures, as well as the number of cities in the lists of Hierocles and George of Cyprus, the total number of inhabitants in the early sixth-century (pre-Justinianic plague) diocese, including Cyprus, was approximately 6 million.

The Late Antique Settlement Peak in Oriens

Archaeological work in cities such as Caesarea has demonstrated the number and wealth of eastern Mediterranean urban settlements during Late Antiquity.[52] In those many centres in the heart of Greater Syria that remain unexcavated, there is strong textual evidence demonstrating continued

Alexandria, Antioch, and Apamea, there would have been a lower percentage of urbanization. Perhaps under 20 per cent of the population of the territory of Cyrrhus would have lived in the town itself. On the rate of urbanization, see Scheidel, *Death on the Nile*, 247–8.

[49] Fuller, *Abila of the Decapolis*, 239–53.

[50] J. Shereshevski, *Byzantine Urban Settlements in the Negev Desert* (Beer-Sheva, Israel, 1991), 213.

[51] B. De Vries, 'The Umm el-Jimal Project 1972–1977', *ADAJ* 26 (1982), 109.

[52] See, for example K. G. Holum, *King Herod's Dream: Caesarea on the Sea* (New York and London, 1988) for Caesarea.

strength of urban life up to and beyond the Arab conquest.[53] While the fortunes of individual cities waxed and waned during Late Antiquity, urban life seems to have been generally buoyant. That this dynamism was mirrored in the countryside (or more likely, anchored there) is evident from the results of three decades or so of archaeological survey. As it stands, the data are suggestive of unprecedented settlement on the land in the fourth through seventh centuries in most areas of Oriens and, generally, throughout the Near East: the results of survey indicate that the Diyala Basin supported unprecedented settlement levels during the late Sasanian period.[54]

Twenty-seven published surveys conducted by the Israel Antiquities Authority, covering portions of former Palestine I, II, and III, concur that the 'Byzantine' period (the fourth through early seventh centuries) represents a

[53] Emesa (Homs) was a major arena for the final resistance to the Muslim advance; Epiphaneia still maintained civic status and Byzantine material culture: see P. Pentz, *The Medieval Citadel and its Architecture* (Aarhus, 1998). Edessa was, through the sixth and seventh centuries, a lively Byzantine metropolis, M. Whittow, 'Ruling the Late Roman and Early Byzantine City: A Continuous History', *PastPres* 129 (November 1990). The persistence of lesser cities, such as Aila, is known from the Conquest accounts, in the case of the latter al-Balādhurī's *Futūḥ al-buldān*, trans. P. K. Hitti, '*The Origins of the Islamic State*', 92. The continuing wealth and economic activity of the hinterlands of Apamea and Antioch are well expressed in coin hoards and silver treasure. On the former, see C. Morrisson, 'Le trésor byzantin de Nikertai', *RBN* 118 (1972); on the latter, M. M. Mango, *Silver from Early Byzantium: The Kaper Koraon and Related Treasures* (Walters Art Gallery, 1986); M. M. Mango, 'The Origins of the Syrian Ecclesiastical Silver Treasures of the Sixth–Seventh Centuries', in N. Duval et al. (eds.), *Argenterie romaine et byzantine: actes de la table ronde, Paris 11–13 octobre 1983* (Paris, 1988), 163–84. On continuity in general, see P. Pentz, *The Invisible Conquest: The Ontogenesis of Sixth and Seventh Century Syria* (Copenhagen, 1992), J. Magness, *The Archaeology of the Early Islamic Settlement in Palestine* (2003), and now also A. Walmsley, *Early Islamic Syria: An Archaeological Assessment* (London, 2007).

[54] The collected data are limited in detail and often not backed by published excavation results so that the view of past settlement and land-use presented herein must be regarded as provisional. There is certainly much work to be done on evolving standards of prospecting and recording of data in field survey. Site identification and chronology remain particularly thorny areas. Site magnitude across time remains extremely difficult to detect through field survey, and ceramic chronologies for the Byzantine–Umayyad transition remain problematic, though progress is being made in this area: Magness, *Early Islamic Settlement*; E. Villeneuve and P. Watson (eds.), *La céramique byzantine et proto-islamique en Syrie-Jordanie: IVe–VIIIe siècles apr. J.-C.: actes du colloque tenu à Amman les 3, 4 et 5 décembre 1994* (Beirut, 2001) and Walmsley, *Early Islamic Syria*. Sasanian settlement: the sixth century was the zenith of the Sasanians' political and economic power and the peak of their irrigation agriculture. R. McC. Adams, *Land Behind Baghdad: A History of Settlement on the Diyala Plains* (Chicago, 1965), 69–83; R. McC. Adams, *Heartland of Cities: Surveys of Ancient Settlement and Land Use on the Central Floodplain of the Euphrates* (Chicago, 1981), 179–83. See P. Christensen, *The Decline of Iranshahr: Irrigation and Environments in the History of the Middle East, 500 B.C. to A.D. 1500* (Copenhagen, 1993), 67–72 for a similar view on proposed Sasanian demographic expansion, though M. Morony disagrees: M. Morony, 'Land Use and Settlement Patterns in Late Sassanian and Early Islamic Iraq', in G. R. D. King and A. Cameron (eds.), *The Byzantine and Early Islamic Near East II: Land Use and Settlement Patterns* (Princeton, 1994).

time of very dense settlement.[55] In Palaestina II, in particular, agricultural communities expanded into the Golan—where a total of 143 Roman sites are known, increasing to 173 in the Byzantine period.[56] Similarly, for Mount Hermon, the 'Late Roman–Byzantine' (second to seventh centuries) period was one of 'unprecedented prosperity' and escalating settlement.[57]

A host of survey work from the Transjordan yields similar results, though qualification is required. In the south of Palestine III, for example, the Byzantine administrative presence had apparently faded by the time of the Persian Wars, a situation that did not change by the time of the first strikes of Muslim marauders. The lack of interest of the Byzantine government seems to be reflected in a slow decline of settlement through south-east Palestine and Provincia Arabia from the fourth century onward.[58] There are indications, though, that settlement did prosper in many areas of this region through the period of the early Muslim conquests. Although we learn nothing from the conquest accounts concerning the size or importance of the settlements, several in the Transjordan are mentioned as eventually capitulating to the

[55] G. Avni, *Map of Har Saggi Northeast (225)* (Jerusalem, 1991); Y. A. Baumgarten, *Mapat Shivṭah (166)* (Jerusalem, 2004); I. Beit-Arieh, *Map of Tel Malhatah (144)* (Jerusalem, 2003); A. Berman, H. Stark, and L. Barda, *Map of Ziqim (91)* (Jerusalem, 2004); A. Berman, L. Barda, and H. Stark, *Map of Ashdod (84)* (Jerusalem, 2005); A. Berman, *Map of Nizzanim: West (87)* (Jerusalem, 2005); *Map of Nizzanim: East (88)* (Jerusalem, 2005); R. Cohen, *Map of Sede-Boqer—East (168)* (Jerusalem, 1981); R. Cohen, *Map of Sede-Boqer—West (167)* (Jerusalem, 1985); Y. Dagan, *Map of Lakhish (98)* (Jerusalem, 1992); R. Frankel and N. Getzov, *Map of Akhziv; Map of Hanitah* (Jerusalem, 1997); Z. Gal, *Map of Gazit (19–22)* (Jerusalem, 1991); Z. Gal, *Map of Har Tavor (41) and Map of En Dor (45)* (Jerusalem, 1998); D. Gazit, *Map of Urim (125)* (Jerusalem, 1996); R. Gophna, *Map of Herziliyya (69)* (Jerusalem, 1998); Y. Govrin, *Map of Nahal Yattir (139)* (Jerusalem, 1991); M. Haiman, *Map of Har Hamran—Southwest (198)* (Jerusalem, 1986); M. Haiman, *Map of Mizpé Ramon—Southwest (200)* (Jerusalem, 1991); M. Haiman, *Map of Har Hamran—Southeast (199)* (Jerusalem, 1993); M. Haiman, *Map of Har Ramon (203)* (Jerusalem, 1999); Y. Hirschfeld, *Map of Herodium (108/2)* (Jerusalem, 1985); M. Kochavi and I. Beit-Arieh, *Map of Rosh ha-Ayin (78)* (Jerusalem, 1994); Y. Ne'eman, *Map of Ma'anit (54)* (Jerusalem, 1990); Y. Ne"eman, S. Sender, and E. Oren, *Map of Mikhmoret (52): Map of Hadera (53)* (Jerusalem, 2000); Y. A. Olami, *Map of Daliya (31)* (Jerusalem, 1981); Y. A. Olami, A. Ronen, and A. Romano, *Map of Haifa—West (22)* (Jerusalem, 2003); Y. A. Olami and Z. Gal, *Map of Shefar'am (24)* (Jerusalem, 2003); Y. A. Olami, S. Sender, and E. Oren, *Map of Yagur (27)* (Jerusalem, 2004); J. Patrich, *Map of Deir Mar Saba (109/7)* (Jerusalem, 1995); A. Raban, *Map of Nahalal (28)* (Jerusalem, 1982); A. Raban, *Map of Mishmar Ha-'Emeq (32)* (Jerusalem, 1999); A. Ronen and Y. A. Olami, *Map of Atlit (1)* (Jerusalem, 1978); A. Ronen and Y. A. Olami, *Map of Haifa—East (23)* (Jerusalem, 1983); S. A. Rosen, *Map of Makhtesh Ramon (204)* (Jerusalem, 1994); D. Weiss, B. Zissu, and G. Solimany, *Mapat Nes Harim (104)* (Jerusalem, 2004). In Israel and Jordan (and in the discussion that follows), the "Late Roman" period is 2nd–4th centuries, whilst the Byzantine period ends with the seventh century.

[56] Z. U. Ma'oz, "Golan", in E. Stern (ed.), *NEAEHL* (Jerusalem, New York, and London, 1993), II.545.

[57] S. Dar, *Settlements and Cult Sites on Mount Hermon, Israel* (Oxford, 1993), 26.

[58] Fiema, *Economics, Administration and Demography*, 242–3. Urban life apparently continued in cities such as Petra, Adroa (Urdruh) and Areopolis. Petra now seems to have been

Muslims. Archaeologically the data are more plentiful, with investigations, conducted in the Jordan Rift Valley, Moab, the Kerak Plateau, and Wadi al-Hasa, suggesting that the late antique settlement peak extended to portions of southern Oriens as well.[59]

An absence of published regional surveys renders the data for the rest of Oriens more impressionistic. The nineteenth- and early twentieth-century architectural surveys of Melchior De Vogüé, Butler, and Lassus already

doing much better than has recently been allowed, whilst too little is known to hypothesize concerning other cities in southern Transjordan. In the countryside, there is a striking contrast between the arid margins of the Transjordanian landscape and the Negev region, with its massive investment in agriculture and vibrant community life centred on substantial villages.

[59] As much of the information comes from work done in modern Israel and Jordan, in the immediately following discussion of survey data, the 'Byzantine' period is 4th–7th centuries, whilst 'Late Roman' is 2nd–4th centuries. In the Central Moab survey, 47 per cent of sites yielded Byzantine pottery, an increase over the 25 per cent evidenced in the Late Roman period. On the Kerak Plateau, of 442 recorded sites, 282 yielded Byzantine pottery, the most of any period, see J. M. Miller, *Archaeological Survey of the Kerak Plateau: Conducted During 1978–1982 Under the Direction of J. Maxwell Miller and Jack M. Pinkerton* (Atlanta, Ga., 1991). Parker's *Limes Arabicus* project documented settlement recovery in the Byzantine period (*c.* AD 284–500): S. T. Parker, *The Roman Frontier in Central Jordan: Interim Report on the Limes Arabicus Project, 1980–1985* (Oxford, 1987); S. T. Parker, 'The Limes Arabicus Project: The 1987 Campaign', *ADAJ* 32 (1988); S. T. Parker, 'The *Limes* and Settlement Patterns in Central Jordan in the Roman and Byzantine Periods', *SHAJ* 4 (1992), 321–5; S. T. Parker and J. W. Betlyon, *The Roman Frontier in Central Jordan: Final Report on the Limes Arabicus Project, 1980–1989* (Washington, DC, 2006), 46–8. Similarly, the Wadi el-Hasa survey noted an Early Roman peak of 55 per cent of sites occupied, and Late Roman occupation on only 20 per cent of sites, increasing to 32 per cent of sites by the Byzantine period. B. Macdonald, *The Wadi el Hasa Archaeological Survey, 1979–1983, West-Central Jordan* (Waterloo, Ont., Canada, 1988), 97–112. The Hesban Survey: R. D. Ibach, *Archaeological Survey of the Hesban Region: Catalogue of Sites and Characterization of Periods* (Berrien Springs, Mich., 1987), 183, offers evidence of dramatic Byzantine settlement expansion. Whereas only 30 per cent of sites exhibited Late Roman occupation, 85 per cent of sites yielded Byzantine ceramics. Andrews University's survey of the limits of the Madaba plain at the junction of the *territoria* of ancient Madaba and Philadelphia, concluded that the region's peak occupation occurred in the Byzantine period: L. G. Herr, *Madaba Plains Project: The 1989 Season at Tell el-Umeiri and Vicinity and Subsequent Studies* (Berrien Springs, Mich., 1997), 187–8; A. Northedge, C. M. Bennett, and J. Bowsher, *Studies on Roman and Islamic Amman: The Excavations of Mrs C-M Bennett and Other Investigations* (Oxford, 1992), 40. Villeneuve's survey around Iraq al-Amir in the Jordan Valley demonstrated that 63 per cent of sites were inhabited in the Roman-Byzantine period, F. Villeneuve, 'Prospection archéologique et géographie historique: la région d'Iraq al-Amir (Jordanie)', in P.-L. Gatier, B. Helly, and J.-P. Rey-Coquais (eds.), *Géographie historique au Proche-Orient (Syrie, Phénicie, Arabie, grecques, romaines, byzantines): Actes de la Table Ronde de Vallebonne 16–18 septembre 1985* (Paris, 1988), 282–3. A more recent survey of the same area, C. J. Chang-Ho and J. K. Lee, 'Archaeological Survey and Settlement Patterns in the Region of 'Iraq al-'Amir: A Preliminary Report', *ADAJ* 42 (1998), 587–606; C. J. Chang-Ho and J. K. Lee, 'The 1998 Season of Archaeological Survey in the Regions of 'Iraq al-'Amir and Wadi al-Kafraya: A Preliminary Report', *ADAJ* 43 (1999), 521–39, concludes that the Byzantine period was one of densest settlement. Finally, the survey of the Wadi az-Zarqa, in northern Palestine II, noted an increase in site numbers between the Hellenistic and Late Roman/Byzantine periods, G. Palumbo et al., 'The Wadi Az-Zarqa/Wadi Ad-Dulayl Excavations and Survey Project: Report on the October–November 1993 Fieldwork Season', *ADAJ* 40 (1996).

suggested a settlement boom in late antique Syria. More recent survey work in the Hauran has revealed dense Roman settlement and continued Byzantine settlement activity.[60] In the Massyas valley (al-Biqāᶜ), little work has been done, but survey results suggest a Byzantine decline, with settlement in the marginal valley fringes abandoned. Although they are less numerous, Byzantine sites, however, tend to be larger than their Roman predecessors.[61]

Recent survey work in the Chalcis region at Androna (modern al-Andarin) region paints the same picture of a late antique settlement boom as found over much of former Oriens.[62] Further north, the Balikh River survey yielded 23 securely dated Byzantine settlements around Callinicum (modern al-Raqqa), as opposed to 55 Early Islamic period sites (eighth to tenth centuries).[63] On the Middle Euphrates, Islamic-period sites predominate; 'Abbasid and Ayyubid agricultural investment, in the form of a massive canal project, are especially apparent.[64] By contrast, in the northern Jazira, Wilkinson's survey around Edessa (modern Urfa) noted a 4th–7th-centuries settlement peak.[65] Surveys conducted in the region around Zeugma, on the Euphrates and extending to the Tigris, indicate that the Late Roman/Early Byzantine period 'represented a substantial peak' in occupation.[66]

60 J.-M. Dentzer (ed.), *Hauran I: Recherches archéologiques sur la Syrie du sud à l'époque héllenistique et romaine* (1; Paris, 1985); J.-M. Dentzer (ed.), *Hauran II: Recherches archéologiques sur la Syrie du sud à l'époque héllenistique et romaine* (Paris, 1986); J.-M. Dentzer, 'Les recherches archéologiques françaises dans la perspective de l'exploration de la Syrie du Sud basaltique', *AAAS* 42 (1997), 297–318.

61 L. Marfoe, 'Between Qadesh and Kumidi: A History of Frontier Settlement and Land Use in the Biqa, Lebanon', Ph.D. diss. (Chicago, 1978).

62 H. C. Butler, *Architecture and Other Arts* (New York, 1904); H. C. Butler et al., *Publications of the Princeton University Archaeological Expedition to Syria in 1904–1905 and 1909* (Leiden, 1907–49); B. Geyer and M.-O. Rousset, 'Les steppes arides de la Syrie du Nord à l'époque byzantine ou la rouée vers l'est', in B. Geyer (ed.), *Conquête de la steppe et appropriation des terres sur les marges arides du Croissant fertile* (Lyon, 2001), 111–21; J. Lassus, *Inventaire archéologique de la région au nord-est de Hama* (Damascus, 1935); M. M. Mango, 'Excavations and Survey at Androna, Syria: The Oxford Team 1999', *DOP* 56 (2002), 307–15; M. M. Mango, 'Excavations and Survey at Androna, Syria: The Oxford Team 2000', *DOP* 57 (2003), 293–7; A. Musil, *Palmyrena, A Topographical Itinerary* (New York, 1928).

63 K. Bartl and S. R. Hauser, *Continuity and Change in Northern Mesopotamia from the Hellenistic to the Early Islamic Period: Proceedings of a Colloquium Held at the Seminar für Vorderasiatische Altertumskunde, Freie Universität Berlin, 6th–9th April, 1994* (Berlin, 1996), 336.

64 B. Geyer and J. Y. Monchambert, 'Prospection dans la basse vallée de l'Euphrate Syrien', *AAAS* 33 (1983); this preliminary report offered little detail on other period sites. The surface finds from the prosperous Ayyubid agrarian expansion may well mask other material.

65 T. J. Wilkinson, *Town and Country in Southeastern Anatolia* (Chicago, 1990), 117.

66 Algaze et al., 'Tigris–Euphrates Archaeological Reconnaissance Project 1989–1990', 205–6; G. Algaze, R. Breuninger, and J. Knudstad, 'The Tigris–Euphrates Archaeological Reconnaissance Project: Final Report of the Birecik and Carchemish Dam Survey Areas', *Anatolica* 20 (1994), 22.

Cilicia and Cyprus exhibit a similar trend toward peak site numbers in Late Antiquity. Survey in Cilicia has been extremely limited and focused primarily on urban architectural remains. An exception is Blanton's preliminary work in Rough Cilicia, formerly part of Isauria. According to Blanton, the 'Late Roman' period (fourth through seventh centuries) settlement in the survey area increased from an estimated 18,000 to 24,500 people (from a pre-Roman population of approximately 1,400).[67] Papacostas' collection of Cypriot surveys shows marked continuity and settlement on the land during the Roman–Byzantine period.[68] On the island's north coast, for example, the rugged and agriculturally marginal Kormatiki peninsula yielded evidence of dense occupation during the 6th–7th century.[69] These results are repeated throughout the north coast, as well as in the Mesaoria, the north-eastern foothills of the Troodos massif, and in the south coast hinterland of Amathus, Paleopaphos, and in the Vasilikos Valley.[70]

These data accord with evidence from elsewhere in the late antique world, such as Lycia, where settlement also reached a high point.[71] The same is true for several regions in Greece. Intensive surveys from the Argolid, Boiotia, Keos, Laconia, Megalopolis, Methana, the Nemea Valley, Euboea, the Skourta Plain, the Gulf of Corinth area, Kenchreai, Akra Sophia, Messenia, Nichoria, Phokis, and Lokris provide evidence of expansion and prosperity of rural settlement during the Byzantine period.[72]

Archaeological survey is still a relatively young discipline with an evolving set of methodologies. In addition, there remain substantial areas not covered by any survey work. Limited as we are, our current knowledge suggests a broad increase in the size and number of cities and a corresponding growth of

[67] R. E. Blanton, *Hellenistic, Roman and Byzantine Settlement Patterns of the Coast Lands of the Western Rough Cilicia* (Oxford, 2000), 60 tbl. 4.1 These figures are preliminary and to be treated with caution.

[68] T. Papacostas, 'The Economy of Late Antique Cyprus', in S. A. Kingsley and M. Decker (eds.), *Economy and Exchange in the East Mediterranean During Late Antiquity* (Oxford, 2001), 109.

[69] H. Catling, 'An Early Byzantine Pottery Factory at Dhiorios in Cyprus', *Levant* 4 (1972).

[70] P. Aupert, *Guide d'Amathonte* (Paris, 1996), 178; H. W. Catling, 'The Ancient Topography of the Yialias Valley', *RDAC* (1982); S. Hadjisavvas, 'Perforated Monoliths: Myths and Reality', in M.-C. Amouretti, J.-P. Brun, and D. Eitam (eds.), *Oil and Wine Production in the Mediterranean Area from the Bronze Age to the End of the XVIth Century* (Paris, 1991); Papacostas, 'Economy of Late Antique Cyprus', 109; M. Rautman, 'The Busy Countryside of Late Roman Cyprus', *RDAC* (2000); M. L. Rautman, M. C. McClellan, and L. Benson, *A Cypriot Village of Late Antiquity: Kalavasos-Kopetra in the Vasilikos Valley* (Portsmouth, RI, 2003), 236–42; L. W. Sørensen, D. W. Rupp, and P. G. Bilde, *The Land of the Paphian Aphrodite* (Göteborg, 1993), 277.

[71] C. Foss, 'The Lycian Coast', *DOP* 48 (1994), 2.

[72] C. Kosso, 'Public Policy and Agricultural Practice: An Archaeological and Literary Study of Late Roman Greece', Ph.D. diss. (Chicago, 1993); C. Kosso, *The Archaeology of Public Policy in Late Roman Greece* (Oxford, 2003).

the number and size of many rural settlements throughout much of the Roman–Byzantine East. Rather than a dilapidated group of impoverished eastern provinces falling to the Muslim conquerors, one may reasonably suppose that rural prosperity continued in many areas well into the sixth century, and perhaps into the seventh century and later. For the bulk of our period of study, therefore, the eastern countryside was as crowded and as vibrant as it had ever been. It was also a changing landscape, the transformation of which was an important part of the momentous developments in the final stages of the ancient world.

2

The Countryside in Late Antiquity

As in most pre-industrial societies, in Late Antiquity the wealth of those with social and political power was ultimately tethered to the land. Trade and industry, although bolstered by the far-reaching connections and deep monetization of the eastern empire, never vied to replace land ownership as an anchor for aristocratic wealth. Land continued to be the vehicle that generated most wealth.

The political reforms of the third and early fourth centuries engendered tremendous social and economic change. Among the major shifts was a rise in the number of imperial bureaucrats, and with it the attendant rewards spawned by imperial office-holding. The tax system of Diocletian and Constantine fostered a civil administration that was larger, more focused, and more efficient than the decentralized, city-based formula. The creation of a stable gold coinage—first as the *aureus* of Diocletian, and then the slightly reduced *solidus* of Constantine—also sent waves through the wider sea of the ancient economy; the ripples continued well into the sixth century. One result of monetary reform was the manipulation of the gold : copper relationship by the state and the elites to their own advantage and to the detriment of the poor whose lives were often spent handling copper, not gold. These factors, among others, were responsible for the growth of large estates and affected the manner in which estates were managed and wealth generated. Recent scholars, therefore, largely confirm a fundamental assumption of past generations: Late Antiquity was a time when 'new men' came to dominate the empire, and these new men acquired land fairly rapidly. Their gains were often at the expense of their small and medium-scale landowning neighbours.

The full extent of the trend toward the centrally managed, 'manorial' estate and its social and political impact remain, however, a matter of debate.[1] A much simplified version of the generally accepted view of Late Antiquity may be summarized as follows: the land was dominated by large estate owners, often imperial bureaucrats (or closely connected with them) who

[1] Sarris, 'Origins of the Manorial Economy'; Sarris, *Economy and Society* effectively revives the notion that the estate customs prevalent in the early medieval west are to be traced back to Roman practice adapted in Late Antiquity. For the degree to which Byzantine society was monetized in late antique Egypt, see Banaji, *Agrarian Change*.

could better shield themselves from the burden of taxation. Those on the bottom rungs of the ladder of wealth probably paid an undue proportion of taxes; under such strain, these farmers fled the land or shed their burdens by giving their holdings to wealthy patrons who already possessed the lion's share. In other instances, the cultivators' role as wage earners, whose pay depended on sophisticated account-keeping and bullion ratios of which they would have been largely ignorant, again played into the hands of the wealthy. From a purely fiscal perspective, this process of elite aggrandizement eroded the tax base upon which the stability of the state depended.

According to this model, the flow of land into the hands of the aristocracy encouraged further concentration and centralization of resources. As estates became more specialized with a view to generating cash, in order to maintain and extend the owner and the owner's family's social standing, they also tended to centralize. They made rational use of labour pools and invested capital in more land and labour-saving machinery. They employed aggressive approaches to land improvement and cash crops. Sharply attuned to the monetized market and sensitive to the opportunities offered by overseas trade, the aristocracy tapped into an export trade from which those lacking sufficient resources and personal networks were effectively barred. Within such a milieu, landowners exerted downward pressure on wages. So long as manpower remained plentiful, income for the majority of rural people remained static or degraded. The loss of buying power by the majority of workers on the land meant that they had little access to the market and were particularly vulnerable to fluctuations in the price of goods. They were thus starved of capital and overexposed to the risks of daily life, such as food shortage.

Scholars have also highlighted the weaknesses of a society with a hyper-concentration of wealth. As a small cadre of those with social power cordoned off the avenues to economic and social advancement and further entrenched themselves, they continued to act at the expense of the majority of the citizens, leading to disaffection and social unrest. Furthermore, a decline in the number of those with wealth in favour of the superlatively wealthy reduced access to literacy and other skills required to manage private and public affairs, leading ultimately to a lack of social and political talent. By contrast, a more socially broad economy, with a deeper mix of rich and poor, would offer, if not an antithesis to the above generalization, the possibility of a more durable and responsive society. The possibility therefore arises that conditions in the countryside, and, in particular, an extensive concentration of wealth and power, contributed to the failure of the Roman state in the East, first in the Sasanian Persian occupation, and in the final knockout blow of the armies of Islam.

In order to explore this issue, landownership and the physical manifestations of landownership and husbandry—namely the estate centre and the village—deserve careful attention. At stake is our understanding of the fiscal, economic, and socio-political foundations of the empire. The following analysis explores the question of large-estate dominance, both by surveying the textual evidence for the presence of this type of landholding in Oriens, but also by examining the archaeological data for the existence of different kinds of estates within the diocese. Alongside the great estate and imperial territories, with their complex web of dependencies and tenancies, there undoubtedly survived a number of independent farmers of middling or modest means: the evidence for this survival in the written and archaeological record holds the promise of a more nuanced view of the shifting landscape in the late antique east. The variety presented by the constructed landscape—the nodes of estate centre, village, isolated farmstead, or fortified farm—offer us glimpses of the rich tapestry of the eastern landscape at a high point of its Roman exploitation.

Imperial Estates

Many of the lands administered by the Roman state during the Late Empire belonged to the period of the Republic's takeover of the Hellenistic kingdoms. Confiscations from rebels and the appropriation of the estates of those who died without heirs also augmented these estates. Imperial landholdings greatly increased when the state confiscated all temple lands (*CTh* 5.13.3), though many of these were subsequently given to the Christian church. Most cities in Late Antiquity probably had imperial land within their territories under the supervision of imperial estate managers. The emperor, for example, owned 16 per cent of the whole of Cyrrhus' *territorium*; an inscription (*SEG* XX.455: AD 305 or 311) from Scythopolis (Palestine II) refers to the 'crown lands' (*despotai chorai*).[2]

The organization and administration of these estates were complex. The holdings were supervised by curators (*κουράτωρες*). These administrators, answerable to the praetorian prefect, oversaw a number of different categories of imperial lands. The *res privata* were crown lands reserved, in theory, for the maintenance of the imperial household, but in practice they were used more often for general public expenditure.[3] The *patrimonium* (*sacrum patrimonium*)

[2] D. A. Fiensy, *The Social History of Palestine in the Herodian Period: The Land is Mine* (Lewiston and Queenston and Lampeter, 1991), 59; B. H. Isaac and I. Roll, *Roman Roads in Judaea I: The Legio-Scythopolis Road* (Oxford, 1982), 105; A. H. M. Jones, *The Later Roman Empire, 284–602: A Social, Economic and Administrative Survey* (Oxford, 1964), 415–16.

[3] M. Kaplan, 'Les grands propriétaires de Cappadoce (VIe–XIe siècles)', in C. Fonseca (ed.), *Le aree omogenee della civiltà rupestre nell'ambito dell'impero bizantino: la Cappadocia* (Lecce, 1981), 126.

comprised lands detached from the *res privata* specifically to support state expenditure. Finally, the Divine House (*domus divina, basilikos oikos, theios oikos*) was part of the *res privata* under Justinian, but later formed an independent entity possessing special privileges and prone to managerial abuse, which Tiberius II (578–82) attempted to temper.[4]

Agglomerations of imperial estates were organized for administrative purposes into *salti* (cάλτοι), *regiones* (ρεγέωνες), and *tractus* (κλίμα).[5] The salti often lay in rugged country, and several are known from Oriens. The city list of George of Cyprus in the seventh century includes a number of imperial domains throughout the eastern provinces, demonstrating that the crown had an interest in immense tracts of land throughout the whole of Oriens. From this list we know that there were imperial estates (*klima*, pl. *klimata*) in the highlands of Isauria.[6] Around the Euphrates, in the Anti-Taurus Mountains, and in the northern Jazira, large portions of the province of Mesopotamia were incorporated into imperial domains. In the vicinity of Martyropolis, by the Tigris in eastern Mesopotamia, lay the Klima Arzanene. Further west, on the Anti-Taurus scarp and in the Arsanias river valley were the expansive state lands of the city of Dadima. Territory there was organized into a number of administrative units, foremost among them the Klima Sophanene, the *klima* of Ialimbanōn, but a number of others are also known, indicating a high concentration of imperial ownership.[7]

North Oriens consisted of other units of state land. The Saltus Eragiza probably lay at the great bend in the Euphrates, astride the territory of the city of Barbalissus (Meskene), and also probably included the semi-arid lands of the basalt hills of the Jebels Šbeit and Hass. Broad tracts of imperial land lay around in Chalcis where Julian brought grain to relieve the shortages at Antioch.[8] In Late Antiquity, this previously underdeveloped area was a landscape dotted with important villages, amongst them Tarutia (in the territory of Apamea), Androna, and Zebed. The settlements there were relatively young: Gabbula and Anasartha (also ancient Theodoropolis; today

[4] M. Kaplan, 'Nouvelle de Tibère II sur les « 'maisons divines' »', *TM* 8 (1981), 237–45.

[5] Jones, *Later Roman Empire*, 413. On estates in the east generally see D. J. Crawford, 'Imperial Estates', in M. I. Finley (ed.), *Studies in Roman Property* (Cambridge, 1976), 64–5.

[6] George of Cyprus, *Georgii Cyprii Descriptio orbis Romani*, ed. Gelzer (*Georg. Cyp.*), 854–9, which lists the *klimata* of Casae, Banaba, Cotrada, Bolbosus. Jones, *Cities of the Eastern Roman Provinces*, 139–40, suggests that these were tribal territories confiscated by Servilius Isauricus. Much of the area of these estates must have been exploited for timber and pasturing rather than sedentary agriculture.

[7] These were lands that probably belonged originally to the Armenian kingdom before the partition of 387. *Georg. Cyp.* 958–62, noted the *klima* of Sophene, called Ialimbanōn and the estates (*klima*) which make up the *klimata* of Anzētinēs, Digēsinēs, Garinēs, Bilabētinēs, Palinēs, Orzianinēs, Astianikēs, and Mouzourōn.

[8] Julian, *Misopogon*, ed. Wright, 369a.

Khanasir) first appear in the sixth-century *Notitia Antiochena*.[9] At Bab al-Hawa, in North Syria, impressive remains of a palace and boundary stone survive. The boundary stone inscription mentions the *kouritoros* (curator) Magnus.[10] Magnus was active in the later sixth century during the reigns of Justin II and Maurice. His seals from the latter reign reveal his dual functions as *kouritoros* and *kommerkiarios*.[11] Evidence suggests that the Anasartha region possessed significant imperial lands. An inscription, dated to AD 344 at al-Mshrefe, mentions the *pragmateutes* (probably an imperial estate agent in this context) Agapetus.[12] Another inscription from the same village mentions the curator Lazarus.[13] These inscriptions, alongside Julian's importation of grain from the region, suggest that there was an imperial estate centre at Anasartha.

In central and southern Oriens, the imperial Klima Anatolikon stretched from the Chalkidike and the region of Salamias to the Red Sea and comprised the eastern *limes* where steppe lands and a semi-desert landscape gave way to the true desert of north-western Arabia. The Klima Anatolikon straddled both Phoenice Libanensis and Arabia. The former province also contained the *klimata* of Iabruda and Maglula (*Georg. Cyp.* 990, 993) that lay mainly on the eastern flank of the Anti-Lebanon mountains north of Damascus.[14] Also within Phoenice, the Saltus Gonaiticus was situated in the southern Massyas (al-Biqaᶜ) Valley, which comprised the territory of the former petty kingdom of Chalcis. Further south was the Saltus Constantinianus in the Negev region of Palaestina Prima and the Saltus Hieraticus in Palaestina Tertia.[15] According to George of Cyprus (*Georg. Cyp.* 1075), the Saltus Bataneōs lay in Arabia. The Klima Dysmōn covered a large portion of the Sinai. Several other old domains in Palestine had been organized into *regiones* after the Roman takeover of

[9] As reconstructed by E. Honigmann, 'Studien zur Notitia Antiochena', *BZ* 25 (1925), 73, from later eastern sources.

[10] M. M. Mango, 'Artistic Patronage in the Roman Diocese of Oriens, 313–641 AD', D.Phil. thesis (Oxford, 1984), 272; G. Tchalenko, *Villages antiques de la Syrie du Nord: le massif du Bélus à l'époque romaine* (Paris, 1953), I.116–18; II.41–3. Magnus is also known from Evagrius, *Hist. Eccl.* ed. J. Bidez and L. Parmentier, II.206, 31 = *Ecclesiastical History*, trans. Whitby, p. 269, where it is stated that he was given charge of an imperial estate by Justin II.

[11] G. Zacos, A. Veglery, and J. W. Nesbitt, *Byzantine Lead Seals* (Basel, 1972), nos. 130–1. On Magnus, see D. Feissel, 'Magnus, Mégas et les curateurs des "maisons divines" de Justin II à Maurice', *TM* 9 (1985).

[12] Butler et al., *Syria*, vol. III, no. 881.

[13] Ibid., no. 885.

[14] Jones, *Cities of the Eastern Roman Provinces*, 256. They had formed part of the possessions of the Ituraean kingdom before its absorption. Jones saw as probably corrupt Georgius' inclusion of Gonaiticus in both the provinces of Phoenice and Phoenice Libanensis, but I see no reason why this extraterritorial administrative unit could not have crossed provincial boundaries.

[15] *Georg. Cyp.* 1026–7. For the attribution of these salti to the land behind Gaza, see Y. Tsafrir, L. Di Segni, and J. Green, *Tabula Imperii Romani: Iudea Palaestina* (Jerusalem, 1994), 220.

Herodian lands in the first century: Josephus (*Ant.* 9.7.11) refers to the royal farms around Jericho that produced balsam, some of which continued to do so into the late empire.[16]

This brief survey has provided some notion of the range and extent of imperial landholdings. Their vast expanse made the state the largest landowner in Oriens; this fact has great implications for the settlement and exploitation of the eastern landscape.

Villages and Farms

Most farmers in the Roman–Byzantine East lived in villages.[17] Even a cursory look at the Madaba Map or the Limestone Massif village settlement patterns, convinces one that the Late Roman East was studded with rural settlements; it was a landscape of villages. In the fourth century, Eusebius distinguished between the *polichnē* (πολίχνη) (*Onom.* 20) a small city, and the *polis.*[18] This may simply have been a way to separate the major cities, like Scythopolis, from smaller urban entities like Zoara, which had not long before been villages, but also may indicate differences in the administrative capacities of such settlements. Since the early Roman period, there had been large villages with significant populations and economic importance, but they lacked civic status. In some areas like the Hauran, where cities were relatively rare, large villages were a vital part of the countryside. Rural settlements called μητροκομίαι (mother-villages) apparently had administrative responsibilities over the komai in their vicinity, much as poleis were usually responsible for the villages in their surrounding region (*chorion*).

16 These balsam groves were known to Jerome, who noted them at Engeddi, which was a large Jewish village in Late Antiquity. See Tsafrir, Di Segni and Green, *Tabula Imperii Romani*, 121; J. Wilkinson, *Jerusalem Pilgrims Before the Crusades* (Warminster, 1977), 50. The *regiones* listed in *Georg. Cyp.* 1016–19 are Amathus, Jericho, Livias, and Gadara.

17 Farmers lived not only in hamlets and villages but also in or nearby cities. In AD 542, the fortress of Callinicum fell to the Persians, who thus captured a number of farmers who had sought protection there during the invasion: N. Pollard, *Soldiers, Cities, and Civilians in Roman Syria* (Ann Arbor, 2000), 294. Some of these cultivators probably made their home within the town walls, as did some at the North African town of Syllektos. According to Theophanes (*Chron.* ed. de Boor, p. 191 = Mango and Scott, p. 289), Belisarius captured that place when he followed the peasants into the city after the day's work. In the Egyptian city of Hermopolis, tillers lived in the town and worked the surrounding fields. On this see Bagnall, *Egypt in Late Antiquity*, 50. The physical separation between city and country was variable. In one instance the transition might be gradual, from acropolis and built-urban space, to *proasteion*, to orchard and vineyard, and thence to open fields. In other instances, market gardens ran up to many city walls. Gardens were tended within most cities, and within even the greatest metropolis there was constant contact between the urban denizen and the rustic with all his or her produce, animals, and manners.

18 M. Piccirillo, 'Rural Settlements in Byzantine Jordan', in A. Hadidi (ed.), *SHAJ II* (Amman, 1985), 260.

Villages varied considerably in size. Eusebius also talks of 'large villages' (κῶμαι μεγίσται/μεγάλαι), such as Beersheva (ancient Bērsabe) (*Onom.* 50), which at certain times and places undoubtedly functioned as the *mētrokōmiai*, known epigraphically from south Syria.[19] Such places might have a lively social life and even administrative machinery. In their *epoikia*, the wealthy of Antioch built taverns and churches.[20] In the Hauran and Negev villages, written evidence attests to the presence of village officials involved in the daily management of local government affairs.[21] Minor local administrative functions proved durable and continued widely beyond the Islamic conquests, especially in less urbanized regions, like the Hauran, the Chalcidike, and the Jazira. A case in point is known from Nessana, where, around AD 680, George, the *dioiketēs*, wrote to a friend instructing him to provide chaff to a certain Abraham.[22]

We can discern changes in population and political organization through the life cycle of these villages. Far from stagnant, these settlements sometimes grew into cities. It was not unusual for a *megakomē* to make the leap to city status during Late Antiquity: such was the case for Maximianopolis in the Hauran and Zoara in Palestine III as well as Anasartha in eastern Syria I. The elevation of Maximianopolis was perhaps artificial, dependent on the whims of an emperor, but the cases of Zoara and Anasartha certainly reflected a more enduring socio-economic and demographic prosperity.

Large villages have left impressive archaeological traces, as at Korazim in the Golan and Kaper Barada (Brad) in North Syria. Brad's remains presently cover more than 25 hectares. The Negev villages[23] belong to this category, as well as many settlements in the Hauran. In addition to sizeable populations, these centres might contain other urban amenities, such as churches, bath houses, or fortifications.[24] Rural centres of this kind generally possessed a

[19] H. I. Macadam, 'Cities, Villages and Veteran Settlements: Roman Administration of the Syrian Hawran', in D. Panzac (ed.), *Histoire économique et sociale de l'Empire ottoman et de la turquie (1326–1960): Actes du sixième congrès international, Aix-en-Provence, 1–4 juillet 1992)* (Paris, 1995), 647.

[20] John Chrysostom, *In Acta Apost.* 18: *PG* 60.149ff.

[21] Local councils seem to have been a commonplace. For an overview of the Hauran epigraphic evidence and village administration, see H. I. Macadam, 'Epigraphy and Village Life in Southern Syria During the Roman and Byzantine Periods', *Berytus* 31 (1983) and S. M. Moors, 'De Decapolis: steden en dorpen in de Romeinse provincies Syria en Arabia', Ph.D. diss. (Leiden, 1992), 376–419.

[22] He was διοικι[τῇ] Ν[εςτάνων in P.Ness. III.68.7.

[23] Often called cities or towns, e.g. A. Segal, *The Byzantine City of Shivta (Esbeita), Negev Desert, Israel* (Oxford, 1983) we restrict our use of 'city' and 'town' to settlements with civic status in the eyes of the central government. Elusa and Mampsis appear in the city lists of George of Cyprus, *Desc. orb. rom.* 1049–50 and of Hierocles, *Synecdemus*, ed. Mommsen, 721.10, 721.8.

[24] Nessana, for example, is thought to have been home to about 3,000 people at its sixth–seventh-century peak. See Shereshevski, *Byzantine Urban Settlements*, table 19. Amongst the

larger population than the simple *komē* (κώμη) or *agridion* (ἀγρίδιον), which comprised the majority of rural settlements. Most of the latter were modest in scale: the Jewish village of Sumaqa, covering 3–6 hectares, is probably representative in size of most villages in Oriens.[25]

Villages, both large and small, were in plan either nucleated or dispersed.[26] The nucleated village, where houses with adjoining walls are situated around a focal point, often a house or open area, are a common feature of the Byzantine Near East. Nucleated villages frequently developed in a relatively short period of time around a family dwelling; as the family grew, new houses were built and additional relatives and dependants built clusters of settlement around the original focal point of the village. Many of the villages of northern Syria are of this type, while Ḥorvat Kanaf (Figure 2.1) in Galilee and Qatzrin in the Golan are good examples in Palestine-Phoenice.[27]

Dispersed villages were those that grew up without a central focus, such as Kafr Naffakh (Figure 2.2) in the Golan. Such settlements are not defined by houses scattered far and wide, but rather by a lack of clear nucleus or convergence of routes in obvious centres within the villages. Typically dispersed villages began as a loose collection of houses with some distance between them, perhaps reflecting the dwellings of unrelated or less tightly allied families. Over time the empty space was filled in by buildings, and these were often cramped due to limited building space available. As they grew and changed over time, dispersed villages came to resemble a closed settlement, with houses and courtyards pressed against one another and houses accessed through narrow alleys.

Negev settlements Oboda (Avdat) and Rehovot-in-the-Negev (Khirbat Ruheibeh) each had public baths. See A. Musil, *Arabia Petraea* (Vienna, 1907), II.75 fig. 46, 79, 95 fig. 66, 106; A. Negev, *The Architecture of Oboda* (Jerusalem, 1997), 170–6.

25 S. Dar, *Sumaqa: A Roman and Byzantine Jewish village on Mount Carmel, Israel* (Oxford, 1999), 11.

26 The categories and village-discussion are based on those of Y. Hirschfeld, 'Farms and Villages in Byzantine Palestine', *DOP* 57 (1997), 33–77, whom I follow for Palestine, especially the Golan Heights. They are serviceable generally across Greater Syria. G. Tate, 'Les campagnes de la Syrie du nord à l'époque proto-byzantine', in C. Abadie-Reynal (ed.), *Hommes et richesses dans l'Empire byzantin* (Paris, 1989), offers more detail for the north Syria region. A general overview of the post-Roman world is offered by C. Wickham, *Framing the Early Middle Ages: Europe and the Mediterranean 400–800* (Oxford, 2005), 442–518. For valuable modern comparanda of (pre-industrial) village life see S. S. Reich, *Études sur les villages araméens de l'Anti-Liban. Avec 32 planches en phototypie et 33 figures* (Damascus, 1937); R. L. Thoumin, *La maison syrienne dans la plaine hauranaise, le bassin du Barada et sur les plateaux du Qalamun* (Paris, 1932); R. L. Thoumin, *Géographie humaine de la Syrie centrale* (Paris, 1936); J. Weulersse, *Le pays des Alaouites* (Tours, 1940); J. Weulersse, *Paysans de Syrie et du Proche-Orient* (Paris, 1944).

27 Hirschfeld, 'Farms and Villages', 41; A. Killebrew, 'Qasrin: The Village', in E. Stern (ed.), *NEAEHL* (Jerusalem, New York, and London, 1993), 1222–4.

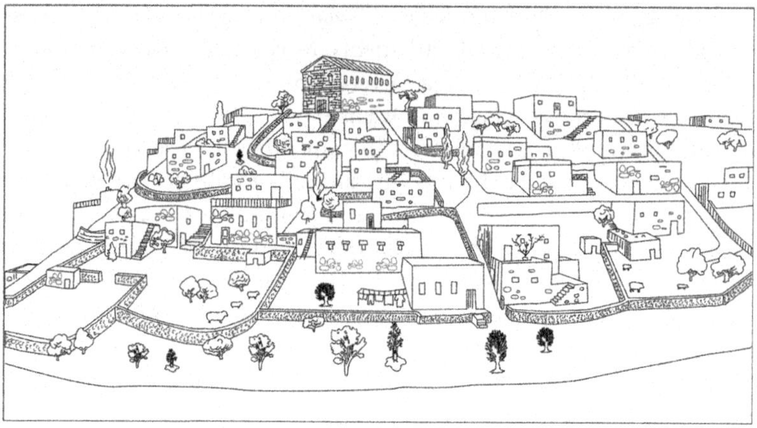

Figure 2.1. Horvat Kanaf village (reconstruction by Laurie Gross on the basis of Ritmeyer, from archives of Z.U. Ma'oz in Hirschfeld, 'Water Supply', fig. 10).

The farmers, traders, craftspeople, and others, who belonged to the small settlements (numbering in the hundreds throughout the Levant), most often lived in single-family units (monks providing a rare exception). The village house was the basic 'building-block' in terms of the physical and social structure of the village as it contained the fundamental unit of population, the nuclear family. There is thus a great deal to be learnt about the eastern landscape from a closer look at the physical remains of these houses.

The Village House (oikos)[28]

Most farmers lived in the same building in which their tools and livestock were kept and their crops processed and stored. Often these houses were in villages, but sometimes they were in cities or isolated in the countryside. Modern scholarship has made several attempts to divide village houses into broad types. For the Late Roman western Mediterranean, Lewit classified dwellings according to the area they occupied, while Hirschfeld, in his study of the Palestinian house, used morphological characteristics to distinguish various sorts of domestic space.[29] Since late antique dwellings exhibit rich variety, I have combined these two approaches to attempt a more refined view.

[28] The term '*oikos*' (οἶκος) is applied in its broadest sense: a dwelling. Y. Hirschfeld, *The Palestinian Dwelling in the Roman–Byzantine Period* (Jerusalem, 1995), 21–3.

[29] T. Lewit, *Agricultural Production in the Roman Economy A.D. 200–400* (Oxford, 1991), 22–3 and Hirschfeld, *Palestinian Dwelling*. Lewit's study of the archaeological data of the western

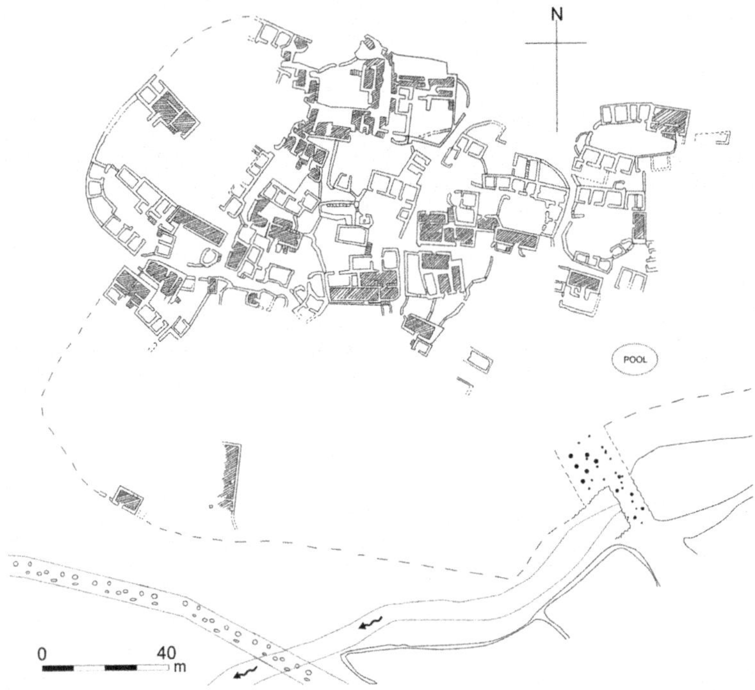

Figure 2.2. Kafr Naffakh (after Dauphin, 'Seventh Century Measuring Rod', p. 12, fig. 2).

Various types of rural houses will be described and related, and their agricultural features discussed. I have tried to use the difficult terms 'farmhouse' and 'farmstead' sparingly. The former I use to describe a rural dwelling (including within villages) with identifiable agricultural components. Obviously, such a definition would probably encompass the majority of houses in rural Oriens, which often provided shelter and agricultural workspace simultaneously. Isolated farmhouses or agricultural complexes I have termed 'farmstead' and describe below. In the secondary literature, large farmsteads

empire attempted to rationalize the classifications of farm buildings by placing them into the categories of 'small' (under 500 m^2), 'medium' (500–2,500 m^2) and 'large' (over 2,500 m^2). Since I do not factor in courtyard space, my categories occupy slightly smaller areas. The categories in Hirschfeld, *Palestinian Dwelling*, 21–3, are: (1) The simple house, the most basic and common dwelling in the Roman–Byzantine period in Palestine, which is comprised of a one-room core dwelling area and a courtyard attached to front or back. (2) The complex house, which is generally a simple house expanded through the addition of dwelling space on three or more sides of the courtyard. (3) The courtyard house, which is characterized by a courtyard surrounded on all four sides by domestic quarters.

often have been called 'villas', but this designation is problematic and I have tried to avoid it.[30]

Domestic structures are a reliable indicator of rural wealth and as such they reflect the status and the tastes of their builder. Dwellings also offer surrogate data regarding land tenure and farming practice. The aims of the following outline of small, medium, and large houses are to illustrate the range of characteristics prevailing in the houses of Oriens and to discuss some of those features within the context of their agricultural function.[31]

A fifth–sixth-century house in the village of Ḥorvat Kanaf, 5 km north-east of the Sea of Galilee, typifies simple small-house architecture. The house is irregular in plan and possesses a front courtyard and two living rooms for domestic activities. A 'window wall', an inner wall pierced by low openings in its base, separated the front room from the second, larger room. Sleeping quarters were in a loft above the latter. Two small rooms to the rear of the house served for storage.[32] Such half-open dwellings were common throughout the East, both in villages and in isolation. A good example of the latter is the small farmstead from Nahal Govta on Mount Hermon in Phoenicia (Figure 2.3). The house is rectangular in shape and built of large, crudely dressed stones. The total interior space is 45 m^2, divided into two sections comprising a main living area and a smaller mosaic-paved room, possibly a winepress. Fronting the house entrance on the west is a courtyard, on the northern side of which is a sheepfold. Slightly east of the dwelling is a small tower, probably used for storage of agricultural produce of the small farm, covering an area of 0.2–0.3 ha.[33] Other good examples of small rural dwellings (ranging in floor area from 30 to 150 m^2) are found at

[30] 'Villa' seems to best describe a planned complex exhibiting elaborate architectural components, such as *impluvia*, bath houses, and pleasure gardens in addition to other decorative features. Tchalenko referred to the larger dwellings in the Limestone Massif as 'villas': Tchalenko, *Villages*, I.356 n. 6. Tate pointed out the inadequacy of such a designation, and refers to such buildings simply as 'house', too vague to be of use in the present study, see G. Tate, *Les campagnes de la Syrie du Nord du IIe au VIIe siècle* (Paris, 1992), 15. The use of the terms 'farmhouse' and 'farmstead' are not without problems, but this terminology seems the best English classification of certain dwelling types. While certainly not all of the houses known from survey in Oriens are those of farmers, the vast majority undoubtedly were. The sample presented here will provide a useful indicator of general types and demonstrate how I have sought to control closely the application of these terms.

[31] Village houses often preserve traces of agricultural use, such as troughs and stables on the ground floor.

[32] Hirschfeld, 'Farms and Villages', 27–8.

[33] S. Dar, 'The History of the Hermon Settlements', *PEQ* 120 (1988), 140–4; Hirschfeld, 'Farms and Villages', 40.

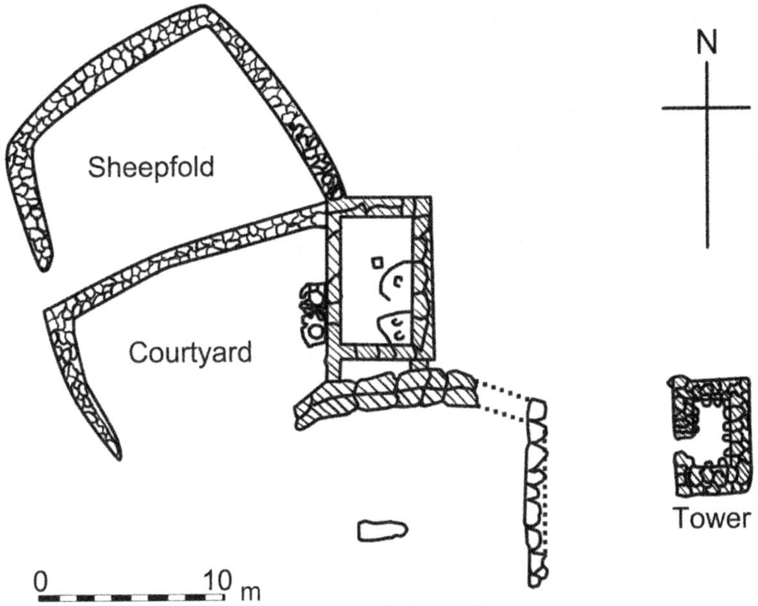

Figure 2.3. Nahal Govta (after Dar, *Settlements*, p. 141).

Öküzlü (Cilicia II), Mylai (Cilicia I), Wakim (Arabia), and Umm Rihan (Palestine I).[34]

An example of a medium house in Oriens is the tower-house at al-Safiyeh in the Hauran (Figure 2.4). This dwelling was constructed of large, finely squared basalt with a total area of approximately 288 m^2.[35] The house itself is of two storeys; the tower rose to a third storey. The latter served for storage and as a refuge and lookout. The rear of the house's ground floor contained a stable with mangers incorporated into the wall. Another medium-sized farmhouse, that at Qirqbize in Syria I, is a towered house with open courtyard and winepress. This dwelling, though relatively modest in dimension (its total estimated area of about 156 m^2 is on the smaller side for a 'medium' house) possesses architectural features often found in much larger *proasteia* (described below) including decorated stone panels and a portico. Other houses representative of this size are found in Palestine II at Pella (House G—an urban example) and in Arabia at Luben.[36] Houses of this size exist in an array

[34] Cilicia: S. Eyice, 'Un site byzantine de la Cilicie: Öküzlü et ses basiliques', in L. Hadermann-Misguich, G. Raepsaet, and G. Cambier (eds.), *Rayonnement Grec: Hommages à Charles Delvoye* (Brussels, 1982), 360–1; Wakim: Butler, et al., *Syria*, II.A.417–18; Umm Rihan: Hirschfeld, *Palestinian Dwelling*, 34–5.

[35] Butler et al., *Syria*, II.A.126–7.

[36] B. Hennessy, A. McNicoll, and R. H. Smith, *Pella in Jordan* (Canberra, 1982), 130–9; Hirschfeld, *Palestinian Dwelling*, 69–70.

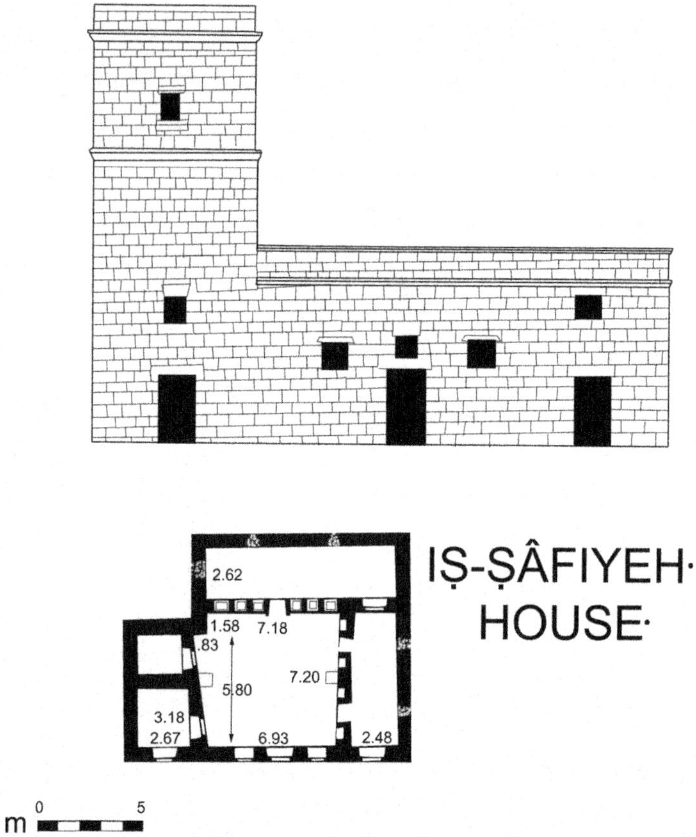

Figure 2.4. Medium-sized house at as-Safiya (after Butler et al., *Syria*, II.A ill. 100).

of settlement and geographic contexts. That they differ from one another in layout and construction method underscores the variety apparent even within a relatively narrow group of houses in late antique Oriens. This indicates substantial local flavour and individual taste in the country villages of the Roman–Byzantine East.

Some of the better examples of large houses within Levantine villages are preserved at Umm el-Jimal. House XVIII is a large dwelling (in excess of 650 m^2 roofed space) representative of Nabataean-Syrian architecture of the sixth century.[37] Apartments surround a central rectangular courtyard (24×19 m), a typical arrangement of houses in the Hauran.[38] Separate

[37] In its present form, the village of Umm el-Jimal still preserves a great deal of the sixth-century village: see B. De Vries, 'Location, Plan and Design of the Late-Antique Town', in B. De Vries and J. W. Betlyon (eds.), *Umm el-Jimal: A Frontier Town and its Landscape in Northern Jordan* (Portsmouth, RI, 1998). The date of House XVIII is probably Early Byzantine. Occupation continued into the Early Islamic period (*c.*AD 661–800).

[38] M. R. Brown, 'A Large Residence (House XVIII)', in De Vries and Betlyon, *Umm el-Jimal*, is the basis for the description.

apartment blocks constitute the main dwelling areas; these were upper-storey living spaces that were entered through external, cantilevered stone staircases, another hallmark of Hauran houses. The lack of mangers, typical of many ground-floor rooms of Nabataean and later Hauran type houses, indicates that the ground floors may have been used for human living space or for storage. The House XVIII courtyard preserves remnants of mangers where livestock were kept. A relatively modest-sized internal cistern ($1.25 \times 1.7 \times 1.4$ m) was supplemented by the external reservoir recorded on plans of the building.[39]

Village Houses and Village Wealth

Houses similar in scale to House XVIII at Umm el-Jimal, such as one at Jemerrin in Arabia (> 1100 m^2) and House 10 at Serjilla in Syria II (on the lowest end of the large house, about 615 m^2), provide further examples of larger-scale dwellings in the villages of late antique Oriens.[40] Big dwellings like House XVIII stand in marked contrast to the small dwelling found at Horvat Shema' in Upper Galilee (Phoenice), whose two-storey dwelling area measures a modest 40 m^2.[41] Such variation in size can be both internal to specific settlements, and regional: local building choices might have depended on the overall wealth of a given area, the available building materials, or local architectural traditions.[42]

House architecture is generally a reliable indicator of wealth. Few would argue that the original owner(s) of House XVIII at Umm el-Jimal had greater economic resources than those who built the dwelling at Horvat Shema'. Within a given village context, contemporary houses can be compared with one another to give some indication of the differences in wealth of their owners. At least one writer in Late Antiquity thought that one's house should be appropriate to one's station and landed wealth. Palladius (1.7) stated that the value of a property should be reflected in its buildings, as the expense of its maintenance could far exceed initial construction costs.[43] Palladius thus indicates that there existed a social attitude throughout Late Antiquity in which one's dwelling at least partly reflected one's means.

[39] Ibid., fig. 139.

[40] Butler et al., *Syria*, II.B.127–9; I.A.300–2.

[41] Cohen, *Map of Sede-Boqer—East (168)*, 41; Hirschfeld, *Palestinian Dwelling*, 34, fig 10.

[42] Social choices determined by the wants and needs of individual families or economic arrangements are less readily detected in the archaeological data. Does the division of dwelling space in House XVIII represent the segregated quarters of a single, wealthy owner's family; or is it the dwelling of separate groups within a single extended family; or does it represent a block of apartments rented to tenants by a landlord, who may or may not have also lived in the house? Material evidence alone often cannot answer these questions.

[43] Palladius, *Opus Agriculturae*, ed. Rodgers.

A desire for a degree of display in village architecture is obvious in the more than 700 villages of the Limestone Massif of North Syria. There one finds the remains of thousands of houses, the majority of which are built of fine limestone ashlars. Many dwellings are of such a durable quality that today many of the ruins still stand two-storeys high. The Limestone Massif domiciles often have architectural decoration, consisting of carved mouldings and medallions and ornate capitals.[44] It is possible that these buildings were held in tenancy, but the quality of construction and their decorative features argue for private ownership, or at the least for long-term leases.

The chief building element of the houses—the ubiquitous, attractive limestone—was easily available in regional quarries, but the houses were nonetheless expensive to build and represent significant outlays of capital. Illustrative of this is the church at Khirbet Hassan (on the Jebel Bariša in Syria I), which is recorded to have been built at a cost of 580 solidi, plus 430 modii of grain, beans and lentils.[45] In its form, scale, and manner of construction, Khirbet Hassan does not differ substantially from many houses in the Limestone Massif. Its interior space, about 220 m^2, is similar to medium-sized houses in the region, such as House 2 at Refade (204 m^2) and a house noted by De Vogüé at Mujeleia (264 m^2).[46] Many houses in the region were several times larger. It does not seem far from the mark, then, to suggest that the fine limestone houses of modest size, hundreds of which remain in the Limestone Massif, represent sizeable capital investments of well over 200 solidi in cost.

Even stone buildings of a kind humbler than most in the Limestone Massif were not inexpensive. A private rural tower at Imtan (ancient Mothanae) was constructed in AD 485/486 at a cost of 40 solidi. A tower like this probably took a team of builders 300–400 days to complete.[47] This specific one is known to have required the services of a professional architect (*οἰκοδόμος*). In the mid-to-late second century, the Rabbinic literature noted (TB Ket. 91b)

[44] Tate, *Campagnes*, 98–107.

[45] J.-P. Sodini, 'Les églises de la Syria du Nord', in J. M. Dentzer and W. Orthmann (eds.), *Archéologie et histoire de la Syrie*, II: *La Syrie de l'époque achéménide à l'avènement de l'Islam* (Saarbrücken, 1989), 348. The value of the crops is fairly insignificant when compared with the value of the coin. Assuming an equal division of the three, the grain (assuming wheat) would have been 9.5 solidi and about the same for the lentils, with the beans worth a bit under 6 solidi, for a rough total of 25 solidi.

[46] Admittedly, a church building price may well have included the equipping of the church with precious silver furnishings, and this might account for as much as half of the building costs at Khirbet Hassan. Yet even if we allow that the furnishings of the church at Khirbet Hassan were uncommonly rich, their cost is unlikely to have exceeded half the value of the whole, Mango, *Silver from Early Byzantium*, 7–13.

[47] *SEG* VII.1141. Since such towers often were an integral part of many medium and large houses and since many possessed multiple towers, this inscription is valuable because it provides evidence of the cost of a typical portion of a rural house. Man hours are calculated by comparison with similar towers in Samaria. S. Dar and S. Applebaum, *Landscape and Pattern: An Archaeological Survey of Samaria 800 B.C.E.–636 C.E.* (Oxford, 1986), 95.

that a house in Palestine cost 10 gold aurei.[48] What these figures demonstrate is that anything more than the most austere, mud-brick houses could represent a significant capital outlay. In the fifth century, for example, an artisan might make just 1/4 solidus per month. A common dwelling in the Limestone Massif, therefore, commanded a value that exceeded many times over a skilled worker's annual salary.

Differences in scale and form in house architecture tell us something about the degree of economic differentiation amongst village inhabitants in North Syria in Late Antiquity. That some small-scale farmers were better off than others is certain; by the standards of the day some were apparently wealthy and able to afford substantial dwellings, or oil- and winepresses. As noted above, they also participated in local community life, assisted in church foundations, and offered precious liturgical fittings to sanctuaries. Whatever the legal status of these farmers (i.e. whether *coloni* or emphyteutic leaseholders or freeholders), village architecture, and the accounts of Libanius' problems with his recalcitrant dependants, speak of a resilient and remarkably independent rural milieu in Late Antiquity.

The material data add considerably to the textual evidence. Village houses, churches, and public buildings, such as bath houses, are markers of sizeable capital outlay by small communities. Many rural sites, such as Nessana and Androna, had public baths and multiple, substantially built churches. It is unlikely that much of the money needed to build and service these structures came from outside the local community. The presence of substantial wealth in villages is nowhere more evident than in North Syria. Excavations of 'Site 13', located north-east of Antioch on the plain below the Jebel Zawiye, uncovered a hoard of 534 Byzantine gold coins ranging in date from Maurice (582–602) through Constantine IV (668–85), with Heraclian issues being the most numerous.[49] Better known is the impressive collection of more than fifty-six ecclesiastical vessels of the sixth–seventh centuries. The treasure, weighing 80 lb, was found in Kaper Koraon in northern Syria and belonged to the village church of St Sergius. Some donations, such as that by the imperial

[48] A modius of grain cost about 1/25 of an aureus, comparable to the later gold/grain ratio of wheat costing 1/15–1/40 solidus per modius. For the sake of simplicity, the purchasing power of an aureus and solidus are considered more or less equivalent, but this is a rule of thumb and is, of course, speculative. D. Sperber, *Roman Palestine, 200–400: Money and Prices* (Ramat-Gan, Israel, 1974), 108. Comparative prices for houses in Late Antiquity: a house in Hermopolis was sold in AD 266 for 2,000 silver drachmas while a 'mansion' in Babylonia in AD 320–50 cost 500 drachmas. An indication of relative value can be found at Theadelphia in Egypt, where in AD 256, one *artaba* of wheat fetched 9/10 of a drachma. In fourth-century Babylonia, a *griva* (roughly a modius) of wheat sold for about 1/4 of a drachma. Sperber, *Money and Prices*, 146. The Babylonian house was thus probably worth in the order of 133 solidi.

[49] C. Foss, 'Syria in Transition, A.D. 550–750: An Archaeological Approach', *DOP* 51 (1997), 227; Morrisson, 'Le trésor byzantin'.

curator Megas, overseer of imperial estates of the nearby domain of Hormisdas, were from men with outside interests. But there were other, less prominent donors who were certainly natives of the villages they favoured with their gifts.[50] These manifestations of widespread and varying wealth contrast with the conventional picture of a small elite, monopolizing rural resources, with which I opened this chapter.

Isolated Farmsteads

A sixth-century traveller who journeyed overland from Bostra to Antioch would have traversed a patchwork landscape. The traveller would have passed through many of the close-knit, black basalt villages, their houses looking in on one another rather than out to the fertile plain. Beyond the villages lay the carefully tended gardens and orchards that broke into the open grain fields. Our traveller might be in lonelier country, but not in an empty one: outside the village lay scattered farmsteads, military posts, way-stations, estate houses, and monasteries. As in medieval Europe, small-scale farmers in the late antique East lived most often in villages. However, the isolated farm (*αὐλή*)[51] was a common feature of the landscape of some areas of late antique Greater Syria.[52]

Besides *aulē*, *monagria* (*μοναγρία*) was also used to describe an isolated farm. In the fourth century the word appears in Alciphron (4.17.7). The latter used *monagria* in paraphrasing Philo when describing the haunts of monks which were found in 'isolated farms' and gardens.[53] In his sixth-century *vita*

[50] The economics of silver donations to Syrian churches are detailed by Mango, *Silver from Early Byzantium*, 7–13, who estimates that the surviving articles of Megas' contribution at Kaper Koraon were worth 75 solidi. Contributions of silver elsewhere in Oriens were much larger: in AD 437–8, a consul offered 2,880 solidi worth of silver to a church at Edessa.

[51] Originally the word for the courtyard of the home, *αὐλή* appears as such in the Nessana Papyri (P.Ness. III.22, 29, 31). Most often, the word refers to palatial dwellings or their central halls, especially the imperial palace (Themistius, *Or.* 15.269.4; 16.290.18; Procopius, *De Bell.* 1.24.9; 3.3.5). In late antique Egypt, *aulē* often referred to an urban courtyard home, such as would have been common throughout the ancient east Mediterranean: see G. Husson, *Oikia: le vocabulaire de la maison privée en Egypte d'après les papyrus grecs* (Paris, 1983), 45–54. In rural contexts the word is used for sheepfolds: Ps.-Oppian, *Cyn.* 4.397; Synesius, *Ep.* 66.53; Sozomen, *Hist. Eccl.* 1.11.2. In the pastoral novel *Daphnis and Chloe*, written after the second century, the rural dwelling of Chloe's family is consistently called *aulē*, which underscored both the rusticity and its situation beyond the village. Stephanos, *Ethn.* ed. Meineke, 146.9–11: *Αὐλή, ἡ ἔπαυλις 'χραύσῃ μέν τ' αὐλῆς ὑπεράλμενον'. λέγεται καὶ αὖλις, καὶ σύνθετον δύσαυλις, ὁ οἰκήτωρ αὐλίτης, αἱ δ' ἐν τοῖς ἀγροῖς οἰκήσεις αὔλια.* Zosimus (*Hist. Nov.*, ed. Paschoud, 5.8.1) used the word in reference to Eutropius' confiscation of Stilicho's rural estate (*Εὐτρόπιος δὲ πρὸς πάντα Στελίχωνι συνεργήσας τὰ κατὰ τούτου βεβουλευμένα τῶν ἐν τῇ αὐλῇ πραττομένων κύριος ἦν*).

[52] There is no set of features, beyond their situation, that would differentiate these dwellings from any other rural oikos and no description of the rural aulē as a type.

[53] Benner and Fobes (Loeb edition) render *aulē* in Alciphron 4.17.7 as 'preserve', thus intellectualizing the sense of an isolated holding in the context of a lover's quarrel. Philo of

of Simeon the Mountaineer, John of Ephesus expressed astonishment at the scattered farmsteads sprinkled around upper Mesopotamia:

> and he [Simeon] saw that, though these mountains were rugged and towering, houses and domestic animals [or cattle sheds] were scattered over them all, even though they were some five or seven miles, or even as much as eight and ten miles from the village; because the boundary of that village is so extended...[54]

John's comments make it clear that, in his view, farmers lived in villages; when they did not, this was worthy of comment.

The phenomenon of isolated rural farmsteads in the landscape is indeed one of the more striking aspects of the late antique period, since such settlements are generally held to have been uncommon in other times. In Palestine, isolated rural dwellings are prominent in the late antique archaeological records.[55] A good example of such a late antique farm is that at Nahal Govta (Figure 2.3) on Mount Hermon (Phoenice).[56] This modest complex is built on a small clearing on a hill slope, and the buildings cover an area of 20×40 m. The main dwelling is quite small (external dimensions 10.5×5.3 m) and opens to a yard (9×14 m). Abutting the farmyard on the east is another enclosed workyard, while to the west of the dwelling is an area identified as an animal pen (9×13.5 m) constructed in rough-hewn local stones. The last feature of the farmstead is a small *pyrgos* (external dimensions 4×5.8 m) with walls 1 m thick. Rectangular houses of this type (termed 'Ituraean' by Shimon Dar) were common throughout the foothills of Mount Hermon and attest to the practice there of mixed arable agriculture and herding.

The farmstead of Ramat Hanadiv (Figures 2.5, 2.6, 2.7) in the foothills of the Carmel (Phoenice) is focused on a central courtyard. The surrounding ground floor is devoted to storage and agricultural activities, with vaulted chambers one storey high on the western side of the building. East of the courtyard are a series of three chambers consisting of stables and a cellar. To the west of the courtyard lie the main entrance and store-rooms, whilst directly north is a second stable area and to the north-west a chamber (4×12.8 m) which was interpreted as a wine cellar by the excavator. Living

Alexandria mentions the *monagria* (presumably he was thinking of those in Hellenistic Egypt) on two occasions (*De Abrahamo* 23.3; *De vita contemplativa* 20.1) and Eusebius juxtaposed them with gardens (ἐν μοναγρίοις καὶ κήποις) as places beyond the social life of people, that is, at the margins of town and village settlement.

[54] John of Ephesus, *Lives of the Eastern Saints*, trans. Brooks, 232. Translation slightly adapted.

[55] Hirschfeld, 'Farms and Villages', 48–9.

[56] Dar, 'Hermon Settlements', 140–4, on which the description is based.

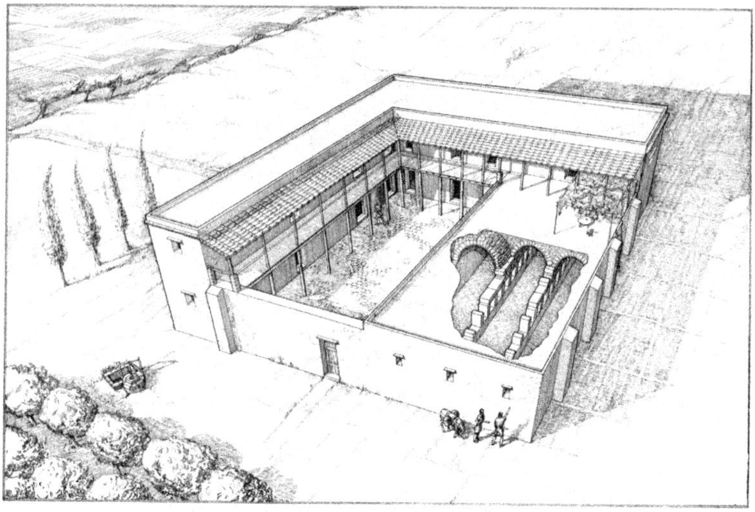

Figure 2.5. Ramat Hanadiv: reconstruction of the Byzantine farmhouse (by Erez Cohen from Hirschfeld, *Ramat Hanadiv Excavations*, fig. 124. Reproduced by permission of The Israel Exploration Society, Jerusalem.)

quarters were on upper floors along the two wings flanking the northern and eastern side of the courtyard.[57]

In the early sixth century, important changes occurred at Ramat Hanadiv. The southern parts of the complex were damaged and rebuilt in masonry of a lower quality than that of the original structure. Dwelling units were built within the central courtyard that effectively limited the open space. A second courtyard, probably intended to replace the first, was then built to the south of the building.[58] Thus, whilst the farm at Ramat Hanadiv started out as a relatively open, well-constructed farm, later expansion phases of an inferior quality probably reflect *ad hoc* measures to accommodate family members.

Ramat Hanadiv is important for several reasons. It has been completely excavated and its findings well published by Hirschfeld. While finely built and relatively large (528 m^2 in total area), the house at Ramat Hanadiv is not ostentatious and lacks the trappings we expect in rich oikoi, such as bath houses and mosaics. The remains of a suburban villa (*proasteion*) discovered at nearby Caesarea apparently did have such features.[59] Ramat Hanadiv thus provides a southern parallel for the well-built stone houses of the Limestone Massif with which it shares a general functionality, an architectural quality,

[57] Hirschfeld, 'Farms and Villages', 44.

[58] Y. Hirschfeld, 'Ramat Ha-Nadiv', in E. Stern (ed.), *NEAEHL* (Jerusalem, New York, and London, 1993), 1260.

[59] Holum, *King Herod's Dream*, 181–4; Y. Hirschfeld and R. Birgercalderon, 'Early Roman and Byzantine Estates Near Caesarea', *IEJ* 41/1–3 (1991).

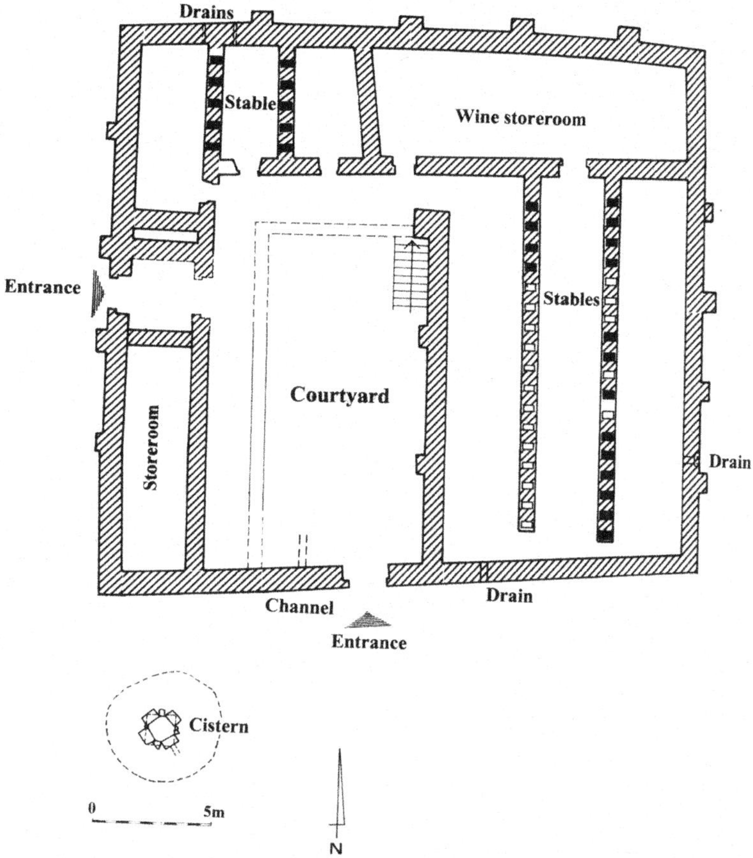

Figure 2.6. Ramat Hanadiv: plan of the ground floor of the Byzantine farmhouse (from Hirschfeld, *Ramat Hanadiv Excavations*, fig. 125. Reproduced by permission of The Israel Exploration Society, Jerusalem.)

and a somewhat marginal agricultural situation. Its owner, though prosperous, was not likely to have been amongst the richest of Caesarea. The building presents a picture of a working farmstead and family home of those without great wealth, but surviving well above subsistence.

Of what size were the farms worked by the small and middling farmers of Late Antiquity? At Qawaret beni Hassan, in Samaria, Dar estimates that the average family farm was 17.5–20 iugera (4.5–5 ha).[60] The farm at Wadi Kurnub, investigated by Evenari, was 40–48 iugera in size (10–12 ha).[61] However, the village lands of Qirqbize, if distributed equally, yield an average

[60] Dar and Applebaum, *Landscape and Pattern*, 247.
[61] Evenari, Shanan, and Tadmor, *The Negev*, 210.

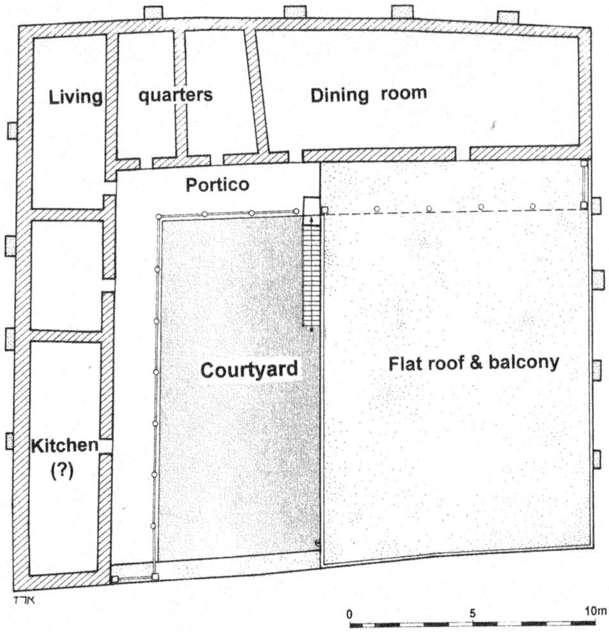

Figure 2.7. Ramat Hanadiv: plan of the upper storey of the Byzantine farmhouse (from Hirschfeld, *Ramat Hanadiv Excavations*, fig. 125. Reproduced by permission of The Israel Exploration Society, Jerusalem.)

holding of only 7.2 iugera (1.8 ha) per household,[62] which is considerably smaller than the average size of farms granted to discharged veterans, of about 50 iugera (12.6 ha).[63] Many farms were much smaller still; for example the reconstructed area of the site called Michael's Farm in the Negev is only 2.8 iugera (0.7 ha).[64] How such small farms operated and survived will be discussed when exploring subsistence (Chapter 3).

Ecclesiastical and Monastic Properties

Ecclesiastical and monastic properties were widespread and often substantial, probably second in extent only to those of the imperial government. From the time of Constantine I (324–37), the church benefited considerably from

[62] Based on 217 people in the village, with 6 per family, exploiting an arable surface of 65 ha.

[63] According to *CTh* 7.20.3, veterans were granted 100 *modii* of seed corn, which implies an area of 50 *iugera* (2 *modii* sown per *iugerum*).

[64] Evenari, Shanan, and Tadmor, *The Negev*, 105–8. Papyri from third-century Philadelphia in the Egyptian Fayum record holdings from 1–87 *arourae* in size (0.272–23.9 ha), A. K. Bowman, 'Landholding in the Hermopolite Nome in the Fourth Century A.D.', *JRS* 75 (1985), 151.

imperial largess, and a major economic boon to the church was that it paid no tax on its properties (*CTh* 11.1). Like the emperor, ecclesiastical authorities rented out their lands in emphyteusis (*Novella* 120). According to Zachariah of Mytilene, Anastasius (AD 491–518) bought the land from the church at Amida on which the fortress-city of Dara was built. The earlier village of Dara, with its surrounding fields and pasture, had thus formed an ecclesiastical *epoikion*. Following the state's purchase of the village, the emperor provided houses and land for cultivators of the village.[65] In AD 410–20, during the time of the founding of Martyropolis (later capital of Armenia IV), Theodosius II sent gold, silver, and skilled workers to the bishop, Marutha, and granted villages, farms, vineyards, and olive-yards to churches there.[66] The Roman church also owned large areas of lands in Oriens, including lands in the region of Tyre that produced annual revenue of about 1,500 solidi. According to the *Liber Pontificalis*, it also had property in Antioch that rendered an annual total of 572 solidi.[67]

The ecclesiastical estate in the province of Phoenice, at Shelomi in western Galilee, has been well excavated, with findings published by Dauphin.[68] The estate centre, founded in the fifth or early sixth century, reached peak prosperity in the early seventh century, but was destroyed by fire soon afterwards around the time of the Sasanian invasions of Palestine in the 610s.[69] The farm's surviving structures cover an area of approximately 500 m^2 and include an enclosure with a courtyard and fine mosaic pavements bearing grapevine and amphorae motifs. The farm produced a melange of crops, typical of Mediterranean farming systems, including grain, olives, grapes, apples, and pears. The find of an iron *threinax* (winnowing fork) attests to cereal production on the estate, which was probably geared toward self-sufficiency and perhaps cattle ranching. The Shelomi farm took part in the

65 Ps.-Zachariah Rhetor, *Chron.*, trans. Hamilton, p. 165: 'And he [Anastasius] liberated all the serfs who were in it, and granted to each of them his land and his house.'

66 Mango, *Artistic Patronage*, 414.

67 In addition to two gardens near Antioch. These produced smaller sums, of 10 and 11 solidi. Constantine presented to the see of Peter the entire island of Cordionon near Tarsus, along with its revenue of 800 solidi. On these papal properties, see R. Davis, *The Book of Pontiffs (Liber pontificalis): The Ancient Biographies of the First Ninety Roman Bishops to AD 715* (Liverpool, 1989), 19–20.

68 See C. Dauphin, 'A Seventh Century Measuring Rod from the Ecclesiastical Farm at Shelomi in Western Galilee [Israel]', *JÖB* 32/3 (1982); C. Dauphin, 'A Byzantine Ecclesiastical Farm at Shelomi', in Y. Tsafrir (ed.), *Ancient Churches Revealed* (Jerusalem, 1993); C. Dauphin and S. Kingsley, 'Ceramic Evidence for the Rise and Fall of a Late Antique Ecclesiastical Estate at Shelomi in Phoenicia Maritima', in G. C. Bottini, L. D. Segni, and L. D. Chrupcala (eds.), *One Land—Many Cultures: Archaeological Studies in Honour of S. Loffreda* (Jerusalm, 2003), who interprets this estate as having been managed under emphyteusis. Whilst this arrangement is certainly a strong possibility, the data are ambiguous as to the exact form of tenancy at Shelomi.

69 Dauphin and Kingsley, 'Ceramic Evidence', 65.

Palestinian wine trade, but as a consumer of local vintages, not as a producer, as no winepress was found in excavation.[70] Church officials, *oikonomoi*, oversaw such estates. The epigraphic record contains evidence of probable ecclesiastical estates under oikonomoi at Namara, Imtan, ʿAmra, al-Hit, Jebab, and Selemiya (Salamias) near Apamea.[71]

The increase in monastic foundations from the fourth through seventh centuries made a significant impact on cultivation and production in Oriens. The efforts of monks in the semi-arid fringes of the empire led to extensive agricultural development there. The rapid growth of monastic houses in northern Mesopotamia during Late Antiquity, especially in the Tur ʿAbdin, is a good illustration of this. Procopius of Caesarea (*Buildings* 2.10.8–3.1) catalogued Justinian's repairs and additions to the monasteries of seven monastic complexes around Dara alone.

Most monasteries participated in common agricultural functions, even if only to provide food for a community of monks who aimed for self-sufficiency. In the marginal regions of the Limestone Massif and the Judaean Desert, monasteries invested substantially in creating an agricultural infrastructure that converted sterile tracts into productive landscapes. Unfortunately, for our period and place of study, the size of an average monastic holding is generally unknown. What is certain is that monks were actively engaged in farming marginal areas. There, sterile land in theory belonged to the emperor, but in practice the brothers cultivated without hindrance. The holy men of the Judaean Desert exerted great effort to develop the landscape's agricultural and hydraulic potential. Near the Dead Sea, the monastery at Khirbet al-Quneitra is known from excavation to have possessed a reservoir, gardens, and orchards, in addition to an oil-press capable of processing a sizeable surplus of olive oil for the market.[72] The fifth- or sixth-century *wadi* terrace systems of the Chariton Monastery created arable land in the Judaean Desert, and canals carried rainwater from the short desert cloudbursts to fields and gardens.[73] Chariton was equipped with terraced fields, an aqueduct, a series of reservoirs, and cisterns, which were linked to a sophisticated irrigation system.[74] Similar irrigation schemes and orchards operated at Khirbet ed-Deir.[75] Monastic houses frequently possessed granaries, such as that at

[70] Ibid., 65 (animals), 68–9 (wine). The site thus serves as a good reminder that not all properties within known surplus-producing economic units were geared to serve the opportunities offered by overseas markets (in this case for Palestinian wines).

[71] See Waddington, nos. 2173a (Namara), 2038 (Imtan), 2091 (ʿAmra), 2124 (El Hit, dated 354), 2413m (Jebab), 2633 Selemiya.

[72] Y. Hirschfeld, *The Judean Desert Monasteries in the Byzantine Period* (New Haven, 1992), 46–7.

[73] Ibid., 150.

[74] Y. Hirschfeld, 'The Water Supply of the Monastery of Chariton', in D. Amit, J. Patrich, and Y. Hirschfeld (eds.), *The Aqueducts of Israel* (Portsmouth, RI, 2002).

[75] Y. Hirschfeld and R. Birger, 'Khirbet ed-Deir (Judean-Desert) 1981–1984', *RB* 93/2 (1986); Hirschfeld, *Judean Desert Monasteries*, 41–2, 150–1.

Khirbet ed-Deir, which could hold 5.5 m^2 of grain. Stables also were frequent components of the complexes. Stables at the monastery of Theoctistus, for example, were built into a cave 17 m long, 9 m wide, and 5 m high.[76]

The remains of the North Syrian monastery at Deir Dehes indicate that the monastery produced agricultural products on a commercial scale.[77] It housed twin oil-press with a winepress located outside. The winepress is a small, simple installation, approximately 2.8 m on each side. The oil-press however is a twin installation, with each press capable of processing approximately 5,400 litres of oil per season. The capacity of the oilery far exceeded the needs of a relatively small community, a habitation area of approximately 360 m^2 that housed a maximum of about thirty monks, and suggests surplus production for market.[78]

Estates and Their Owners

Late Antiquity was a time of intense investment in property. The old rank of landowners faced challenges from a newer eastern bureaucratic class whose wealth and influence were drawn initially from the power centres of administration, both in the capital and in the provinces. By the sixth and seventh centuries, the mandarins of the house of Flavius were in their prime, and they invested their wealth in these provinces in what Romans had always invested in: land.[79] The rhetor Evangelos, whom we learn about from Procopius of Caesarea (*Anecdota* 30.19), was not an old-money aristocrat with serious political clout: he made a fortune in the capital and invested his wealth back home. Cassianus Bassus, the author of the *Geoponica*, was another such character. A lowly *scholasticus*, Cassianus nevertheless had an estate on the rich corn-lands at the foot of the Jebel Riha in Syria II, and probably owned others as well. There also apparently were opportunities for new, local men. The *kastron* at Stabl Antar, the bath house and *kastron* at Androna, and the *horion* at al-Tuba were built by men who possessed no official titles.[80]

[76] Hirschfeld, *Judean Desert Monasteries*, 35.

[77] J.-L. Biscop, D. Orssaud, and M. M. Mango, *Deir Déhès, Monastère d'Antiochène: étude architecturale* (Beirut, 1997), 20–5.

[78] I. Peña, P. Castellana, and R. Fernandez, *Inventaire du Jebel El-A'la: recherches archéologiques dans la région des villes mortes de la Syrie du nord* (Milan, 1990), 33.

[79] This aristocracy was formed in part through incorporation of the old, landed families, and in part through their displacement. See P. J. Heather, 'New Men for New Constantines? Creating an Imperial Elite in the Eastern Mediterranean', in P. Magdalino (ed.), *New Constantines: The Rhythm of Imperial Renewal in Byzantium, 4th–13th Centuries: Papers from the Twenty-sixth Spring Symposium of Byzantine Studies, St Andrews, March 1992* (Aldershot, and Brookfield, VT, 1994), for the origins of these changes, and Banaji, *Agrarian Change* for their actions and impact in Egypt.

[80] See Butler et al., *Syria*, vol. III, nos. 915, 19.

Many local elites, it seems, found room for opportunity in the marginal regions: in hills of the Limestone Massif, Rough Cilicia, and the Golan, or in the wide open steppe south of the Euphrates. In the fourth century, Aurelius Bellicus established an agricultural stronghold in this steppe, one that offered refuge to his tenants in times of trouble, but at its heart his fortified compound was a farm, a centre from which the land could be profitably worked.[81]

Like their early Roman predecessors, the great late antique landowners were urbanites. Though there were many people in villages with lands, agricultural surplus, and portable wealth, the real economic centre of gravity in the late antique East was always the cities. This is nowhere more apparent than in Egypt, where recent work has highlighted the concentration of wealth in the hands of city dwellers.[82] In addition to the concentration of wealth in cities, it often was, at least in Egypt, gathered into a few hands. Bowman's study of data from the mid-fourth-century Hermopolite land register indicates a 'very high degree of inequality in land distribution', while Gascou and MacCoull's analysis of the sixth-century Aphrodito cadastre has produced an image of a landscape in which wealthy private individuals and monastic or ecclesiastical establishments controlled much of the city's territory.[83] Similarly, Jones concluded in his study of cadastral inscriptions from Magnesia-on-Meander in Asia Minor that this landscape was controlled mostly by large-scale landowners.[84] In later work, Duncan-Jones calculated that the wealthiest 7.5 per cent of the owners in the Magnesia survey owned 49.7 per cent of the land, with the largest estate holding a dizzying 21.6 per cent of the total.[85]

It is hard to know whether this picture, derived from neighbouring regions, should also be applied to the landscape of Oriens, but to suggest a broadly similar situation seems fair based on available sources. Men and women of

[81] R. Mouterde and A. Poidebard, *Le Limes de Chalcis: organisation de la steppe en haute Syrie romaine* (Paris, 1945), 199–200.

[82] R. S. Bagnall, 'Landholding in Late Roman Egypt: The Distribution of Wealth', *JRS* 82 (1992); Banaji, *Agrarian Change*; Sarris, 'Origins of the Manorial Economy'; Sarris, 'Rehabilitating the Great Estate'.

[83] Bowman, 'Landholding in the Hermopolite Nome'; J. Gascou and L. Maccoull, 'Le cadastre d'Aphroditô', *TM* 10 (1987), 118. The deep inequalities of wealth and attendant status dominated late antique society throughout the lands of the Roman world. See A. Marcone, 'Late Roman Social Relations', in A. Cameron and P. Garnsey (eds.), *The Cambridge Ancient History*, vol. xiii: *The Late Empire, A.D. 337–425* (Cambridge, 1998).

[84] A. H. M. Jones, 'Census Records of the Later Roman Empire', *JRS* 49 (1953).

[85] R. Duncan-Jones, 'Some Configurations of Landholding in the Roman Empire', in M. I. Finley (ed.), *Studies in Roman Property* (Cambridge, 1976), 19. While I make much of these figures here in order to demonstrate the concentration of wealth in the hands of a few rich owners, it is also imperative to note that the inscription shows that a little more than half the land was probably *not* in the hands of the greatest landholders.

Libanius' class were denounced by John Chrysostom, whose homilies make clear the tension between urban and rural, and rich and poor. As noted above, great landowners, like Libanius and the church, often owned entire villages (*epoikia*) populated by tenants who worked their estates. In addition to the village(s) populated by Jewish tenants noted above, Libanius also possessed the *Zezous chorion*, which had been purchased by his uncle Phasganius.[86] He may also have owned estates around Apamea.[87] Other large landowners are known only incidentally, like the wealthy man in John of Ephesus' *Life of Habib* who possessed large vineyards. Similarly, in the region of Amida-Martyropolis, Thomas the Armenian possessed a walled estate with irrigated gardens, orchards, and a park (*paradeisos*);[88] while an unnamed magistrate in the vicinity of Kalesh in Mesopotamia possessed a farm with many stone buildings, large tracts of land, vineyards, orchards, and many slaves.[89] Justin II (565–78) granted tax immunities to an estate at Aina, near Beroia (Aleppo).[90] Not all estates were on such a grand scale: that of 'Marinus' at Raqit in the Carmel, which functioned in the fourth and fifth centuries, and possibly in the sixth and seventh as well, comprised an area of 250 ha, among the larger holdings certainly, but not of the order of great estates.[91]

The wealthy also maintained dwellings in suburban or rural holdings that were often integral to large working farms. More often, estate business was conducted through a manager (*pragmateutes, oikodomos, ephestōs*) on widely scattered holdings.[92] A large estate, known from a boundary marker, mentions the pragmateutes Lazarus, who lived probably in around AD 550.[93] If the reading of pragmateutes is correct in the inscriptions at Tarutia in the territory of Apamea, then John, the estate manager, assisted in building a tower there in AD 509/510.[94] The pragmateutes Leontius in AD 488/9 paid for a synagogue or church pavement at Emesa, indicating the presence of a large estate in the territory of that city.[95] A seal in the museum of Adana belonged

86 J. H. W. G. Liebeschuetz, *Antioch: City and Imperial Administration in the Later Roman Empire* (Oxford, 1972), 44.

87 Ibid., 45 n. 6.

88 John of Ephesus, *Lives*, trans. Brooks, 12–13, 289.

89 Ibid., 75–6.

90 Mango, *Artistic Patronage*, 272; *IGLS* 2.262.

91 S. Dar, *Raqit: Marinus' Estate on the Carmel, Israel* (Oxford, 2004). The area is known from archaeological survey; of course we cannot tell if the owners of these estates also held lands outside their immediate estate centre, a common practice.

92 The fact that the largest landowners were urbanites helps to explain the general lack of rural dwellings akin to the 'villas' of the Roman West. The large farms in the countryside probably reflect more the managerial centres of these estates, not the primary dwellings of the owners.

93 Mango, *Artistic Patronage*, 305; *IGLS* 4.1905.

94 Mango, ibid., 312; *IGLS* 4.1630.

95 *IGLS* 4.2246.

to a certain Gregory (*pragmateutes*), who probably managed an estate in the territory of Tarsus.[96]

Estate Centres

Suburban and rural estates (sg. *proasteion*, *προάστειον*) were an important feature of late antique aristocratic life. They were frequently homes to which the wealthy might retire and where they could enjoy some seclusion within reach of city life. Such were the *proasteia* that surrounded Theodoret's Cyrrhus and the retreats of the wealthy in Libanius' Antioch, some of which probably had a dual pleasurable-economic function.[97] Proasteia, like grand rural oikoi, might have features such as corner towers (*pyrgoi*) or gateways. In the thinking of Late Antiquity, however, defence was probably not the characteristic that defined the majority of them; they were generally 'open' structures. In the lush Antiochene suburb at Daphne, the fifth-century house of Ardaburius provides an excellent archaeological example of the lavish proasteion. The great palace of the emperor Zeno's rebellious general Illus at Akkale in Cilicia is another example of a large proasteion, massively built, replete with a substantial reservoir and bath.[98] The house of Ardaburius Daphne possessed elaborate architecture and sumptuous mosaics, as did the sixth-century proasteion at Jenah outside Beirut and the (fifth–sixth-century) villa of Awza'i on the coast south of Beirut.[99]

In the eastern provinces outside of Oriens, rural proasteia are occasionally mentioned in literary sources, though not as often as their suburban counterparts.[100] We have brief notices of such grand dwellings at the centre of farms: the sixth-century *vita* of Peter the Iberian (p. 100) mentions the presence of a villa at Migdal Tutha in Mesopotamia. A proasteion is suggested by the finds of rich mosaic pavements in the countryside of Hierapolis (Mambij) in northern Syria.[101] Such references however are rare, as is the archaeological evidence.

Perhaps the best example of a proasteion with a strong agricultural function in Oriens is the Third Mile Estate outside Ashkelon (ancient Ascalon). The Third

[96] J.-C. Cheynet, 'Sceaux byzantines des Musées d'Antioche et de Tarse', *TM* 12 (1994), 469 no. 140.

[97] Theodoret, *Hist. Eccl.*, *PG* 82.1185. I thank Jairus Banaji for this reference.

[98] F. Hild and H. Hellenkemper, *Kilikien und Isaurien* (Vienna, 1990), 165–6, pls. 19–26.

[99] W. A. Campbell and R. Stillwell, *The Excavations, 1933–1936* (Princeton, NJ, 1938); M. H. Chéhab, *Mosaiques du Liban* (Paris, 1957), 59–69, 127–39.

[100] J. J. Rossiter, 'Roman Villas of the Greek East and the Villa of Gregory of Nyssa *Ep.* 20', *JRA* 2 (1989), 105.

[101] J. Balty, *La Mosaïque de Sarrîn (Osrhoène)* (Paris, 1990).

Mile Estate (Figure 2.7) was so named by its excavators because of its position on the Via Maris, the major route from Palestine to Syria, 4.5 km north-east of Ascalon and 2.3 km east of the sea. The building complex covers an area of 0.8 ha.[102] This proasteion is the only thoroughly excavated example (though only partially published) of a rural building in the late antique Near East that has both the architectural elements of the grand country estate and the industrial installations of a working farm, thereby linking a wealthy resident owner to the source of his income, in this case, surplus production of wine and oil.

The Third Mile Estate (Figure 2.8) was a short journey away from the markets and port of Ascalon and thus well-placed to transport goods. As advised for estates by the fifth-century agronomist Palladius (1.7) it was situated on a road. According to the excavator, the farm began its existence as a bath house, olive press, and winepress; other installations were built in later phases.[103] The upper storeys of both the great central storage warehouses and the halls to the east and west of the storehouses constituted the living quarters. The ground floor held winepresses. The storehouses, four or five in number, occupied an area of over 2,000 m^2. The walls (0.8 m thick and of local sandstone) were coated in plaster and whitewashed, as prescribed by Palladius. The large eastern winepress hall occupied an area of 220 m^2. Mosaic finds from the winepresses reveal something of the decoration of the upper, residential levels. To the north-east of the hall lies a bath house (14 × 21 m) with *caldarium*, *sudatorium*, *tepidarium*, and *frigidarium* and a furnace to the south. The latrine and sink lay adjacent to the baths to the north.[104] Overflow from the bath house was utilized for fish breeding, again as suggested by Palladius (1.42). A series of stone and concrete pools covered an area of 200 m^2, trapezoidal in shape and plastered with *opus signinum*. The western pool, measuring 4.5×5.0 m and 1 m in depth, was reinforced with piers and received piped water. These supply pipes were connected to another, sunken, pool, *c.*4.0×7.0 m, which lay at a lower level. Both pools had storage jars embedded within them to serve for fish breeding.[105] A third pool served as a collecting point for harvesting fish.

In addition to fish, the Third Mile Estate produced other commodities: a series of winepresses in the centre of the site, measuring a total of 380 m^2, were clearly for large-scale surplus production of wine. South-west of the centre of the Third Mile Estate was an amphora kiln complex that produced Gaza/Ascalon jars (LR4/LR5), which were principally used for the storage and transport of Palestinian wine during the Byzantine era. To the north of the

102 Hirschfeld, 'Farms and Villages', 46.

103 Y. Israel, 'Ashqelon', *ESI* 13 (1993), 101.

104 Ibid., 104. 105 Ibid., 105.

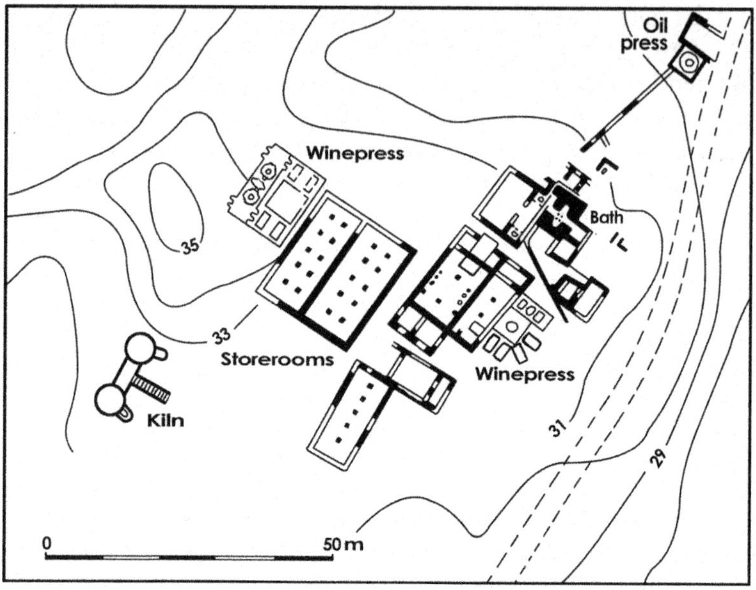

Figure 2.8. Plan of the Third Mile Estate (after Israel, 'Ashqelon', p. 100).

large winepress lay a substantial (*c*.290 m^2) olive oil press room, which was also used for the estate's grain milling as evidenced by the discovery of a basalt Pompeian-type rotary mill.

Although this complex—better than any other set of late antique buildings from Oriens—fits the model of the villa as a centre of pleasure as well as labour, caution is required. The vast majority of space is in fact occupied by agricultural and industrial installations, whilst the bath house is of modest proportions and the mosaics are plain. Conspicuously absent is a large oikos where the master of the house and his family would have been domiciled. This may await discovery in a nearby location, but it is possible that the Third Mile Estate was strictly a working farm or home of an overseer in the service of an absentee owner, probably residing in Ascalon or Gaza. In the meantime, we await the fuller publication of the site, and other estate sites of late antique Palestine, where estate houses have been recorded at places like Caesarea, Bet Shearim, and Tell Abu Shoshah.[106]

[106] See Z. Safrai, *The Economy of Roman Palestine* (London, 1994), 91, for an extensive list that does not establish the exact character of these buildings: they are all apparently large rural dwellings, some displaying luxurious features, others exhibiting agricultural functions, but the two meeting in one building is not, to my knowledge, well attested.

The Agricultural Components of a Large Farmstead

Palladius, relying on Vitruvius, provides an account of the construction and component parts of an idealized large farmhouse.[107] Palladius also recommended that dwellings should be proportional in size to the value of the farm, since the maintenance costs of buildings would be much higher than their initial construction costs (1.7). The family farmhouse was to be placed on elevated ground and to have foundations a foot and a half wider than the wall built upon them. Winter habitations should face south, whilst summer quarters should face west. In order to allow for ample lighting (1.12), the *triclinium* (dining area) and bedrooms were to be built to a height equal to half the sum of their length and width. Building materials should include timber, brick, and lime mortar. These materials seem to be used less often in Oriens, where stone was the favoured building component for erecting large dwellings. Buildings were to be whitewashed (1.14). The Palladian villa was further equipped with cisterns, wine cellars, granaries, oil-presses, storage areas, stables, courtyards, and aviaries. Other buildings included space for storing hay, straw, and wood (1.32), installations such as threshing floors (1.36), and an apiary (1.37). Bath houses (1.40) were an important part of the farmhouse, with excess water used to run watermills and supply fishponds.

A core component of many ancient and late antique houses, regardless of wealth and social standing, was the courtyard, which continues to be a prominent feature of houses in south-west Asia today. The courtyard provided a working area for processing cereal crops, vegetables, and fruits. The late antique basalt dwelling at Qatzrin in the Golan has an oven flanking the courtyard, where grain was probably prepared.[108] Raisins were likely dried in the courtyard, where the produce from vineyard and field could be sheltered from vermin whilst the grapes dried in the sun. Traditionally, some fruits and vegetables were grown within the sheltered environment of the courtyard. As attested in Mishnah Ma'as. 3.8–10 (200 BC), the practice of using the courtyard as a kitchen garden was centuries old by Late Antiquity. The Mishnah further suggests that the courtyard of such houses often housed implements and sheltered fowl and other animals.[109] At Qatzrin, the courtyard's palm-thatched roof offered humans and animals protection from the elements. House XVIII at Umm el-Jimal in the Hauran lacks the ground-floor stables

[107] V. Rose and W. H. Plommer, *Vitruvius and Later Roman Building Manuals* (Cambridge, 1973), 2–37. The *Geoponica* mentions such structures only when describing activities related to them, and there is no discussion of the overall construction and layout of the farm.

[108] A. Killebrew and S. Fine, 'Qatzrin: Reconstructing Village Life in Talmudic Times', *BARev* 17/3 (1991), 50.

[109] Ibid., 50–1.

common to the region's dwellings, but the courtyard mangers suggest livestock husbandry there.[110]

Cisterns, described in greater detail in Chapter 4, provided water for household use, for animals stabled within the house, and probably for small-scale agricultural irrigation and handicrafts (as well as for processing oil). Often these water storage features, most frequently filled by the collection of rainfall from rooftops, were placed within courtyards.

In two-storey houses, storage space was typically on the ground floor.[111] According to Palladius (1.18), wine store-rooms were to be placed to the north of the estate (where the wine would remain cooler and thus better preserved), away from the bath and other possible contaminating elements. The *Geoponica* (6.2), the bulk of which belongs to the fourth through sixth centuries (see Appendix), recommended placing the wine cellar toward the east and north. According to Palladius (1.20), oil store-rooms should face the south, where the oil could benefit from the winter sun.

At Oboda, a house excavated below the acropolis included a modified cave (Room 16) that its excavator, Negev, interpreted as a wine cellar (*οἰνῶν*).[112] A bench within Room 16 accommodated the flat-bottomed jars (LR5) associated with late antique Palestinian wine production. A narrow passage, 0.6 m wide and 1.5 m long, connected Room 16 to another cellar, Room 17. Additional shelves within the latter held the LR5, apparently during final fermentation. Deep within the complex, Room 17 was admirably sheltered and well-suited to fermentation. The cellar had a 3,750 litre capacity. Typically, store-rooms for wine and oil would contain sunken *dolia*, like those contained in the basements of many Mediterranean houses up to the present day.

According to Palladius (*Op. Ag.* 1.19) granaries should face north and be built with a brick floor with separate bins for each type of grain. The walls should be covered with clay and sprinkled with olive waste (*amurca*). The *horion* built by Aurelius Bellicus at al-Tuba near Anasartha in north central Syria resembles the Palladian form. It is an 'elongated rectangle' measuring approximately 27×8.5 m (229.5 m²). A door gave access to a central room, 5.3×4.5 m, with an arch to the ceiling constructed of voussoirs 0.60 m on a side. To the north and south, the hall opened into flanking storage rooms that adjoined the towers.[113] Field surveys have recorded numerous stone grain silos, such as those at Har Saggi and Nahal Sirpad in the Negev. Rock-cut

110 Brown, 'A Large Residence', 194; Butler et al., *Syria*, II.A.199–200.
111 Hirschfeld, *Palestinian Dwelling*, 264.
112 Negev, *The Architecture of Oboda*, 165–6.
113 Mouterde and Poidebard, *Limes de Chalcis*, 198–9.

granaries, like those found in southern Palestine at Nahal Minan and Nahal 'Avedat, were also common.[114]

According to Palladius (1.21), stables were to be situated near the kitchen, where, during the winter months, animals could benefit from the hearth's warmth. Numerous surviving Late Roman houses, such as 'House G' at Pella, Giv'at Orha, and Korazin, show that the small and medium house frequently incorporated stables. The courtyards of most houses sheltered animals of various kinds, especially poultry.[115] One of the defining characteristics of the large farmhouse appears to be the expense taken to ensure shelter for animals. Butler noted that the stables of the Limestone Massif were as impressive in quality as the houses to which they belonged.[116] At Mampsis in the Negev, Building XII, a large private house, with a floor area of more than 2,000 m², possessed sizeable stables on its eastern side.[117] These stables were vaulted structures measuring approximately 10 m on a side, with an open area in the centre where provender and foodstuffs were stored.[118]

Non-integrated, purpose-built stables are known from the Hauran, where stocking was a major economic pursuit.[119] Butler published two important examples. The first, Subhiyeh, is a sizeable (289 m²) structure with mangers on three sides of the first floor and fodder storage or habitation above.[120] The second is from Kara'ah, a village in which Butler noted that nearly all the buildings were stables. The published example is built of roughly quadrate

[114] Avni, *Map of Har Saggi Northeast (225)*, 66; Haiman, *Map of Har Hamran—Southwest (198)*, 44. These structures generally are constructed of small fieldstones and covered with stone slabs, and are between 0.8 and 1 m in diameter, though the one at Har Saggi is 3 m in diameter and 1 m in height. Haiman, *Map of Har Hamran—Southwest (198)*, 21; Haiman, *Map of Har Hamran—Southeast (199)*, 29. That at Nahal Sirpad measured 3×4 m, was 1 m deep and was lined with fieldstones. Nahal Minan: Haiman, *Map of Har Hamran—Southwest (198)*, 50; Nahal 'Avedat: Lender, *Map of Har Nafhah (196) 12–01* (Jerusalem, 1990), 4. The ancient practice of cereal storage in pits was still current in nineteenth-century Arabia, see S. Merrill, *East of the Jordan: A Record of Travel and Observation in the Countries of Moab, Gilead, and Bashan* (London, 1881), 139–40.

[115] Hirschfeld, *Palestinian Dwelling*, 69–72.

[116] Butler, *Architecture and Other Arts*, 125. At House 119 at Umm el-Jimal the living space inhabited by people does not greatly exceed that of the animal population housed there, indicative of the care taken to ensure the welfare of the animals upon which the household depended, and the importance of animals in the agricultural economy. B. De Vries and J. W. Betlyon, *Umm el-Jimal* (Portsmouth, RI, 1998), 108, fig. 57.

[117] Hirschfeld, *Palestinian Dwelling*, 76; Shereshevski, *Byzantine Urban Settlements*, 25–8.

[118] De Vries, 'Location, Plan and Design of the Late-Antique Town', 108, fig. 57.

[119] D. Miller, 'From the Archives of Architecture: Agrarian Settlement and Society in the Roman Hawran', in D. Panzac (ed.), *Histoire économique et sociale de l'Empire ottoman et de la turquie (1326–1960): actes du sixième congrès international tenu à Aix-en-Provence du 1er au 4 juillett 1992* (Paris, 1995), 655.

[120] Butler et al., *Syria*, II.A.

basalt and is unique in design, having an arched central chamber flanked by narrow rooms with mangers along their entire length.[121]

Columbaria (pigeon- or dove-houses) are treated in some detail by Palladius (1.24) and in the *Geoponica*. The latter (14.1) describes two types. The first was built atop the farmhouse with small windows on all sides, which provided ventilation and access to the birds. In rural Syria today, the mud-built, rooftop pigeon-house is a ubiquitous feature of village life and a link to the past. The second type of columbarium described was a separate, free-standing structure. At Ramat Hanadiv, a tower dovecote from the Hellenistic–Early Roman period was excavated; possibly, it continued to function in the site's Late Roman phase.[122] Pigeon-houses also commonly appear in archaeological record as subterranean, rock-cut installations.[123] Pigeon-breeding was extremely important for all levels of agricultural society, as the birds are prolific and easily kept, and dovecotes are cheap to construct; they thus provided a major source of protein in a diet that was generally lacking in meat. These birds had substantial value in the city and village markets and thus provided a way for those of limited means to obtain cash. They also provided a highly prized fertilizer (*Geop.* 14.1).

Fishponds (*piscinae*) were typically an integral part of estates, though not exclusive to them. Farmers placed importance on producing a variety of fish, and provided water installations for their maintenance. Aquaculture was especially important for seaside estates and villages (*Geop.* 20.1). Archaeological remains of seaside *piscinae* are known from Israel.[124] According to Palladius (1.17), eels and river fish were sometimes raised in cisterns. An aquaculture installation has been found at Tell Tanninim, where pools incorporated broken Gaza amphorae for fish breeding.[125] Such pools in

[121] Ibid., II.A.142–3.

[122] B. Zissu, 'The Dovecote at Horvat 'Eleq', in Y. Hirschfeld (ed.), *Ramat Hanadiv Excavations: Final Report of the 1984–1998 Seasons* (Jerusalem, 2000).

[123] Dating is therefore difficult. During survey of the hill country between Jerusalem and Ramallah, a columbarium with Hellenistic and Byzantine period pottery was found. Another such columbarium was discovered in Israel at Khirbet er Rasm. Typically, such rock-cut columbaria in Palestine are dated to the Second Temple period, since doves and pigeons were important sacrificial animals. Many probably continued to be utilized throughout Late Antiquity: Dagan, *Map of Lakhish (98)*, 65.

[124] Such as that from Akhziv. See R. Frankel and N. Getzov, *Map of Akhziv, Map of Hanitah*, 57, fig. 1.6. Raffi Frankel kindly brought this installation to my attention.

[125] R. R. Stieglitz, 'A Late Byzantine Reservoir and *Piscina* at Tell Tanninim', *IEJ* 48/1–2 (1998). At 'Ein Bikura in Judaea, Byzantine-period inhabitants used the pool for fishponds: S. Gibson, B. Ibbs, and A. Kloner, 'The Sataf Project of Landscape Archaeology and the Judean Hills: A Preliminary Report on Four Seasons of Survey and Excavation (1987–89)', *Levant* 23 (1991), 39. The pool at Khirbet Sabiya in the Sharon, measured 9.5×5 m and has walls preserved to a height of 1.7 m with three rows of jars set into the wall. E. Ayalon, 'The Jar Installation of Khirbet Sabiya', *IEJ* 29 (1979), 175.

Palestine were sometimes fed with *saqiya*-driven water-lifting equipment, indicating significant capital investment. The labour and material required to construct fish-breeding pools, such as those at Caesarea and Dor, existed only within estate or communal resources.

Towers (πύργοι)

Ancient towers (*pyrgoi*) occur frequently in the lands of former Oriens. Estate houses nearly always had one or more attached towers to provide a lookout point, a measure of security to the inhabitants, and a place for safe storage—functions they served since at least the Hellenistic period.[126] Some served as dwellings, like the tower house at Serjible, in the Jebel Sim'an in northern Syria.[127] This tower of five storeys is solidly built of limestone ashlars and measures over 17 m in height. On the ground floor, a single arch and parallel wall support the stone-slab roof. The arch-and-slab cover was a common feature of Syrian architecture, with the lower storey serving as either a stable or store-room.[128] The existence of these structures, which await fuller study, does not necessarily imply general insecurity, but they do show a concern for security that ran deep throughout the countryside. In some instances, the presence of a *pyrgos* may signify nothing more than the owner's preoccupation with protecting his crops from his neighbours' poaching. But there may have been greater concerns in some regions; the towers in the Chalcidike, for example, may well reflect a more widespread insecurity posed by bandits in frontier country.[129]

Fortified Farms

This study defines a fortified farm as an enclosed compound generally focused on a tower. In this type of dwelling, defensive characteristics predominate,

126 Rossiter, 'Roman Villas', 105–6. Tchalenko, *Villages*, I.31, called towers 'a general and permanent feature of Mediterranean architecture'. De Vogüé recorded several towers in North Syria, and Butler followed up on his work there and in the Hauran, where he recorded many examples: Butler et al., *Syria*, II.A.126.

127 Butler et al., *Syria*, II.B.230–1.

128 Initially published as a storeroom by Butler, *Architecture and Other Arts*, 255, he later considered it a stable. Butler et al., *Syria*, II.B.230, an interpretation disputed by Peña: see I. Peña, P. Castellana, and R. Fernandez, *Les reclus syriens: recherches sur les anciennes formes de vie solitaire en Syrie* (Jerusalem, 1980). Dar and Applebaum, *Landscape and Pattern*, 110–22, who recorded and studied more than 1,200 rural towers in Samaria and related them to the ancient agricultural landscape. They excavated forty-five examples, and concluded that most were constructed during the Hellenistic period. Like the Byzantine era, this was a time of greatly expanding wine production. Dar's survey and excavation work led him to conclude that these tower structures were intimately linked with working the land. They served as watch-booths, provided cellars for wine, and housed agricultural labourers during peak work times.

129 Lassus, *Inventaire archéologique*.

exceeding those customary in many farmsteads and houses. Whilst the medium and large farmhouse may possess towers as a matter of course—and these towers may have served to provide refuge or a lookout point—the buildings to which they are attached frequently differ from fortified farms in construction, organization, and aspect.

In the written sources, there is apparently no special word that sharply segregates private, fortified dwellings from those of state-military character. Thus, the words *pyrgos*, *tetrapyrgos* (τετραπύργος), and *phrourion* (φρούριον), for which we have examples clearly referring to state fortifications, probably were also used to designate some civilian farms. Perhaps the likeliest of these terms to designate a fortified, private estate centre (apart from βᾶρις, described below) is *phrourion*. In other areas in the East and at different times, the word designated a fortified estate centre.[130] In 509, following the end of the Persian War of Anastasius, John, who was without rank or title, built a *phrourion* at Taroutia.[131] It is possible that John was providing defensive measures on the imperial lands or that he was simply a wealthy local who felt the need to protect his estates from attack.

By the first quarter of the fifth century AD, the process of farm fortification was fully under way. A law of Honorius and Theodosius (*CJ* 8.10) recognized the fact of widespread fortification of rural dwellings.[132] The emperors granted permission for walled estates in all regions where the practice was deemed desirable, and named the following provinces in Oriens: Mesopotamia, Osrhoene, Euphratensis, Syria II, Phoenicia Libanensis, and Cilicia II. All the named provinces within Oriens were on, or very near, the frontier, and it seems certain that concerns about security lay behind the edict. Landholders were apparently building fortified compounds in response to local disorder, including banditry and tribal raids, although the extent and intensity of such activity is difficult to gauge.[133] According to Libanius (*Or.* 50.26), thieves around Antioch operated at night, often in league with innkeepers. Ammianus (28.2.13) described the Marathocupreni bandits who operated around Apamea. Local soldiers were often uninterested or incapable of

[130] C. Schuler, *Ländliche Siedlungen und Gemeinden im hellenistischen und römischen Kleinasien* (Munich, 1998), 73.

[131] F. Trombley, 'War and Society in Rural Syria *c.*502–613 A.D.', *BMGS* 21 (1997), 161–2. This *phourion* functioned as a local refuge in the event of hostilities, a feature that these late antique *phouria* seem to share; perhaps they were more often viewed as places to which the rural population could withdraw than as organized points of military resistance.

[132] A. Demandt, *Die Spätantike: römische Geschichte von Diocletian bis Justinian, 284–565 n. Chr* (Munich, 1989), 335.

[133] B. Isaac, 'Bandits in Judaea and Arabia', *HSCP* 88 (1984); S. T. Parker, 'Peasants, Pastoralists, and Pax-Romana, a Different View', *BASOR* 265 (1987).

dealing with such social disorder, and thus the rich took matters into their own hands.

The complex at al-Tuba, built in northern Syria near Anasartha and Zebed (ancient Kaprozebada), has already been mentioned for its fine fourth-century granary, above which were the living quarters for its elite owner Aurelius Bellicus or his overseer.[134] In addition, the complex contained an enclosure, 139–160 m on its long sides and approximately 90 m on its short ends. Towers flanked the ends of the storehouse and allowed observation to the north and east. The chapel/funerary monument and necropolis are civilian rather than military in character, and the walls, at 1.2 m thick, are stout, but not of the sort to hold off an invading army. The buildings are constructed of local basalt, and the wall is pierced by three portals to the north, facing the direction of Zebed. Despite its fortified aspect, it seems that al-Tuba was a private residence rather than a fort. Aurelius Bellicus gave no title, and it therefore seems unlikely that this is any kind of official building.

Often, the fortified farms of the Levant are interpreted as part of a defensive system planned by the Roman authorities. This may lead to their misidentification as military forts. Zeev Rubin interpreted the farms at Nahal Sa'adon as the settlements of frontier troops, *limitanei*, constructed in the Negev by the central government.[135] However, it is interesting to compare the eastern examples with the fortified farms in Cyrenaica and Tripolitania, which were organized around a central tower. These are referred to, using the Arabic term, as *gasr* (pl. *gsur*), and they often possess ancillary buildings. In many cases, they form the centres of previously existing open farms.[136] Recent scholarship has rejected the notion that the gsur were part of an organized defensive scheme involving garrisons of limitanei. In light of this, we should probably interpret the examples from the Levant as not being purely military in origin.[137]

Similarly, Butler and others have considered Stabl Antar, a massive basalt building about 60 m square, a fort. However, it seems that this structure may well have been a fortified farm, and, as such, the centre of a large domain, as I have suggested elsewhere.[138]

[134] Mouterde and Poidebard, *Limes de Chalcis*, 198.

[135] Hirschfeld, 'Farms and Villages', 59 n. 99.

[136] D. J. Mattingly, 'Libyans and the "Limes": Culture and Society in Roman Tripolitania', *AntAfr* 23 (1987), 84.

[137] D. J. Mattingly, 'Farmers and Frontiers: Exploiting and Defending the Countryside of Roman Tripolitania', *LibSt* 20 (1989), 141–3; D. J. Mattingly and J. W. Hayes, 'Nador and Fortified Farms in North Africa', *JRA* 5 (1992), 415–18; D. J. Mattingly, *Tripolitania* (London, 1995), 194–209.

[138] M. Decker, 'Towers, Refuges, and Fortified Farms in the Late Roman East', *LA* 56 (2006).

Fortified Estates

The *baris* is a fortified estate centre and seems to have been a widespread settlement phenomenon in Hellenistic Asia Minor, where the word is preserved in several place names.[139] Cyril of Alexandria (*c.*AD 378–*c.*444) mentioned the *baris* and equated the structure with a pyrgos.[140] In the first century, Josephus (*BJ* 5.1) used the word to describe the retreat of the Hasmoneans in the vicinity of Amman, Jordan at Iraq al-Amir. From this, Ernest Will argued that the archaeological remains of this complex provide a paradigm for the baris.[141] Iraq al-Amir was an impressive royal residence, a fact that explains the external decoration of the façade, which would have been an unlikely feature of a private house, even a wealthy one. Furthermore, the palace has a walled-in cultivated area, store-rooms, an underground dovecote and baths. The baris was an eastern, palatial rural dwelling, possessing a prominent pleasure component and substantial architectural decoration—in addition to a fortified aspect and agricultural installations. The baris shares features common to fortified farmhouses and proasteia but the baris was also built for pleasure and for aesthetic appeal. As a type, this kind of structure continued into Late Antiquity.

The Byzantine complex at Qasr ibn Wardan, built in the Constantinopolitan style in AD 561–72, strongly resembles the Hellenistic baris at Iraq al-Amir.[142] Situated in the basin formed by a ring of basalt ridges, the complex formed the heart of an agricultural estate in Syria II. The scale and richness of the architecture, which includes a palace (49.4 × 52 m), church (15 × 18.5 m), and barracks, led Butler to speculate that Qasr ibn Wardan was an imperial project.[143] A garden enclosure (*paradeisos*) also survived until Butler's day.[144] Traces of a *qanat* system formed part of an elaborate irrigation system, which provided water to the fields and paradeisos.[145] The ensemble in many ways

[139] L. Robert, *Noms indigènes dans l'Asie-Mineure gréco-romaine* (Paris, 1963), 15; Schuler, *Ländliche Siedlungen*, 72–3, puzzled over why the word is used instead of the more common *πύργος*, *φρούριον* or *ὀχύρωμα*. The distinction appears to be clarified somewhat by the archaeological material discussed below.

[140] Cyril Theol., *Exp. in Ps. PG* 69.1040: *Βάρις μὲν γὰρ καλεῖται πύργος ἅπας*.

[141] Butler et al., *Syria*, II.A, offered considerable detail on Iraq al-Amir. More recently, French archaeologists have undertaken extensive work at the site: see E. Will et al., *Iraq al Amir: le château du Tobiade Hyrcan* (Paris, 1991).

[142] F. De'maffei, 'Il Palazzo di Qasr ibn-Wardan: dopo gli scavi e i restauri', in A. Iacobini and E. Zanini (eds.), *Arte profana e arte sacra a Bisanzio* (Rome, 1995); P. Grossmann, 'Zu den Bogen und Gewölben in dem Wüstenpalast von Qasr ibn Wardān', *DM* 12 (2000).

[143] Butler et al., *Syria*, II.B.27.

[144] Ibid., II.B.26.

[145] D. R. Lightfoot, 'Syrian Qanat Romani: History, Ecology, Abandonment', *JAridEnviron* 33 (1996), 323, noted the presence of a *qanat* system near the site. A. W. Zakariyya, *Jawla athariyya fi ba'd al-balad al shamiyya* (Damascus, 1934) had earlier reported that this was an ancient system,

anticipates the Umayyad desert palaces at Qasr al-Hayr al-Gharbi, Qasr al-Hayr el-Sharqi, and Qasr Hallabat, where walled areas enclosed irrigated gardens.[146] There is then a unifying link between Byzantine management of the steppe landscape, its agricultural development and exploitation, and that of their Umayyad successors.

The Pattern of Land Holding in Oriens

How big were the large estates of Late Antiquity to which the proasteion and baris belonged? At the end of the fourth century, Chrysostom suggested that many of the Antiochene wealthy owned 10–20 houses or more in addition to their great landholdings and other outward trappings of wealth.[147] In return for his handover of Italy, the Goth Theodahad requested of Justinian an estate that would produce 1,200 lb of gold annually.[148] Credible figures for the size of a large estate in Oriens can only be approached obliquely through evidence provided in texts like the *Life of Simeon of the Olives*.[149] In the eighth century, Simeon aggrandized his monastic estates on the Tur 'Abdin in Mesopotamia by planting 12,000 olives, an undertaking that would have required around

cleared out by settlers in the early twentieth century. Personal prospection, carried out with Dr Andrew Wilson during the Oxford Excavations at Androna, discovered a basalt-lined well-shaft (approximately 3×1 m) to the north of the palace. Wilson (personal communication, July 1999) believes this well was fitted with a *saqiya*-driven water-lifting device, as it is similar in construction to *saqiya* wells elsewhere. The paradeisos was a common feature in the eastern landscape, although rarely noted in textual sources. The discovery of a pleasure garden in the centre of Petra is of great interest for garden history and urban history generally; see L.-A. Bedal, 'A *Paradeisos* at Petra: New Light on the "Lower Market"', *ADAJ* 43 (1999), 227–39. Another late antique paradeisos is known from nearby Anasartha (*IGLS* 2.316). The practice of building garden game parks was probably of Persian origin.

146 G. Bisheh, 'Excavations at Qasr al-Hallabat, 1979', *ADAJ* 24 (1980); G. Bisheh, 'The Second Season of Excavation at Hallabat, 1980', *ADAJ* 26 (1982); G. Bisheh, 'Qasr al-Hallabat: An Ummayad Desert Retreat or Farm Land?', *SHAJ* 2 (1985); G. Bisheh, 'Qasr al-Hallabat: A Summary of the 1984 and 1985 Excavations', *Archiv für Orientforschung* 38 (1986), 158–62; O. Grabar, *City in the Desert: Qasr al-Hayr East* (Cambridge, Mass, 1978); D. Schlumberger, 'Les fouilles de Qasr el-Heir el-Gharbi (1936–38), rapport préliminaire', *Syria* 20 (1939); D. Schlumberger et al., *Qasr el-Heir el Gharbi: textes et planches* (Paris, 1986). The gardens may have been influential in plant-breeding and transmission, and in the development and acclimatization of new varieties. See A. M. Watson, *Agricultural Innovation in the Early Islamic World* (Cambridge, 1983), 118–19. It is noteworthy that the first Umayyad caliph of Cordoba, 'Abd al-Rahman (756–88), sent agents to Syria to collect rare plants and seeds for the creation of his impressive gardens.

147 *PG* 58.633. We do not need to accept at face value Chrysostom's suggestion that many of the elite possessed 1,000–2,000 slaves, though in order for his words to have resonated at all with his audience, his figures must have been grounded in reality.

148 Procopius of Caesarea, *Wars*, ed. Haury, 5.6.19.

149 S. Brock, 'The Life of Simeon of the Olives', *OKS* 29 (1979).

480 iugera of land (120 ha). This is a similar figure to the 420 iugera (106 ha), average-sized farms calculated by Duncan-Jones in his study of the epigraphic evidence from Magnesia in Asia Minor.[150] It is far from the 15,132 iugera of the largest holding at Hermopolis in Egypt and even further removed from the largest estate around Magnesia that controlled 28,160 iugera of land (70.9 km^2 or 7,090 ha).[151] Similarly, the *Life of Philaretos the Merciful* (eighth century) presents evidence of another massive estate in Paphlagonia of perhaps 2,100 iugera in arable land, with an additional 21,333 iugera of pasture land.[152] Though the evidence is limited, the range of different sizes of estates is clear. The disparity between the average, great estate owner at Magnesia and the most prosperous of the group was apparently enormous.

Small- and Medium-Scale Landowners

A law of Justin II (*Novella* 138) of AD 566 mentions what Lemerle perceives to be a rising hierarchic order: *geōrgoi* (γεώργοι), *misthōtai* (μισθωταί), *emphyteutai* (ἐμφετευταί), *kektēmenoi* (κεκτήμενοι)—that is, *coloni*, lease-holders, holders of *emphyteusis*, and free-holders, respectively.[153] Though in the specific case of Justin's manuscript, Lemerle's interpretation of the word *geōrgos* (as dependent tenant (*colonus*) as one common meaning of the word) is valid, there are other cases in which the word describes a different set of social realities. In the Egyptian documents of Late Antiquity, *geōrgos* may mean 'peasant', 'farm labourer', or 'landless peasant', or denote a small-holding, independent farmer.[154] The early seventh-century *Life of Theodore of Sykeon* (116.2) uses the word *geōrgos* to describe a small-scale, independent cultivator. *Misthōtai* were, by legal definition, free cultivators who had served thirty years on the land as lessees and were of free status. These men lost their right to leave the land, as did their descendants (*CJ* 11.48.19). Thus, this class must always have been a small one, which Lemerle saw as representing a kind of purgatory between the heaven of free land-holding and the hell of *colonate*

150 Duncan-Jones, 'Some Configurations of Landholding', 21, tbl. 3.

151 Ibid., 23, tbl. 4.

152 Plough-land is based on the *Vita's* statement that Philaretos (ed. Rydén, 1.5–15) owned 100 yoke of oxen, with the assumption of a minimum 21-day ploughing season. Each team of oxen worked one *iugerum* per day. The estimates on the pastureland are Kaplan's: see M. Kaplan, *Les hommes et la terre à Byzance du VI au XI siècle: propriété et exploitation du sol* (Paris, 1992), 79.

153 P. Lemerle, *The Agrarian History of Byzantium: From the Origins to the Twelfth Century, the Sources and Problems* (Galway, 1979), 26. *Geōrgoi* is troublesome: in this official context, *coloni* seems right, but it still appeared in the sixth–seventh centuries as simply 'cultivator' or 'farmer' of independent status.

154 Banaji, *Agrarian Change*, 108, 91–4, 253.

status. As the standard lease cycle was five years (*CJ* 4.65), these men would have had to serve six leases in succession before they lost their rights. In essence, the arrangement was a lifetime lease. *Emphyteutai* were unique in that they were neither owners nor lessees (*CJ* 4.66.1), but possessed certain characteristics of both. Emphyteutic land could be bought and sold as well as undergo improvement by the lessee, and the emphyteutic contract itself could be exchanged.

Legally, independent farmers were 'registered on the tax rolls under their own name and their own place of residence' (*CJ* 11.48.4).[155] They owed their tribute directly to the imperial government, and were *possessores/ktētores*. In other words, they owned land which they probably farmed or actively managed themselves. They therefore stand in contrast to those merchants and craftsmen in town and country who owned no land.[156] Similarly, but without the benefits and risk of freedom and wealth, were slaves, who were common throughout Late Antiquity. Even though some people maintained a hundred slaves or more, unfree labour apparently did not play a fundamental role in the agricultural economy of the eastern empire.[157]

It is possible to further refine our view of landholding. Unlike in the case of Egypt, whose archival sources are unmatched in constructing a picture of Byzantine agrarian activities, the information for Oriens is much poorer. Documents within the sixth–seventh-century Nessana Papyri and the sixth-century Petra Papyri (the latter awaiting full publication) do, however, deserve further examination. As a group, these texts suggest that a number of small- or medium-scale, independent owners persisted in the countryside well into the sixth century. At Nessana, in particular, soldiers or veterans were prominent at this level of agrarian society. The sixth-century documents found in the late antique basilica at Petra detail the activities of lower officials and apparently private citizens who seem to have been owners on the same scale as the citizens of Nessana.

The so-called sixth-century 'soldiers' archive' from the village of Nessana in southern Palestine provides a rare glimpse into the daily life and landholdings of prominent villagers who can be characterized as independent landowners on a modest scale. The men involved in *P.Ness.* III.14–30 are frequently referred to as *stratiōtai* (soldiers). *P.Ness.* III.16 (11 July 512) records a transaction involving John, son of Asad, John's sister, and Zunayn, son of Abraham. Both John and Zunayn were soldiers. *P.Ness.* III.16.10–16 details

[155] Kaplan, *Les hommes et la terre*, 95–9.

[156] Lemerle, *Agrarian History*, 18.

[157] Liebeschuetz, *Antioch*, 64–5, believes, as do I, that slaves were not a major part of the agricultural workforce. However, see *Syr.Rom.Law.* ed. Vööbus, §§21ff. for manumission by those with up to a hundred slaves.

the purchase by John of Zunayn's share of a garden that the two owned jointly, and of an additional plot belonging to Zunayn. John purchased Zunayn's holdings for 4 and 1½ solidi, respectively. Lines 29–32 record that John bought another plot from Zunayn, for 1½ solidi. Some of the land involved was apparently part of the vineyards that lay in the environs of the village. Nessana, like other Negev settlements, was a supplier of the famous Gaza wine, which may shed light on the agricultural activities in which these soldiers were engaged.

In another papyrus, *P.Ness.* III.24.5, two brothers, both soldiers at Nessana, cede to a comrade a piece of land, 5 *kabiaiai* in extent. The value of the land is unknown. The 5 *kabiaiai* in this papyrus are noted as adjoining the property of Zeno, son of Firsan. The latter appears as witness to the completion of a contract in *P.Ness.* III.25. In *P.Ness.* III.26, two soldiers borrowed 4 solidi from a colleague, with part of the collateral for the loan in the form of a threshing floor. In *P.Ness.* III.21, we find the soldier Sergius dividing his property, including his farmland, among his four children. One of those children later is found lending 4 solidi to a local tribesman (*P.Ness.* III.28).

These Nessana documents permit shadowy glimpses into rural life. They document Semitic language speakers, probably Arabs, integrating themselves into Byzantine society through imperial service. Zeno and John both use Greek names, but are the sons of men with Semitic names—Firsan and Abraham, respectively. Sergius may have been Arab; his namesake, St Sergius, was immensely popular among the pre-Islamic Arab residents of the eastern frontier, and Sergius was a common name in Syria.[158] These texts also show a group of imperial servants, albeit soldiers and thus relatively minor ones, who were busy investing in land and other small, speculative business exchanges (money-lending). Their families managed to hold on to a modicum of wealth for some time: the money-lending of Sergius' son hints that the family remained locally prominent throughout the next generation, and perhaps beyond.

A prominent landowner, on a slightly larger scale, emerges from the Petra papyrus, P.Petra I.3–5 (dated August 538), which documents the transfer of taxation from one landowner to another. Patrikios (Flavius Patrikios, *lamprotatos* (I.5.2), possibly a curial, acquired the tax burden on the real estate of Panolbios, also *lamprotatos* (I.3.4).[159] The transaction involved an estate of

[158] E. K. Fowden, *The Barbarian Plain: Saint Sergius Between Rome and Iran* (Berkeley, 1999), 101–4.

[159] *Lamprotatos* was a rank among the lower senatorial orders and was, by the sixth century, quite common and offers little insight into the real wealth or position of its holder. Arjava, 'Zum Gebrauch der griechischen Rangprädikate des Senatorenstandes in den Papyri und Inschriften', *Tyche* 6 (1991), 23.

Panolbios, referred to as *homas* (ὁμάς) ('the whole'), some portion of which was located around the city of Petra itself, other portions of which were located near Augustopolis.[160] Three documents (P.Petra I.3–5) mention land areas and a cash sum. The latter, 12 solidi, was paid by Patrikios to Panolbios 'for his expenses'.[161] In lieu of this annual cash payment, the editors suggest that Patrikios assumed certain tax burdens incumbent upon Panolbios' estate. The extent of these estate holdings cannot be confidently reconstructed, but some interesting details do emerge. For example, Panolbios' possessions included 10 iugera of the *patrimonium*, a term used to describe imperial estates. He had, therefore, become owner of former imperial lands, either through purchase or by other means. We may have something approaching the totality of Panolbios' holdings listed in P.Petra I.5 (again incomplete), in which 8 free (*eleutheroi*) *koriaiai*, 9 *satiaiai* and 4 *kabiaiai* stand alongside 5, one-fourth, and one-sixteenth iugera of the *patrimonium*.[162] The whole area amounts to around 11 ha.

Since many of the documents from Petra remain unpublished, these examples cannot provide a conclusive picture. What bears mentioning, however, is that the sums of money and the area of land mentioned are relatively modest. Whether 'the whole' (*homas*) refers to Panolbios' entire real property—or merely to what he owns in the vicinity of Petra, or simply one estate amongst many that a given owner could possess—is presently beyond certainty. Nevertheless, the sum of money (12 solidi in this case) and the land area involved (8⅓ *koriaiai* approximates to between 10 and 12.5 ha) are far from the order of the largest estate owners discussed below.[163] The figures from Nessana and Petra, albeit limited, apparently belonged to the world of a social group that in Egypt has been termed the 'middle Byzantine bureaucracy'. Members of this group were people like Aurelius Harpokration and his wife, Thaesis, whose modest estate amounted to approximately 2¼ ha.[164]

160 The term ὁμάς is found commonly in later Byzantine land legislation when describing collective units of land: Lemerle, *Agrarian History*, 97–9. In *P.Petr.* I.3–5 it is tempting to see this designation as used to denote both the *autarchic* land of the estate owner, and that worked by an *epoikia* or tenant farmers dependent on the owner.

161 λόγῳ ἀναλωμάτων αὐτοῦ; *P.Petr.* I.3.4.

162 See *P.Petr.* I.5.12–13. These figures are obviously problematic, incomplete as they are. In addition to the land mentioned in *P.Petr.* I.5, for example, are the 10 iugera of the *patrimonium* mentioned in I.4. According to Schilbach, 1 *koriaia* = 30 *satiaia* and 1 *satiaia* = 4 *kabiaia*. The *koriaia* was 12,616–15,140 m² depending on the quality of the land, likewise the *satiaia* 420–504 m² and the *kabiaia* 105–127 m². Using the smaller land areas for the units yields an aggregate of about 11.8 ha: see E. Schilbach, *Byzantinische Metrologie* (Munich, 1970), 76–8.

163 The sixth-century Egyptian aristocrat Sophia held an Arsinoite estate perhaps 10,000 *arourai* (approximately 225 ha) in extent, to cite only one example of the magnitude of such landholdings. See Banaji, *Agrarian Change*, 141.

164 Banaji, ibid., 115–19.

The documents from Petra and Nessana, therefore, probably shed light on the dealings and wealth of the lower orders of the landed class. It is tempting to see Flavius Patrikios as one of the new, bureaucratic aristocracy, akin to those of Byzantine Egypt so adroitly illuminated by Banaji.

The *gēdion* (*γήδιον*), the small farm of independent cultivators, is mentioned in several late antique sources. Without providing much detail, the fourth-century Libanius alludes to small farms around northern Oriens. In his *Antiochikos*, Libanius used the contrast between the small farm and the large estate as a rhetorical device. To Libanius, the gēdion is 'easily worked', implying its centralization. This contrasts with the estate of the rich, which 'stretches far and wide into both background and foreground'. In Libanius' thinking, the small-scale farm was comprised of contiguous plots of no great extent or minor, scattered holdings close to the home; whereas the grand estate possessed fields and orchards spread over vast areas, even across provincial borders.[165] Procopius of Gaza offers evidence of the intensive exploitation of the gēdion; he advises that it should be planted with vines, which promised an abundant return.[166] Theophylact Simocatta mentions the gēdion several times in his letters.[167] He was thinking of his native Egypt and thus offers a comparative perspective from other eastern regions, as do the fourth-century Basil of Caesarea and Gregory of Nyssa in Asia Minor. The former uses the word gēdion within the context of (exaggerated) poverty; to his mind the word denotes an amount of land from which a humble person could derive his or her food supply.[168] Gregory's *Contra Eunomius* provides rare insight into the life of the small-scale farmer when he commented on the heretic Eunomius' father, a man who worked his own gēdion 'stooped over the plough'.[169]

In the textual and epigraphic record, there is further evidence of a lower stratum of landowners in Oriens. The independent farmers (*akkara*) of Aramaic-speaking Mesopotamia persisted throughout Late Antiquity.[170] More than half of all building inscriptions in Syria I/II, Mesopotamia, Phoenice Maritima /Libanensis, Palaestina Tertia, and Arabia come from villages,

165 A. F. Norman, *Antioch as a Centre of Hellenic Culture as Observed by Libanius* (Liverpool, 2000), 49 = *Or.* 11.210 *ὥσπερ γὰρ ἐν ταῖς γεωργίαις τὸ μὲν ὀλίγον γήδιον ῥᾳδίως, οἷον ἐν ζωγράφων πίναξιν, ἤσκηται, αἱ δὲ τῶν εὐδαιμόνων ἄρουραι τοῦτο οὐχ ὑπομένουσιν.* The term also appears in the fourth century in Sopater Rhetor's *Διαίρεσις ζητημάτων*, ed. Waltz, 8.308.21.

166 Procopius of Gaza, 'Un'epistola inedita', ed. Maltese, line 11.

167 See Theophylact Simocatta, *Epistulae*, ed. Zanetto, 5.3 and 8.2.

168 Basil of Caesarea, *In proph. Is.*, ed. P. Trevisan, 3.122.6. *Ὅποιος ἦν πτωχὸς Ἰωάννης, μὴ οἶκον ἔχων, μὴ οἰκέτην, μὴ βοῦν ἀροτῆρα, μὴ γήδιον,μὴ κλίνην, μὴ τράπεζαν, μὴ ἄρτον.*

169 Gregory of Nyssa, *Opera*, ed. W. Jaeger, 1.1.49.9: ... *γεωργὸς γάρ τις ἦν ἐπικεκυφὼς τῷ ἀρότρῳ καὶ πολὺν πόνον περὶ τὸ βραχὺ γήδιον ἔχων ...*

170 J. B. Segal, *Edessa: 'The Blessed City'* (Oxford, 1970), 140–2.

which seldom contain aristocratic houses. Many of them must thus represent the work of local small- and medium-scale proprietors.[171] The farmers who lived near Idjaz, and set up an inscription there, were 'ploughmen' who worked to 'furrow the hateful earth beneath yokes [of oxen]'.[172] Likewise, as presciently pointed out by MacAdam, the Hauran apparently continued to have a healthy community of farmers, traces of which survive in the epigraphic evidence.[173] At Ghariyyah ash-Sharkiyya near Bostra in Arabia, an inscribed late antique epitaph honours Diomedes as 'wealthy from farming'.[174] In 385, the farmer of the village of Nahita, Masalem son of Rab, erected a building using wealth 'from his own farming labours'.[175] Slightly earlier is the inscription paid for by the farmers of Zorava (*Zoraouē geōrgoi*), who had erected a statue of Nike during the reign of Marcus Aurelius or Caracalla.

On numerous other occasions, farmers throughout Oriens were sufficiently well-off to erect village churches or to make donations of silver plate to sanctuaries within their villages.[176] At Qabr Hiram near Tyre, a group of cultivators (*geōrgoi*) paid for the church mosaic pavement. At Zahrani in the territory of Sidon, farmers donated money for the marble revetment of the church ambulatory.[177] Much of the silver plate dedicated to the village churches of Syria during Late Antiquity came from places such as Kaper Barada (Brad), a village that was, as the material remains demonstrate, a prosperous place throughout the last four centuries of Roman rule.

In addition to their contributions to religious buildings, mosaic pavements, and silver plate, some independent farmers could also afford well-built houses. In the Golan, the substantial growth in the number of sites from the Roman–Byzantine periods reflects shifts in the Jewish settlement within the provinces of Palestine. It also suggests the growth of independent farming communities, or perhaps communities of tenant farmers holding leases in emphyteusis. In the case of the Limestone Massif of North Syria, the late antique settlement boom and the wealth expended on private homes is striking. Ancient place names that

171 Mango, *Artistic Patronage*, 6–10.

172 Butler et al., *Syria*, vol. III, no. 1023, though there is no certainty these cultivators did not belong to an *epoikion*. Trombley, 'War and Society in Rural Syria', 158, favours the interpretation of independent farmers as responsible for the inscription.

173 H. I. MacAdam, 'Some Aspects of Land Tenure and Social Development in the Roman Near East: Arabia, Phoenicia and Syria', in T. Khalidi (ed.), *Land Tenure and Social Transformation in the Middle East* (Beirut, 1984), 53. Diomedes: R. Dussaud and F. Macler, *Voyage archéologique au Safâ et dans Djebel ed Drûz* (Paris, 1901), 203 no. 88.

174 MacAdam, 'Epigraphy and Village Life', 113; MacAdam, 'Some Aspects of Land Tenure'.

175 Waddington, no. 2412*l*.

176 Mango, *Silver from Early Byzantium*, 6–13.

177 Chéhab, *Mosaïques du Liban*, 101, no. 3.

endure in this stony but fertile upland silently witness the persistence of independent late antique farming communities, until the abandonment of the sites sometime in the early medieval period. The Semitic *kefr/kapr* of Kafr Nabo, Kafr Lata, Kaper Barada (Brad), and Kapropera (now El Bara) attest to these settlements having been *kōmai*, that is, fiscally independent villages, that were responsible for their own taxation liability.[178] These kōmai were some of the greatest settlements of the so-called 'Dead Cities': Kapropera, and Kaper Barada were several times larger than a number of upland villages (a number of which were *epoikia*—estate-owned villages), and Kaprozebada (Zebed), their counterpart in the eastern basalt hills, rivals some late antique cities in the extent of its ruins.

In aggregate, the data from textual references, epitaphs, building inscriptions, silver donations, and place names suggest that independent farmers flourished throughout late antique Oriens.[179] The presence all over the region of finely decorated houses also points in the same direction. It is unlikely that farmers who did not have some security of tenure would choose to invest in substantial, decorated dwellings. It is probable that emphyteutic leases provided stability akin to ownership and thus encouraged such investment. Such long-term lease-holders may have built fine homes, safe in the knowledge that the dwellings could not be easily taken without compensation.[180] However, it is also true that villages in the Limestone Massif, with names suggesting that they were estate villages (epoikia), such as Bamuqqa, and Batūta, had fine houses. Farmers who worked as labourers or tenants on great estates apparently built, or were provided with, houses of good quality.[181]

[178] Place names: Tchalenko, *Villages*, I.312; III.8c, 9, *kefr/kaper* denote independent communities, and contrast with Bet/Beit which may indicate villages belonging to an estate. Obviously, the usefulness of place names in determining land tenure is limited: some villages carried names older than our period or had names that bore no reflection of their make-up. In favour of taking these names seriously as evidence for Late Antiquity is the fact that the foundation and heyday of most of these villages was restricted to the fourth–seventh centuries. Place names may offer some insight into the social composition of various settlements. An interesting example presents itself in Kaper (Kefr) Koraon: the place is mentioned as *ἐποικίου Κοραων*; it is later called *κώμη Καπερκοραον*. This may indicate that Kaper Koraon began its life as an estate village sometime in the second or third centuries, but by the sixth century had evolved into a free village peopled largely by independent farmers (*SEG* XLIV.1322); its evolution, with a dependent settlement becoming a village of independent farmers, is therefore in exactly the opposite direction to that posited by a model of increasing late antique oppression.

[179] We must be careful not to ignore other possible explanations for the wealth of the countryside: the rich of Antioch were known to have owned ten or even twenty houses. The investment in churches, inns, and other public buildings in these villages may therefore have come not from within the villages, but from the patron's funds.

[180] Given the fact that the landowners would have been legally bound to pay for the improvements on the land when rescinding a lease, the building of such structures may have deterred the landlord from such action.

[181] Tchalenko, *Villages*, I.384–5, 400. Such villages might even be embellished by the landlord: John Chrysostom, *In Acta Apost.* = *PG* 60.149.

The refinement and implementation of the system of long-term leasing was one the most important changes in land-tenure during Late Antiquity. Emphyteusis is prominent in a series of initiatives taken by the emperors transferring imperial land to private cultivators and was institutionalized by the late fifth century.[182] At the outset, the authorities found it difficult to define the precise relationship between owner and lessee and among these two parties and the land itself. As the law of Zeno defined it (*CJ* 4.66.1), emphyteusis was a privilege distinct from lease and ownership, while possessing certain characteristics inherent to both.[183] A fixed annual fee—or in lieu of this, public service—was due for the right to the perpetual lease. Only the failure to pay the annual amount could lead to eviction. So long as the tenant retained the contract by paying the annual fees or labour, the holding was alienable (though the landlord had a right to intervene) and could be inherited by the tenant's descendants. Emphyteutic land could also be transferred through sale; in this case, the property owner had the first right to buy and was entitled to a small portion of the purchase price.

Tchalenko saw the institution of emphyteusis as a key social catalyst for the development of the Limestone Massif villages, and the benefits are immediately obvious.[184] The prominence of the system within the legal codes argues for its evolution and importance. Emphyteusis apparently offered security and generally modest terms, as well as a lack of liability on the part of lessee (if the land suffered harm, it was the owner, not the lessee, who had to absorb the loss). The owners themselves, however, also had much to gain from such

182 Zeno had to refine the legal codes to reflect the prominence of emphyteusis, a lease system that had evolved from previous Greek practice. A. J. Cappel, 'Emphyteusis', in A. P. Kazhdan et al. (eds.), *ODB* (Oxford, 1991), 693, describes these leases as 'perpetual'. Emphyteutic leases were let out for periods of 25–29 years, but they could be passed down to relatives of the original leaseholder for up to three generations; see M. Kaplan, *Les propriétés de la couronne et de l'Église dans l'empire byzantin, Ve–VIe siècles: documents* (Paris, 1976), 46–7. After the transfer through three generations, the lease could be renewed once again, but the lessor was not required to let out the land again in emphyteusis. In actual fact, these leases probably amounted to perpetual access to the land, with the limitations described above (such as the inability to arbitrarily evict) being used to safeguard the liberties of the owner to transform and dispose of the holding. In this way, the lessee was protected from sudden termination of the arrangement and the lessor still maintained long-term rights intrinsic to ownership. Leases offered between the emperor and the church could be perpetual (*Nov.* 17 c.1 and 2 (535)), and Justinian (*Nov.* 120 c.1 and 6 (544)) allowed the churches, apart from that of Constantinople, to grant perpetual leases, whilst the latter was permitted to lease ruined house property in perpetuity.

183 Cappel, 'Emphyteusis', 693. In practice, of course, the situation of one who was a perpetual leaseholder for a distant or uninterested landowner would have been little different from that of outright ownership, and there certainly is evidence to suggest that there was confusion in the function of these leases and the definitions of their holders. Zeno and Justinian constantly juxtapose *emphyteusis* and *dominus*. E. Levy, *West Roman Vulgar Law: The Law of Property* (Philadelphia, 1951), 78–9.

184 Tchalenko, *Villages*, II. 415–17.

arrangements. Private and ecclesiastical authorities could obtain permanent occupancy on their estates and an uninterrupted supply of income. They also freed themselves of the managerial burden of such estates, allowing the lessees to work, produce, and improve the land as they liked. From a fiscal standpoint, the arrangement was a beneficial one: tenants operating under such a lease agreement were less mobile than the landless and thus easier to record and tax.

At least in part, the late antique concern with emphyteutic lease probably reflects a desire by the emperors to increase tax revenues through an increase in productive land. Previously unproductive land was brought under cultivation in this manner and thus taxed. It is probable that in the landscape of the Syrian Chalcidike and the Golan we see the fruits of the central government's efforts to extend cultivation by offering generous lease terms. By the sixth century, there are frequent references to the institution of emphyteusis in the legislation of Justinian.[185] This official recognition of the evolution of eastern land tenure throughout Late Antiquity indicates that the arrangement was deeply embedded within the agricultural landscape of Byzantium. If this is correct, then it is possible the movement to such lease arrangements reflected real changes in the management and possession of land; old tenancies and ownership norms were eroded, and a new group of practices and relationships developed around emphyteutic leaseholders. On imperial land, in particular, the effect may have been marked, with sizeable settlement shifts, intensification, and a reduction of previously uncultivated land.

In terms of wealth, the degree of separation between independent farmers and emphyteutoi was in many instances probably slight. Many emphyteutic leaseholders were not poor to begin with; one had to pay for the lease contract upon inception (*CJ* 4.66.2–3) and probably the annual payment or labour as well.[186] The *Theodosian Code* (10.31–7) deals with land let out in this manner, primarily land belonging to cities, much of it apparently leased at the time by curials. Evidence for such leases is known from Petra; there, an emphyteutic lease was contracted in perpetuity by a certain Gessios during the reign of Justinian or Justin II.[187]

Emphyteusis may also have been the arrangement in operation between Libanius and his recalcitrant Jewish tenants near Antioch. For generations, these lessees had been on the same soil and in the service of Libanius and his ancestors. They bribed soldiers and other officials in an attempt to gain greater independence, using money that Libanius claimed was due him. In

185 See, for example, *CJ* 1.2.24.14; 1.2.17.3; 11.62.12.1.

186 Levy, *West Roman Vulgar Law*, 77–8.

187 T. Gagos and J. Frösén, 'Petra Papyri', *ADAJ* 42 (1998), Inv. 86.

civil court, the wealthy Antiochene philosopher won his case against his tenants, who were subsequently jailed. As a result of this, their relatives and friends immediately presented gifts of barley, wheat, ducks (which may indicate that Libanius' property lay near the Orontes or Lake Antioch), and horse fodder to the local military commander, who released the prisoners (*Or.* 37.13).[188] Whilst emphyteutic lands were often imperial lands let out by the central government, by the fifth century, ecclesiastical and private individuals had also entered into such agreements.[189]

There were numerous other forms of rental and lease agreements made in private treaties between landowners and tenants. The Demai tractate of the Jerusalem Talmud displays a wide range of lease arrangements in action, including leases, rents, and sharecropping. According to this Rabbinic text, renters paid in coin and lessees paid in produce, while the sharecropper paid in labour, taking as payment one-half, one-third, or a quarter of the produce.[190] The same document also describes sharecropping arrangements in which the owner of the property receives a portion of the field's produce as rent. In the case of a vineyard, the share was rendered in wine, that of the olive grove in oil, and so on. Leases and rental agreements apparently were often quite variable in their contractual arrangements, involving handing over produce, labour, money, or a combination thereof. The Rabbinic sources further illuminate the range of leases and contractual labour agreements among Jewish communities in Mesopotamia and Palestine.[191] A form of arrangement, best described as a type of *métayage* in which the landlord provided the seed and the tenant paid half the crop in return, was common. Tenants usually bore short-term leases (annual) on which they paid a fixed rent in produce or coin. Sometimes the crop demanded as payment by the landlord was not the same as that which the tenant had planted in the rental plot. A tiller might lease ground and sow barley, but the landlord might demand dates as the rental price. Another form of leasing was the lifelong sharecropping agreement, in which the master or tenant could terminate the contract at any time. In such circumstances, the landlord might owe money

[188] The evidence for the type of lease arrangement is inconclusive. We cannot assume that these Jewish tenants were coloni. They certainly do not behave like powerless dependent tenants; on the contrary they marshal substantial wealth in defence of their interests. Libanius' belief that his Jewish tenants used his own money to bribe soldiers does not prove that he viewed the *whole* of the farmers' property as his own (as legally he could have done had they been coloni). It is most likely that the farmers used the portion of rental produce or money due Libanius to bribe soldiers and thus switch to more immediately powerful patrons.

[189] Cappel, 'Emphyteusis', 693.

[190] TJ Dem., *passim.*

[191] J. Newman, *The Agricultural Life of the Jews in Babylonia Between 200 C.E. and 500 C.E.* (London, 1932), 49–61, is the basis for the following discussion.

for capital improvements made by the permanent planter who wished to terminate the lease.[192] Upon the death of the tenant, the holding might be leased in turn to his family, but the owner of the plot could also choose to pay the family for the previous tenant's capital improvements and resume control over the tract. Such an agreement is strongly reminiscent of many of the features distinctive of emphyteutic holdings. These contracts perhaps reflect age-old Near Eastern land tenure practices, aspects of which very likely underlie Graeco-Roman and Byzantine systems of organization.

Also worth noting are the regulations governing usufruct. This differs from leases and emphyteusis in that the beneficiary was not responsible for bad harvests and had certain powers over the land, but apparently not as extensive as those of people holding land in emphyteusis (*Digest* 7.1.9; 7.1.10). A variety of leases thus existed side by side, dependent upon local custom and the wishes of individual landlords.

Despite extensive evidence of rural prosperity, conditions for cultivators—whether they were tenants or hired labourers—could be harsh, and many lived at, or slightly above, subsistence. Appalling conditions for agricultural workers, and entrapment through debt, were common.[193] The comments of John Chrysostom (*In. Matth.* 63.3) on conditions of labourers around Antioch are well known.[194] The saint renders a picture filled with oppression: the workers (geōrgoi, in this case probably coloni) on the estates of the wealthy of Antioch owed heavy rents and labour services. These labour obligations are apparently private obligations to the landlord and involved tending fields in addition to those which the geōrgoi worked for themselves. These obligations are insisted on even when the harvest is bad. Additionally, the farmers shouldered heavy debt, were oppressed by overseers (*epitropoi*), and were imprisoned. Later in the same passage, Chrysostom addresses the fate of the farmers who worked in wine production, one of the chief sources of wealth for the rich of Antioch. Forbidden to take any of the produce of the full wine vats over which they toiled, the workers received little money. This picture accords well with that noted by Sarris for the labourers on Egyptian great estates. There workers were expected to husband their own plots, but were also required to till the owner's own in-hand land.[195]

Active landowners, who often employed managers to direct their tenants and paid labourers, contrast with landowners like Libanius. Libanius' style of estate management was apparently more detached; he seemed to care for the

[192] Ibid., 60.

[193] See Bagnall, 'Landholding', 75–8; J. B. Segal, 'Mesopotamian Communities from Julian to the Rise of Islam', *ProcBritAc* 41 (1955), 117–19.

[194] *PG* 58.591.38ff.

[195] Sarris, 'Origins of the Manorial Economy', 285–9.

rent, not the precise mechanisms of estate production. His picture of the renters around Antioch is also very different from that of his pupil Chrysostom. Libanius' lessees used money to seek the protection of the powerful, rather than having patronage imposed upon them. As to the identity of their new patrons, Libanius is purposely vague, but he probably is speaking of military officers or (less likely) wealthy patrons within Antioch. Once secured by their new protectors, the lessees are able to endure the owner's lawsuits against them and defied his attempts to manage their work.[196] The tenants' sunset in Chrysostom melds into the tenants' sunrise in Libanius. Chrysostom's subjugated farmers may have overcome Libanius' overseers, but how often?

The above discussion has outlined the textual evidence for the continued presence in the East of small- and medium-scale landowners. We can be certain of the continued persistence of this group of landowners throughout Late Antiquity. What eludes us is a firm grasp on the proportion of the population who were landowners and how property was distributed among them. At Hermopolis, the metropolis of the eponymous nome in Egypt, Bagnall suggests that perhaps 20 per cent of urban dwellers owned land and that throughout the whole of the population of the nome, a 35 per cent rate of land ownership was likely.[197] Despite a concentration of land in the hands of a few individuals and institutions, a surprisingly wide array of people owned small plots of land. At third-century Philadelphia in the Fayyum, the owners included soldiers, veterans, women, doctors, and town craftsmen.[198]

In a well-known passage, John Chrysostom informs us that 10 per cent of Antioch's population was rich, 10 per cent poor, and the other 80 per cent lived somewhere in between.[199] Were the middle 80 per cent landowners or were they craftsmen and merchants ultimately dependent upon the plutocrats for their patronage and wealth? Again, comparison with contemporary Egypt is instructive. A mere three Hermopolite families held more than half the city's lands.[200] Nevertheless, we also are assured of the survival of a large

[196] The reference to the failure of the courts to do justice for the owner is almost certainly based on his own experience, in which Libanius brought some of his own tenants to trial and obtained a conviction, only to see the *dux* overturn the ruling. His elliptical language in this passage was thus almost certainly due to his desire to avoid further conflict with the military men responsible.

[197] Bagnall, 'Landholding', 139–40.

[198] A. K. Bowman, 'Landholding in the Hermopolite Nome in the Fourth Century A.D.', *JRS* 75 (1985), 151.

[199] *PG* 58.630.14 ff.

[200] Bagnall, *Egypt in Late Antiquity*, 68–9. I arrive at the total of the wealthiest holdings (56.3 per cent) from the following statement: 'In Hermopolis . . . it appears that 38.8 percent of the *total* land is owned by six members of a single family, and another 17.5 percent is owned by two additional families.' (Original emphasis.)

number of modest landowners: 53 per cent of the Hermopolite owners possessed quantities of land unlikely to provide them with their subsistence, whilst another 35 per cent belonged to what we have termed the medium-sized landholders, those who could live comfortably from their rents but who were far below the city's wealthiest.[201] In the sermons and speeches of the elite of Oriens, we have met the holders of gēdia. These people occupied a position in life that possessed a certain precarious dignity, although Basil equated life on the gēdion with bare subsistence and a minimal holding for survival. Even if this is a rhetorical device designed to denigrate his opponent Eunomius, it must have resonated with his audience.

The case of Eunomius, mentioned earlier, is worthy of further comment. From relatively obscure origins on his family's gēdion, Eunomius came to be a cultured leader of the neo-Arians, a fact that illustrates the pretensions and upward mobility of his peers and argues that life for the small and medium landholder was not always of the meanest sort. The mobility and wealth which this group *could* attain is well-illustrated in the epigraphic record, particularly in the Hauran, where strong village identities and pride exhibit themselves through euergetism drawn from the wealth of the farmers.[202] It is in the villages and the farmsteads and estates that one may further assess systems of ownership, production, and wealth.

The make-up of the landscape of Oriens was thus heterogeneous. Diversity amongst types of farm structures underscores the multiplicity of methods used to tame and exploit the land. It is clear that institutional estates, both ecclesiastical and imperial, were an imposing presence on the land. It is also true that the estates of wealthy, private individuals stand out in the corpus of evidence. Thus there is much to support the contention that there was a widespread class of flourishing agriculturalists who held extensive tracts of land. But though the prevailing orthodoxy amongst historians of Late Antiquity argues for steady elite aggrandizement, to the great detriment of those lower down in the social order, the present evidence from Oriens cannot sustain the notion of a collapse of the ranks of middling and small farmers.[203] The resounding prosperity of the wealthy cannot mute the substantial textual

[201] See Bagnall, ibid., 69, who estimates that the middle group of landowners amounted to 49 per cent of the total landholders.

[202] See *IGRR* III.1187 (= Waddington, no. 2546); *IGRR* III.1213 (= Waddington, no. 2399); see *SEG* XXIX.1592 for a group of fifth-century village elders at Huarte near Apamea who paid for a mosaic pavement in the Michaelion. See also Waddington, nos. 2391, 2405.

[203] The dominant view is articulated by M. Kaplan, 'L'exploitation paysanne byzantine entre l'antiquité et le moyen-âge (VIème–VIIIème siècles). Affirmation d'une structure économique et sociale', in V. Vavrínek (ed.), *The Byzantinological Symposium in the 16th International Eirene Conference* (Prague: Academia Praha, 1985), 101–6; Kaplan, *Les hommes et la terre*; E. Patlagean, *Pauvreté économique et pauvreté sociale à Byzance 4e-7e siècles* (Mouton, 1977), to mention but a few.

and material data that trumpet continued resilience amongst the small and medium-sized landowners.

Large estates did expand throughout the empire; of this there is little doubt. But it is critical that we not assume that this progression of land-gathering into the hands of the wealthy advanced unchecked. For instance, some estates are known to have been broken up through inheritance: despite the overwhelming preference for keeping landed holdings intact, the practice among families would often be otherwise. In the fourth century, one the wealthier landowning couples in all of Cappadocia, the parents of Basil of Caesarea, Macrina, and Gregory of Nyssa, divided their estates amongst their nine children.[204] Estates were likewise altered through confiscations, death without heirs, and donations to the church. Wars and plague smote the landowning aristocracy in the same way that they attacked poorer farmers. In any of these cases, estates might be left largely intact in the hands of their new master (state or church), but they might also be leased or sold. As the ranks of the small- and medium-scale landowners were constantly depleted, so they were also replenished. Discharged veterans and serving soldiers, well-to-do artisans, and profit-making merchants wished to invest in land because it was the most stable and desirable venture in the ancient world. This diversity and ebb-and-flow in land tenure amongst rural society persisted into the seventh century. To my mind, the question remains open as to whether or not the Levant, as a whole, made a fundamental tenurial shift and thus transformed itself from an economy in which there were manors to one dominated by them.

What is beyond doubt is that there were essential changes rumbling at the foundations of late antique society. Demographic growth, interrupted only briefly by the Justinianic plague, fuelled a thirst for land. Accompanying the drive to open up new lands was the increasing activity of a wealthy group who had the capital, the knowledge, and, above all, the interest to further enrich themselves.

204 Gregory of Nyssa, *Vita Macrina*, ed. P. Maraval, 20.11–23.

3

Hand to Mouth

Grain in Late Antiquity

Like most pre-industrial people, those in Late Antiquity lived on the razor's edge of hunger. An entire social order relied on slender stalks of wheat and barley. Multiform threats loomed over each crop: war, pestilence, adverse weather, disease of human, beast, or plant, and attacks by insects destroyed crops, ravaged the countryside, or levelled the human or animal population. From the emperor on the Bosphorus to the humble cultivator at Nessana, all looked uneasily towards the harvest. Priests prayed for the produce, and astrologers sought omens in the heavens to predict what to expect from the cornfields. For a landless worker, as much as half the year's wages could be consumed by the cost of food.[1] At its best, grain production was never superabundant, although a good crop meant plentiful supplies in the market and ensured full bellies and seed for next year's sowing. In good years, families could spend some of the money normally reserved for subsistence on other items. Bad years meant shortages, speculation, and sharply rising prices in the market and sometimes urban unrest and political upheaval. Restive city populaces in Antioch rioted in 333 and 384–5 when grain was scarce and prices were high.[2]

The urban poor were most beset by such fluctuations: they were at the mercy of the wealthy landowners of the urban-based aristocracy who controlled large cereal stocks and who could sway market prices. In lean years, taxes could not be collected from farmers who had nothing to eat, and landlords who made no profit would obstinately resist their obligations. The latter framed much of their wealth within the context of control over cereal grains; when crops failed, they were insulated from starvation by their cash reserves and grain stockpiles. With few other avenues to quick gains within a predominantly agrarian society, dearth provided a unique opportunity for landowners to convert commodities into the cash that defined their wealth. During shortages, large-scale landowners in Oriens and throughout the empire made their greatest profits, and thus ensured not only the survival

[1] Banaji, *Agrarian Change*, 66.

[2] D. Stathakopoulos, *Famine and Pestilence in the Late Roman and Early Byzantine Empire: A Systematic Survey of Subsistence Crises and Epidemics* (Aldershot, 2004), 73–4, 182–3, 209–10.

of their families, but the perpetuation and extension of their status. Shortage was therefore a significant profit mechanism within the structures of late antique society, which merchants and large-scale landowners positioned themselves to exploit.

Only in the most extreme cases was there no grain to be had at all, but when it struck, famine killed thousands as it did at Edessa 500–2.[3] Then, as in nearly all such cases, there were two distinct parts to famine, and speculation and price gouging played a role. In the first, the market costs of various food items rose steeply in response to the pressure of demand. In the second, the profit motive induced those who still had stocks of food in their private stores to hoard it in hopes of greater gain; and they did so in a measured way and in response to the market. Sources are not available as to whether or not landowners planned ahead to profit from dearth, but this is likely given that estate owners are known to have planned to profit from seasonal variation in grain supply. As early as Varro's day, of the first century BC, it was stressed that the great estate should hold back its grain until well after the harvest (Varro, *De re rustica* 1.69.1). When grain stocks were at their lowest in the spring immediately before the harvest was the time when maximum profits were obtained. The opportunity to cash in on market fluctuations and difficult times was thus built into an estate structure that was inherently conservative in outlook. For any number of reasons, grain was to be laid up for years (*Geop.* 2.27), but primarily to balance the mercurial harvest cycle. Undoubtedly another consideration was the unspoken, but tremendously important desire to take advantage of local fluctuations in the market.

Although shortages are sometimes noted in the sources, these were generally localized and of short duration.[4] The populous cities of the East—from Alexandria, to Caesarea, to Antioch—testify to plenty; for without a stable, abundant supply of food, the structures of the city would not have been possible. The specialist crafts and trades of the late antique urban world were manifestations of the agrarian subsoil of their society. At the core of civilization stood the farmer on whose humble pursuits depended the most vital component of *civilitas*: a full belly.

HOW MUCH? FAMILY SUBSISTENCE REQUIREMENTS AND URBAN DEMAND

Not since humanity's brief sojourn in the Garden of Eden has there been such a superabundance of food as that which is presently available in many

[3] Ps.-Joshua, *Chronicle*, trans. Trombley, Watt, 38–46.

[4] Recently collected and commented upon in the valuable study by Stathakopoulos, *Famine and Pestilence.*

industrialized countries. Too little food, not too much, was the problem with which the ancients were most familiar. In this, they shared a lot in common with all pre-modern societies. Calculating how much food is required by any group at any time is highly variable and fraught with difficulty. In the case of late antique Oriens, a sufficiently nuanced study is presently impossible, though some calculations can be made on the basis of the minimum quantity of food needed to keep a physically active person alive and healthy, for which all strove equally. Foxhall and Forbes reckon that the total daily calorie requirement of an ancient family of six was about 15,495.[5] As much as 70 per cent of the ancient diet consisted of cereal grains, which would therefore amount to 10,847 calories per day. A kilogram of coarse, ground durum wheat flour contains about 3,390 calories. The annual maximal wheat needs of the family were therefore around 1,168 kilograms of flour, or about 1,226 kilograms of whole wheat.[6] This is about 94 imperial modii of wheat, a convenient figure from which to hypothesize, nothing more.[7]

Fiensy summarized the various estimates, made on the land area, required to feed the ancient family in Palestine.[8] These range from 2.43 ha (9.72 iugera) to 6.8 ha (27.7 iugera).[9] In testing these assumptions, the decisive factor is yield. Ancient yields are very rarely known, particularly from Greater Syria. In any case, the harvest would never have been uniform across the empire for myriad reasons, ranging from soil quality to the type of grain, seed selection, the timing of the harvest, and the impact of disease and pests, to name but a handful.

A critical piece of data does survive. *P.Ness* III.82 offers, to my knowledge, the only fourth–seventh-century quantified data from Oriens from which we can estimate late antique grain yields.[10] There, the harvested grain-to-grain sown ration is 7:1 for wheat and 8–9:1 for barley. It has been argued that the documents from Nessana represent especially high returns based on irrigated

[5] Based on the study of L. Foxhall and H. A. Forbes, 'ΣΙΤΟΜΕΤΡΙΑ· The Role of Grain as a Staple Food in Classical Antiquity', *Chiron* 12 (1982), 48 n. 24, whose cautionary notices are highly relevant; it is hoped that their care is reflected here.

[6] Assuming 5 per cent losses in milling.

[7] Throughout this chapter I have used the 40 litra imperial modius and a litra reckoned at the weight of the *λογαρική λίτρα* of 325 grams. Schilbach, *Byzantinische Metrologie*, 162 ff.

[8] Fiensy, *Social History of Palestine in the Herodian Period*, 94.

[9] Dar and Applebaum, *Landscape and Pattern*, 74, provides data from the Roman–Byzantine site of Buraq in Samaria, where an average area a small-farmer worked for cereals was about 0.67 ha, a little over 2 iugera; D. E. Oakman, *Jesus and the Economic Questions of His Day* (Lewiston/Queenstown, 1986), 64; K. D. White, *Roman Farming* (Ithaca, 1970), 336.

[10] C. Kraemer, *Excavations at Nessana*, iii: *The Non-literary Papyri* (Princeton, 1958), 237–43. Sitos: *P.Ness.* III.40.1, 8, 11–23; 60.13, 15, 16; 61.13, 15, 18; 62.14, 16, 19; 63.9, 11, 14; 64.10, 12, 16, 21, 22; 65.[9], 10, [14], 18; 66.8, 10, 12; 67.11; 69.2, 7, 8, 11, 12; 80.2–7; 81.2–5, 7; 82.1–6, 10–13; 83.1, 2, 4–9, 11; 84.4; 89.4, 5; 89.27; 94.8; 95.41; 164.3.

farming, and therefore cannot be used to generalize.[11] Such doubts are however easily overcome. The Negev is so ill-favoured agriculturally that floodwater farm installations were required to produce any kind of cereal crop. At best, the late antique water management systems there would have done nothing more than place the Negev floodwater-farmed landscape on a par with the coastal plain. From a historical perspective, account records provide primary evidence of the utmost importance. Farmers may be deceptive about the amount of grain sown and reaped in order to evade taxation or rent, but to have successfully done so, their fabrications must have been within the limits of reason. We can therefore be assured that the document we possess reflects a shade of reality, not only for Nessana, but for Oriens in general.[12]

Charting Demand in the Late Antique City: the Case of Antioch

In 362 at Antioch, the emperor Julian (AD 361–3) faced a food crisis created by a number of circumstances. Low rainfall in the winter and spring led to a poor harvest. Most of the blame, though, may be assigned to the imperial court and army, whose presence strained the already taut infrastructure of the city. The emperor responded to the crisis by fixing prices. In so doing, Julian severed a major avenue of the Antiochene wealthy grain producers and dealers towards (what they considered) acceptable profits. Thus affronted, they held their grain back from the market and left the emperor at the centre of a social maelstrom. Faced with this intransigence, the emperor provided 420,000 modii of wheat from the Chalcidike plains surrounding Chalcis and Epiphania and from upper Syria around Hierapolis (*Mis.* 368; 369A). To appreciate the scale of this undertaking, consider the following: transportation of this grain shipment would have required 28,000 camels. The supply was sufficient to feed approximately 262,500 adult males for a month or 4,468 families for a year.

From this incident, we have useful indicators of the size of late antique Antioch, of the transport capabilities of the imperial government in times of food crisis, and a compelling reminder of the productive capacity of the Syrian landscape in the time when the eastern provinces witnessed a vital phase of expansion and agrarian industry. The one place whence the emperor,

[11] Patlagean, *Pauvreté économique*, 247–8.

[12] And a defensible one; contrary to the assertions of Patlagean, the yields of Nessana are hardly spectacular. In the Early Roman Period, Pliny (18.94) provides several notices of yields which were considered magnificent in his day, such as 150:1 in Byzacium. The produce of the Nessana grain fields is feeble by comparison, and extraordinarily low under any modern farming regimen. They are comparable to, though more conservative than, the Italian yields recorded by Varro (*De re rustica* 1.44.1) of 10:1 or 15:1.

whose inability to persuade and unwillingness to compel is so starkly underscored at Antioch, could demand the required grain was from his own land. The imperial estate lands were in all probability the Saltus Eragiza, probably centred on Hierapolis (Mambij) and those around Anasartha, part of the vast Klima Anatolikon. Assuming each iugerum of grain land was ploughed twice (*Geoponica* 3.3), the total number of iugera worked to provide this shipment would have been 52,500. Over a three-week ploughing season, 2,500 iugera would have required ploughing each day in order to complete the cultivation, though in reality we may doubt that the tenants and slaves on imperial estates could have been so efficient. At a minimum, then, Julian mobilized the produce of more than 26,250 iugera of land and the sweat of more than 2,500 cultivators.

The bread procurements during the reign of Julian clearly were exceptional, but the annual demand for cereals in a city such as Antioch was significant nonetheless. Recent work suggests that the inhabited area of the city was even larger than originally thought, with suburbs strewn in all directions.[13] Although rivalled in scale in Oriens only by Apamea, Antioch was not unique in the Roman–Byzantine world. Alexandria and Constantinople would have required as much grain and more. Smaller cities like Jerusalem, with perhaps 30,000 people, may have needed around 470,000 modii of wheat for their annual sustenance. A host of cities were close to Jerusalem in population: Anazarbus, Tarsus, Seleucia-in-Isauria, Mopsuestia, Berytus, Gaza, Bostra, Laodicaea, Hierapolis, Amida, Theodosiopolis, Edessa, and Constantia probably needed about the same quantity of food. This is to say nothing of scores of lesser cities that combined with the metropoleis to create a large number of predominantly agriculturally unproductive mouths. To meet basic subsistence needs, Antioch, with its urban population of 150,000, would have needed around two 2.5 million modii of wheat.[14] With the suburban and rural inhabitants taken into account, the city and its territory is likely to have supported 250,000–500,000 people.[15]

The physical territorium of Antioch was geographically complex, comprised as it was of riparian lands, mountains, and plains. The best cornlands of the city lay around the foot of the limestone scarp, known collectively

[13] J. Casana, 'The Archaeological Landscape of Late Roman Antioch', in I. Sandwell and J. Huskinson (eds.), *Culture and Society in Later Roman Antioch: Papers from a Colloquium, London, 15th December 2001* (Oxford, 2004).

[14] 2.35 million modii to be more precise. The total population is divided into 25,000 of our hypothetical families of six for the purposes of these calculations. The figure of 150,000 people *intra muros* is conservative, based on the estimate of Will, 'Antioch sur l'Oronte: métropole de l'Asie'; Liebeschuetz, *Antioch*, 95, suggested that Antioch possessed 150,000–300,000 people within its walls, while clearly favouring the lower figure, Liebeschuetz, ibid., 41.

[15] Foss, 'Syria in Transition', 201.

as the Massif of Belus (Limestone Massif) and in the river valleys. Recent surveys of the Amuq plain around Antioch have revealed a high concentration of small sites belonging to Late Antiquity.[16] The city, an expanding metropolis, clearly needed farmland. Assuming half the corn-land lay fallow in a given year and yielded at the level documented at Nessana, the maximum proposed population of 500,000 around Antioch constituted slightly more than 83,000 families (assuming about 6 people per family), each of which required 94 modii of corn annually or the equivalent output of about 12 iugera of grain land. Assuming annual cropping, the grain fields of Antioch would cover 996,000 iugera (249,000 ha or 2,490 km^2). The calculated land area is substantial. Translated into the landscape, the area required for cereals alone at Antioch is approximately half the total area of the city territory, of about 6,000–7,000 km^2.[17] In reality, only 25–31 per cent of the total territory land could have been farmed in cereals.[18] The estimated minimum lands (nearly 2,500 km^2) required for cereal cultivation, therefore, far exceeds the probable total arable area of the ancient city territory, which likely ranged between 1,500 and 2,170 km^2.

Apart from the cities of Rome and Constantinople, which certainly depended on imported grain, it is assumed that most cities were organized so as to feed themselves from their own territories. Since the calculations above, though crude, are conservative, the impression is that Antioch's demand outstripped, possibly substantially so, the normal productive capacity of its corn-lands.[19] Indeed the above calculations appear to favour the idea that the landscape around Antioch was intensively farmed, with the fallow suppressed and grain lands planted annually.

The reduction or elimination of the fallow was one of many intensifying strategies used by ancient farmers. The Antiochenes also extended the arable surface of their city territory, thereby enlarging the area of grain lands. This

[16] Casana, 'The Archaeological Landscape'.

[17] Liebeschuetz, *Antioch*, 41.

[18] Assuming the arable area lay around the 25 per cent recorded by a recent environmental survey of the Antakya region. S. Kiliç, S. Senol, and F. Evrendilek, 'Evaluation of Land Use Potential and Suitability of Ecosystems in Antakya for Reforestation, Recreation, Arable Farming, and Residence', *Türk tarim ve ormancilik dergisi* 27 (2003), 19. Grain is most suited to open fields, not to terraces and the rugged hills of Antioch. Though certainly these areas did produce grain in antiquity, their real agricultural value was in other, better adapted, crops.

[19] Three to six pounds of bread were consumed per adult per day. Patlagean, *Pauvreté économique*, 46, 52 (Patlagean's 3–6 lb is a very high figure and would represent an absolute maximum). Annually, this amounts to 1,095–2,190 lb (821–1,643 lb of grain based on loaves comprised of 75 per cent flour), or 21.5–43 modii (40 lb imperial modius, with 5 per cent lost in processing). A 'typical' family of six, in which three adults received a full ration and children half, thus consumed 97–194 modii per year. Kahzdan's figure for the Late Byzantine Empire of 1.5 lb per day would yield a wheat equivalent of about 1.4 lb of grain per day, or only about 13 modii per year.

was accomplished in two ways: first by expanding cultivation into desert areas within the city territory, and second by extending the territory itself. We know that in the mid-fourth century, Antioch had incorporated into its own area the city and territory of Gindarus (modern Jandarīs in Syria), a clear example of horizontal expansion of the farmed landscape.[20] Finally, it is possible that grain played less of a role in the diet of the late antique city than we expect. The latter seems unlikely though, given the universal concern that sources show with the cereal diet and the resultant hunger that occurred when the grain crop failed, which indicates an acute dietary dependence on cereals. Demand per capita was thus relatively inelastic.

Before examining the structures of 'improved' agriculture in Late Antiquity, it pays to examine the foundations of farming praxis. Unencumbered in the first instance by problems of intensification and extension, an analysis of agricultural practice is revealing. In this instance, we delve beyond the matter of how *well* the ancients fed themselves to the infrangible question of how they fed themselves in the first place.

The Agricultural Year

In the cycle of days that formed the agricultural year, the rhythm of grain production set the tempo. For millennia, farmers in the Near East had adapted themselves to their surroundings in the quest for sufficient calories and nutrients to sustain and perpetuate life. Late Antiquity thus inherited an age-old agricultural cycle of planting, growth, and harvest of cereal grains. The ancients groped for assurances that the rains would fall, the heat break, or the rains stop and the heat return. In this quest, they were inclined toward supernatural beliefs and practical observations handed down through the generations. Relics of Hellenistic and possibly Mesopotamian cosmic planting guidelines are preserved in the *Geoponica* (1.8), where we learn that if Sirius rose when the moon was in the house of Aries, there would be problems with one's animals, abundant rain, a bad wheat crop, and a good oil crop. In such cases, we see beliefs probably far more ancient than the Roman empire. The persistence of many traditions is demonstrated by their selection by the tenth-century compilers of the late antique material that forms the bulk of the *Geoponica*. Astrology in agriculture—'planting by the signs'—was part of the vocabulary of every rural person, no matter how Christian he or she professed to be. These beliefs also cut across class groups. Though our notice of them

[20] Jones, *Cities of the Eastern Roman Provinces*, 270 n. 55. Though, in absorbing Gindarus, the Antiochenes would also, of course, have absorbed yet more mouths to feed.

comes from texts written and read by the literary elite, the beliefs themselves belonged to the land, to the rural labourers and villagers whose traditions were oral. In them we possess some of the few direct views into the daily thoughts and concerns of common rural people. That such almanacs were common and of considerable cultural longevity is easily demonstrable. The ninth-century *Nabatean Agriculture*, another medieval compilation that collected late antique material from Mesopotamia, preserves lore similar to that in the *Geoponica*, and in even greater abundance.[21] At least some dependence on ancient Greek sources is fairly certain in the *Nabatean Agriculture*, but the preservation of the sections on weather signs and magic were due to the choices of the compilers and argue for their enduring appeal and social acceptance.

Among the atmospheric and astrological lore are distinct markers which guided the cycle of sowing, care, and harvest. In his *Opus Agriculturae*, the Late Roman agronomer Palladius created a catamenial agricultural guide with a regimen based on the Roman months and anchored within starlore: when Sirius rose (7.9), for example, plants in specially prepared trial beds were sometimes examined. Those which did well could be expected to flourish in the following year, and those that failed were not to be planted. Much of Book 1 of the *Geoponica* is dedicated to the 'air', that is, methods of prognostication of meteorological phenomena and planting signs.[22] Book 3, however, preserves a complete *ephemeris* governing the farm. In this calendar, we see the regimen not only of the great agricultural estate, with its fruit orchards, livestock, vines and olives, and irrigation, but of the small-scale farm as well; the difference was fundamentally a quantitative one, with scale the key variant. Even if one imagines that these precepts and planting guides were somehow a creation of the late antique elite, the knowledge could in no way have been restricted to the elite circles since the size and labour demands of the biggest estates drew tenants and workers from surrounding areas. One should always be mindful of how these handbooks provide a mirror in which we see a reflection of the great estates within the countryside. The same mirror, however, also provides a glimpse of how various segments of rural society interacted within the workings of the large farm. In the course of their work, labourers and small-scale farmers would have seen first-hand the regimen and techniques employed on the centralized estates. Methods judged efficacious and feasible would thus have made it into the repertoire of the free

[21] R. Rodgers, 'Hail, Frost, and Pests in the Vineyard: Anatolius of Berytus as a Source for the Nabataean Agriculture', *JAOS* 80/1 (1980).

[22] One of the four farming cardinals, Palladius, *Op. Ag.* 1.2: *Primo igitur eligendi et bene colendi agri ratio quattuor rebus constat: aere, aqua, terra, industria . . .*

tenants and freeholders. Equally important was the change wrought by the farmer communities on the great estates, whose economic viability and management in the end relied primarily on local workers. The large farms thereby tapped into a vast body of local experience, practice, and opinion. The intellectual curiosity and willingness to experiment exhibited by the handbooks must in large part reflect not only those methods imposed by the richer farmers, but those that they and their managers learned from their experienced tenants, field hands, and neighbours. In any case, with crops and climate dictating their own cycle, differences between richer and poorer in farm management were less significant.

Cultivation

The small-scale farmer had limited control over the type of the land he or she cultivated. Only free tenants possessed some flexibility in the selection of the land they worked, since they, theoretically anyway, possessed some mobility. The vast majority of family holdings were inherited, rented, or purchased around ancestral villages, so most small farmers could not geographically distribute risk factors. The wealthy, by contrast, had amongst their primary aims acquisition of geographically scattered lands suitable for productive agricultural returns under different soil and weather conditions. This focus is traditional in the ancient Mediterranean and sought to mitigate risk through diversity.[23] The small-scale farmer could exercise some control over crop selection, the rate of manure application, and the type of tillage undertaken. Repeated in the third-century treatise of Julius Africanus (*Frag.* 1.19.5) is the standard creed that plains and level ground were best suited to, and reserved for, cereal production, whilst orchards and vines were grown in the uplands. Of course this ideal did not always correspond with reality. Level fields and rich, black earth, commended as best by both the *Geoponica* (1.9) and Palladius (1.6), were not always available for cereals. Grain grown in the hills yielded less, but could be of good quality (Palladius, *Op. Ag.* 1.6.12). Uplands and poor quality soils were indeed especially important in Late Antiquity, when many marginal lands were opened to cultivation. Infertile earth posed a challenge but was looked to with increasing frequency: in Arabia, says the *Geoponica* (2.23), some light soil was not tilled, but, prior to the rains, simply sown with barley—a reference to the sowing of wadis and natural basins, where runoff made cultivation viable.

[23] J. W. Nesbitt, 'The Life of St. Philaretos and its Significance for Byzantine Agriculture', *Greek Orthodox Theological Review* 14 (1969), 155–8.

Preparation of the Soil

The late antique farmer faced the onerous task of preparing the earth for sowing through methods which in his day were already quite ancient. On the coast and in regions where there was sufficient rainfall, the land required attention that differed from that of the semi-arid and arid interior regions. In the mountains, terraces, stacked one upon another, might not be ploughed at all but dug arduously by hand using mattocks or spades. However, in regions with abundant water resources, weeding and crop rotation were probably a major focus, since greenery was more prolific. Compared with the interior, coastal and irrigated areas often possessed an extended growing season, allowing greater crop variety, and, in some instances, more than one crop each year. In more arid areas, dry-farming predominated unless irrigation was widely available.[24]

Dry-farming involved frequent topsoil tillage. This was done in order to destroy weeds that sapped moisture, and to distribute and conserve water reserves by discouraging capillary action that facilitated evaporation.[25] Regions receiving 250–500 mm of rainfall are nowadays defined as semi-arid, and by this definition, nearly the whole of Oriens is semi-arid.[26] These are the lands where dry-farming was most applicable, and the handbooks and comparative evidence from the pre-modern Mediterranean suggest that a large percentage of the agrarian population utilized such practices.[27] Modern dry-farming stresses weed control, timely tillage, maintenance of a compact tilled layer, as well as the use of varieties of crop suitable to the environment.[28] The agricultural handbooks of Late Antiquity address the same concerns and demonstrate that ancient tillage practices were admirably adapted to Mediterranean environmental conditions.[29]

Criticisms from some scholarly quarters regarding the static nature of Roman farm technology remain.[30] Some continue to view Byzantine agriculture as backward because of a misguided view that farmers did not adopt

24 M. S. Spurr, *Arable Cultivation in Roman Italy: c.200 B.C.–c.A.D. 100* (London, 1986), 23–4. For another general view, see Horden and Purcell, *Corrupting Sea*, 231–70.

25 K. G. Brengle, *Principles and Practices of Dryland Farming* (Boulder, Colo., 1982), xi.

26 Ibid., 1.

27 Spurr, *Arable Cultivation*, 23–4.

28 A. Arar, 'The Role of Rainfed Agriculture in the Near East Region: Summary of the Present Situation, Potential and Constraints', in FAO (ed.), *Rainfed Agriculture in the Near East and North Africa* (Rome, 1980), 17.

29 C. E. Stevens, 'Agricultural and Rural Life in the Later Roman Empire', in J. H. Clapham and E. Power (eds.), *The Cambridge Economic History of Europe*, I: *The Agrarian Life of the Middle Ages* (Cambridge, 1949), 94.

30 G. W. Barker, 'The Archaeology of Samnite Settlement in Molise', *Antiquity* 51 (1977), 20–4; G. W. Barker, J. Lloyd, and D. Webley, 'A Classical Landscape in Molise', *PBSR* 46 (1978), 35–51.

certain equipment or agricultural practices, such as dry-farming or changing the design of the plough.[31] Yet much of the success which the cultivators in Late Antiquity achieved, especially on the margins of the previously settled lands, is a tribute to their skills as dryland farmers and their ability to adapt to climatic conditions. Irrigated farming always had its limits, both spatially and productively, and survival therefore often depended on successful techniques of rain-fed cropping.

The farmer's main tool was the scratch plough, what we call today the ard (3.1), the ἄροτρον, Latin *aratrum*. As its name implies, the scratch plough attacked the surface of the ground; it did not turn over the soil, and fields were often criss-crossed to break-up all the earth (Palladius, *Op. Ag.* 1.6.8). Typically, two oxen (*Geoponica* 2.23) drew the ard.[32] The worker applied pressure with his or her foot to keep the implement in the ground to cut a furrow and kept the plough at an angle.[33] Ploughing pulverized the ground, created a tilth in preparation for sowing, and destroyed weeds. In semi-arid and arid landscapes, light ploughing was the prime means of retaining moisture.[34] Modern agricultural trials in northern Iraq have shown that shallow tillage (7.5 cm depth) ensures the highest moisture retention in the cereal root-zone and thus returns the greatest yield.[35] Most of the Mediterranean had no use for the heavy plough, long thought to be a major advancement in early medieval agriculture in northern Europe and superior to the simple ard as it allowed working of heavy soils.[36] In any case, M. S. Spurr has demolished the assertion that the ard could not work heavy soils.[37] All soils of value could be exploited: the *Geoponica* (2.23) describes a wide range. Deep ploughing of wetter lands was possible (2.23). Heavier ards for the ploughing of wet land were drawn by teams of oxen (up to six) and these would have been available only to wealthy farms; the less well-off farmer had to borrow or rent animals and equipment to work challenging land.[38]

Ploughing times varied, depending on location and soil conditions; the authorities of the *Geoponica* (2.23.1) recognized the difference in management

[31] A. Bryer, 'The Estates of the Empire of Trebizond', *Ἀρχ.Πόντ.* 35 (1979), 394.

[32] Kaplan, *Les hommes et la terre*, 50.

[33] Newman, *Agricultural Life of the Jews*, 76.

[34] Y. Y. Klaimi, 'Importance of Winter Cereals in Rainfed Agriculture of the Near East and North African Regions—Present Situation, Potential and Constraints', in FAO (ed.), *Rainfed Agriculture in the Near East and North Africa* (Rome, 1980), 90.

[35] See A. K. El Fakhry, 'Studies in Crop Rotations and Tillage Practices on Wheat Production Under Rainfed Conditions in Northern Iraq', in FAO, ibid., 64, with obvious implications for much of lowland Oriens.

[36] White, *Roman Farming*, 41–5.

[37] Spurr, *Arable Cultivation*, 27–30.

[38] R. A. Horsley, *Galilee: History, Politics, People* (Valley Forge, PA, 1995), 205; A. C. Johnson and L. C. West, *Byzantine Egypt: Economic Studies* (Amsterdam, 1967), 213–14.

of various soil types. To avoid working in the extreme heat of day of the summer months, farmers sometimes ploughed at night (2.23.13). Summer ploughing eradicated troublesome weeds (3.10.5) on the fallow. In preparation for the wheat and barley planting, general tillage most often began in the autumn months (2.14). Over much of Oriens, turning of the soil continued throughout the winter (3.1). Normally a single ploughing was insufficient. At least two workings with the ard were required, with three sometimes recommended (*Geop.* 3.3.10; Palladius, *Op. Ag.* 10.1.1). The purpose of such repeated labour was to create a good seedbed and eradicate weeds. To this end, various types of hoes were also employed. Jerome (*in Isa.* 28.23) mentions a type of hoe (Gr. *cκαλίc*, Lat. *sarculum*), and others, such as the two-pronged drag-hoe (*δίκελλα*, Lat. *rastrum*), were common and are represented on a late antique mosaic in the Great Palace in Constantinople.[39]

Seed Selection, Sowing, and Weeding

Following the ploughing and seedbed preparation, sowing began. The seasons for sowing grain were predominantly autumn and winter, from October through December. Ancient farmers selected seeds for quality, and this resulted in cereal improvements, some quite accidental, across successive production seasons. Seed selection was laborious, even when employing sieves, to select the large quantities of seed required for sowing even the average field.[40] Seeds considered most useful for bread-baking, the largest grains less than two years old, were chosen (*Geop.* 2.16). Farmers produced several concoctions intended to protect the seeds from birds and other pests (2.18) and sometimes coated the grain with manure to increase the rate of germination (2.19).

Grain was planted by hand over the plots or spread using a sieve (*Geop.* 2.19), a slow process but one that provided a more even seed distribution. How much seed was applied to each iugerum varied. Based mainly on Columella, Kraemer proposed a sowing rate of 5 modii of wheat and barley per iugerum. As a general rule, this is preferable to the estimate of 2.5 modii per iugerum of Mayerson, whose figure reflects modern Bedouin practice. Columella's heavier sowing rate was probably valid in the coastal zone, where the earth could support thick plantings, in regions where winter frost meant many plants would not survive and thus heavier seeding was required, or in orchards where cereals were grown between vines or trees (Columella, *De re rustica* 2.9.6).[41] Early Roman seeding figures range from 4 modii of naked

[39] K. D. White, *Agricultural Implements of the Roman World* (Cambridge, 1967), 43–4, 52.
[40] Spurr, *Arable Cultivation*, 40.
[41] Ibid., 56.

wheat per iugerum to 10 modii per iugerum for husked wheat (Columella, *R.R.* 2.9.1; Pliny, *H.N.* 18.55). The latter was the rate for emmer, as emmer was husked and therefore more bulky than naked cereals. In view of the difficulties with Kraemer's and Mayerson's figures and the lack of specific late antique sowing rates for Oriens, Columella's figure of 4 modii of naked grain per iugerum form the foundation of the quantified discussions of this chapter.[42]

Upon sowing, seed was ploughed under, or covered, either by digging or by means of a harrow (*Geop.* 2.24.1). Harrows (ἀγρεῖφνα, *irpex*, *cratis*) were made from brush and wood planking.[43] Oxen or other animals dragged the harrow across the seedbed (2.24), smoothed the tilth, and covered the grain with a light layer of earth that protected it from vermin and promoted germination. Weeding of the young plants (2.24), which improved yields, was probably performed by hand or with the simple hoe (the *sarculum*) or the *dikella*.[44] Once removed, the weeds provided valuable animal fodder.[45] Following seed development, a second hand-weeding often followed (2.24.14).

The Harvest

The reaping season varied across Oriens. Mountainous and cooler districts saw their crops ripen later than those of the coastal plains and southern regions. Barley generally matured earlier than wheat, and, thus, the two cereal harvests generally did not overlap. The grain harvest took place between June and August (*Geop.* 3.6; 3.11) and often began when a portion of the cornfields ripened (2.25.1). Harvesting when not all of the seeds had matured caused problems; dampness in the grain increased its chance of spoilage and lowered flour quality. The reason for such early harvest was the desire to maximize the volume of the harvest; reaping when the whole field ripened meant losing that which had matured earliest, as the brittle ears would separate and fall to the ground before they could be harvested (2.25.1). Reaping was done in the morning (2.25.3), when dew on the grain limited the amount of breakage and dust. Cutting was done with sickles (δρέπανον, Lat. *falx messoria*). The type of Greek *drepanon* depicted in Byzantine manuscript illuminations frequently possesses a serrated cutting edge.[46] This serrated tool is better adapted than

[42] Converted to the imperial modius.

[43] White, *Agricultural Implements*, 149–51. It is uncertain which, if either, predominated, though the limited access to wood in the Levant probably favours those made with brush.

[44] A. Bryer, 'Byzantine Agricultural Implements: The Evidence of Medieval Illustrations of Hesiod's *Works and Days*', *Annual of the British School at Athens* 81 (1986), 70; White, *Agricultural Implements*, 41–52.

[45] G. C. Hillman, 'Traditional Husbandry and Processing of Archaic Cereals in Modern Times: Part 1, the Glume-Wheats', *Bulletin on Sumerian Agriculture* 1 (1984), 117.

[46] Bryer, 'Byzantine Agricultural Implements', 75.

conventional sickles to cope with the tough culm of durum wheat and ensured faster harvesting of all grain types.[47]

The *Geoponica* does not specify whether the grain was headed with the sickle or cut close to the ground.[48] In Oriens, where straw provided fuel and fodder, the stalk was probably cut low in order to retrieve most of the stubble.[49] The serrated edges of the drepanon also support this supposition, since the tool is a pulling sickle used to catch the plants low and draw them toward the harvester. The reaper, sometimes equipped with wooden finger guards or claws,[50] collected the stalks or heads of grain and bound the sheaves.[51] On some occasions, gleaning, in which the poor collected fallen grain, followed reaping.[52]

Workers carried the harvested sheaves or grain heads to the threshing floor, probably at night or early in the morning, when the dampness lowered grain loss, which could be as high as 5 per cent.[53] They then spread the stalks around on the threshing floor and removed the grain heads from the straw. During threshing, animals trampled the grain from the stalks as they pulled the threshing sledge (*τρίβολοϲ* = Lat. *tribulum*).[54] The *tribolos* consists of joined planks studded on the bottom with sharpened flints that would pull the grain from the stalk as the sledge went round the floor. The utility and relative cheapness of the tribolos explain its wide diffusion throughout the Near East till the present day.[55]

Workers sometimes threshed the grain with a flail, a simple wooden rod, known in Iron Age Palestine from the book of Ruth (2.17). The first mention of the jointed flail in the Near East can be found in the writings of Jerome (AD 342–419) (*in Isa.* 19.28).[56] The jointed flail increases efficiency since the hinged piece of wood strikes the grain over its entire length. This tool also spares the worker, as it prevents his or her having to stoop whilst processing the crop.

47 Spurr, *Arable Cultivation*, 70.

48 White, *Roman Farming*, 182.

49 Hillman, 'Glume-Wheats', 119.

50 Bryer, 'Byzantine Agricultural Implements', 76; G. Dalman, *Arbeit und sitte in Palästina* (Hildesheim, 1964), 3.27–9.

51 Hillman, 'Glume-Wheats', 118.

52 Dalman, *Arbeit und sitte*, 3.60–1; Hillman, 'Glume-Wheats', 120. This is especially true in the case of Jewish areas, where gifts to the poor were to be made in such a manner.

53 Hillman, 'Glume-Wheats', 121.

54 White, *Agricultural Implements*, 185. Mentions of the tribolos in the Byzantine East make it fairly certain that the implement was a commonplace: Diocletian's Price Edict (15.41) *τρίβολοϲ ζύλιουϲ*; also mentioned in TB 'A.Z. 24b, and Longus, *Daphnis et Chloe*, ed. Reeve, 3.30.

55 Bryer, 'Byzantine Agricultural Implements", 79. A tribolos cost 200 denarii, only twice the price affixed to a modius of wheat according to Diocletian's Price Edict (15.41).

56 K. D. White, *Farm Equipment of the Roman World* (Cambridge, 1975), 209; Jerome = *PL* 24.326.

Possibly of European origin, this tool was incorrectly thought to be absent from the ancient Levant.[57] Flails were probably most beneficial for those too poor to afford animal-threshing machines or for those who wished to avoid damage to the grain from weather. Rain could ruin a crop on an exposed, animal-trodden threshing floor. Processing under cover with flails could be easily accomplished. The tribolos undoubtedly predominated over most of Oriens, since the summer months, when the harvest was brought in, were scorchingly hot and dry and the need for indoor threshing using flails was probably negligible. However, the appearance of the hinged flail in late antique Palestine, where Jerome lived when he wrote his commentary on Isaiah, is interesting in that it shows the adoption of new technology. It further suggests the dissemination of improved tools amongst poorer farmers.

According to the *Geoponica* (2.26.2), threshing floors had to be situated in windy areas so that the breeze could blow away the winnowed chaff; but the floors had to be at a distance from buildings and orchards which would be damaged by this debris. The threshing area was circular or oblong in shape and levelled and compacted with a stone roller so that it could survive the treading animals and threshing sleds. To keep pests from entering the floor, the area was then treated with olive oil residue (*amurca*) (2.26.5). Threshing floors of the Byzantine period are mentioned in the *Petra Papyri* and survive in Palestine and Arabia, where their remains match the description found in the handbooks; these floors are situated on hillsides and are generally circular with beaten earth surfaces, although rock-cut examples from Samaria are known.[58] Circular threshing floors in the Negev, associated with Byzantine and early Islamic period pottery finds, are usually 8–15 m in diameter, whilst oval examples are around 13 m at their maximum length.[59]

Once threshed, the grain was winnowed. During this phase of processing, workers separated grain from the chaff by tossing the agglomeration into the air so that the breeze would sweep away the lighter husks, leaves, and stalk fragments.[60] Winnowing forks (*θρεῖναξ*, *λικμητήριον*, Lat. *ventilabrum*) were normally used. Winnowing baskets (*λίκνον*, Lat. *vannus*), in which the dirty grain was scooped and shaken until the debris sifted to the surface, were also common.[61]

[57] Hillman, 'Glume-Wheats', 122.

[58] Dar and Applebaum, *Landscape and Pattern*, 17; I. Finkelstein and Y. Magen, *Archaeological Survey of the Hill Country of Benjamin* (Jerusalem, 1993), 49–50, 52; Lender, *Map of Har Nafhah (196) 12–01*, 107. *Petra Papyri*: Avni, *Map of Har Saggi Northeast (225)*, 40, 66; R. W. Daniel, 'P.Petra Inv. 10 and its Arabic', in I. Andorlini (ed.), *Atti del XXII Congresso internazionale di papirologia, Firenze, 23–29 agosto 1998* (Firenze, 2001).

[59] Haiman, *Map of Har Hamran—Southwest (198)*, 21, 57; Haiman, *Map of Har Hamran—Southeast (199)*, 39, 51, 55, 67; Lender, *Map of Har Nafhah (196) 12–01*, 107.

[60] White, *Roman Farming*, 186–7.

[61] White, *Agricultural Implements*, 33.

More than one winnowing may have been necessary.[62] After this, the grain was often sieved to remove insects and unwanted plant particles. The straw and remaining chaff was then collected and the grain heaped on the threshing floor to dry (*Geoponica* 2.25). Finally, the harvest was stored in granaries (*Geop.* 2.37) or silos often situated near the threshing floors.[63] At Har Saggi in the Negev, a Byzantine threshing floor (10 m in diameter) lies adjacent to a circular silo, 3 m in diameter and covered with flat stones.[64] Silos were frequently dug into the ground rather than built above it: a late antique building complex at Be'er Sheva possesses a circular, stone-lined silo dug 2.6 m into the ground.[65]

The labours of the vineyard, orchard, garden, and grain field required implements. The range of tools and their quality greatly affected the efficiency of the labourers and thus their ability to work. Basic equipment for growing cereals included plough, an ox harness, goad, baskets, sickles, threshing sledge, and winnowing fork. But many other chores required additional tools; the care of trees required knives and pruning hooks, while the two-pronged hoe, mattock, or dikella was needed to dig in gardens and vineyards. Estates had well-equipped tool sheds. Palladius (*Op. Ag.* 1.43) noted that a well-stocked estate possessed ploughs, including a double plough for ridging the fields, hooks, hatchets, pruning knives, reaping hooks, scythes, lunate pruning hooks, knives, saws, spades, bramble-bush cutters, axes, rakes, shearing instruments and other tools for livestock rearing. Villagers, on the other hand, might be short on tools and steal them. The thieves that the *Nomos Georgikos* (§22) describes stole a digging fork (λίϲγον), *dikella*, and a plough (§62).[66] The same document talks about the cowherd's staff or crook, called simply ξύλον (§29), and axe (πέλεκυϲ) (§40).

Evidence from the textual sources is incomplete and does not provide a conclusive answer to an important question raised by Anthony Bryer for the Byzantine world, when he wondered to what extent iron tools became rare in the post-Roman centuries. Bryer noted: 'Iron is the key to any advance and, as Hesiod knew, even entirely wooden tools needed iron to fashion them.'[67] Did Byzantium almost drop out of the Iron Age?[68] On the contrary, it seems that

62 Hillman, 'Glume-Wheats', 124.

63 Dar and Applebaum, *Landscape and Pattern*, 193–4.

64 Avni, *Map of Har Saggi Northeast (225)*, 66.

65 N. Negev, 'Be'er Sheva', *ESI* 12 (1992), 95–6.

66 W. Ashburner, 'The Farmer's Law', *JHS* 30 (1910), 100.

67 Bryer, 'Byzantine Agricultural Implements', 47.

68 A. Bryer, 'The Means of Agricultural Production: Muscle and Tools', in A. E. Laiou and C. Bouras (eds.), *The Economic History of Byzantium* (Washington, DC, 2002), 102, states, without citing his source, that 'it has been argued that at times Byzantium almost dropped out of the Iron Age. Does it matter? After all, the polished granite obelisks in the Hippodrome had

iron tools were relatively abundant in Late Antiquity; but this answer is based on an arbitrary and incomplete sample of the evidence for which there has been no fundamental study. A thorough reckoning of this problem, impossible here, is badly needed. For instance, to my knowledge there has been no publication of an extensive cache of tools that could provide insight into an individual farm's equipment. Additionally, we know almost nothing about metal production, the distribution of metalworkers in Late Antiquity, nor their activities and organization. Were they, for example, to be found in ones and twos throughout the bigger villages of Syria, as is the case today within the myriad of small machine shops that dot the towns? Or were they mainly an urban affair, with trips to the cities required to buy and mend tools?

The seventh-century *Life of Theodore of Sykeon* chronicles daily affairs in the village of Mossyna in Asia Minor, where a well-known smith lived, who constructed, from the iron farm tools donated by local farmers, a cage for the saint (§27, 28). Metal and the expertise to fashion it were common but not superabundant: Theodore later acceded to the farmers' entreaties and gave back their metal. They subsequently built him a wooden cage. But the *Life* does not suggest great scarcity of iron, as the saint had metal bonds made for his hands and feet, shackles that weighed 15 lb. He further wore a collar that weighed 18 lb and a belt that weighed 33 lb, in all likelihood also forged in iron. This equipment totals more than 66 lb of iron, all procured in a common village. There is no indication that the villagers were so poor in iron that they could not part with it. On the contrary, it seems that the iron instruments for the saint came from their collective surplus and that they had regular access to a smith within the village community.

In Syria itself, at the village of Androna, probably in the Byzantine or Umayyad period, metalworking was conducted around the reservoirs and in the sixth-century bath which had been replaced by the nearby Umayyad bath.[69] On the other hand, there is also evidence that metalworking might be based in a city, as at Corycus in Cilicia where fifteen epitaphs in the necropolis commemorate smiths. In the metropolis of Antioch, a curial official apparently oversaw the *collegium* of metalworkers.[70] The *Geoponica* (2.40) echoes the warnings of Varro (*De re rustica* 1.16.4) that labourers

originally been cut for the Egyptian pharaoh Thutmoses III (1549–1503 BC) without using iron at all. This should give us pause for thought.' Obviously the issue does matter, as Professor Bryer has raised it on more than one occasion. Bryer, 'Byzantine Agricultural Implements', 47. At stake are fundamental issues of the efficiency with which basic agricultural chores could be conducted, which would have had repercussions throughout the economy.

[69] Mango, 'Androna, 1999', 315; Mango, 'Androna, 2000', 295.

[70] T. R. S. Broughton, 'Roman Asia Minor', in T. Frank (ed.), *An Economic Survey of Ancient Rome*, 3 (Baltimore, 1938), 4.826; Liebeschuetz, *Antioch*, 222.

should not waste time by making trips to the city to have their tools repaired. The responsible estate owner should arrange to have the desired implements manufactured and mended on the estate itself.

An indicator (from a slightly earlier period) of the relative cheapness and therefore abundance of iron in agricultural implements may be found within Diocletian's Price Edict. There, we find that a plough with a yoke was assigned a price of 100 denarii, whilst a *pala*, a small spade used in the garden and orchards, was priced at only 4 denarii. In the same list, one modius of wheat was assigned a value of 100 denarii, which apprises us of the relative cheapness of two implements, the price of which, in all likelihood, included their iron components.[71] Archaeological evidence of the overwhelming volume of small iron objects found in late antique archaeological contexts also supports a generally good supply of metals. Scattered discoveries of tools also clarify the issue, although as in all but the most affluent and wasteful societies, their metal was almost invariably recycled. Often overlooked, but unsurprising, is the fact that many archaeological finds of agricultural tools come from urban contexts, a reminder that many farmers were town dwellers. A *dolabrum* came to light at Anemurium, and vine dressers' knives from Scythopolis (Bet She'an) and Jalame probably indicate that a range of common iron tools and materials awaits a full synthesis.[72]

A Wheat Economy: The Grains of Late Antiquity

Like many people today, those who lived in the Late Roman empire had distinct views of what good bread was. The finest bread was made from wheat, *cιλίγνιον*.[73] Barley was more restricted in the diet of the Romans and their subjects; Rabbinic sources attest to the favouring of wheat over barley,

[71] S. Lauffer, *Diokletians Preisedikt* (Berlin, 1971), §§15.42; 15.45; cf. 15.31a–38 in which the prices of a wagon are rendered with explicit exclusion of its metal components.

[72] Anemurium: J. Russell, 'Byzantine Instrumenta Domestica from Anemurium: The Significance of Context', in R. Hohlfelder (ed.) *City, Town and Countryside in the Early Byzantine Era* (Boulder, Colo., 1982), 136 fig. 1.1. A tool frequently used in general wood cutting tasks, including carpentry and pruning of trees and vines, though it was also used in field work to break up clods after ploughing. The example at Anemurium is double-headed and therefore probably the type which the Greeks called *πέλεκυc*. Bet She'an: G. M. Fitzgerald, *Beth-Shan Excavations 1921–1923: The Arab and Byzantine Levels* (Philadelphia, 1931), pl. XXXVIII.30. Jalame: G. D. Weinberg, *Excavations at Jalame: Site of a Glass Factory in Late Roman Palestine* (Columbia, Mo., 1988), 353, pl. 8.2. Finds of an Early Byzantine metal hoard at Aphrodisias included tongs, a fork, and a hooked rod: R. R. R. Smith and C. Ratté, 'Archaeological Research at Aphrodisias in Caria, 1996', *AJA* 102 (1998), 227, and metal finds from the Saraçhane site in Constantinople, J. W. Hayes, *Excavations at Saraçhane in Istanbul*, vol. II (Princeton and Washington, DC, 1992), are substantial.

[73] J. Teall, 'The Grain Supply of the Byzantine Empire', Ph.D. diss. (New Haven, 1956), 13.

stressing that barley was reserved for the poor and for animals.[74] This still means that barley was an important part of the agricultural landscape throughout Mediterranean history, but the peoples of the late antique Levant were, as we shall see, primarily wheat eaters. They sought to manage these grains and their landscape in order to satisfy their palate. But the range of additional grains grown suggests active adaptation to local conditions, access to a deep cache of cereal types, and considerable variety that fostered both agricultural flexibility and culinary diversity.

The wheat plant, in its many manifestations, has been the object of cultivation in Mesopotamia and the Near East since the Neolithic Period and was grown in Palestine by 7,800 BC.[75] From its place of origin in south-central Asia, wheat became the staple crop of Mesopotamian agriculture. This was true despite the region's hydraulic-based farming and early knowledge of rice, which Mesopotamian farmers largely rejected in favour of barley and wheat.[76] By the Roman period, wheat had superseded barley in the ancient granary of the Mediterranean.[77] There are many reasons why wheat became the dominant plant of human cultivation, amongst them its characteristics of self-pollination and pattern of winter germination. Wheat varieties have unique baking qualities, contain high starch (60–80 per cent) and protein (8–14 per cent), and generally are more nutrient-dense than other cereals.[78]

Modern wheat classification recognizes two groups based on numbers of chromosomes, with further variants determined by the way they are threshed. The first of these is hulled wheat which possesses a tough outer sheath and spikelet glumes encasing the grain; the results of threshing, therefore, is a spikelet–grain combination, not the grain alone.[79] The spikelets may be stored or sold without further processing, but upon consumption the grain must be freed from the husk, usually by pounding with mortar and pestle.[80] The

[74] Safrai, *Economy*, 109; K. D. White, 'Cereals, Bread and Milling in the Roman World', in J. Wilkins, D. Harvey, and M. Dobson (eds.), *Food in Antiquity* (Exeter, 1995), 39–40.

[75] O. Bar-Yosef and R. H. Meadow, 'The Origins of Agriculture in the East', in T. D. Price and A. B. Gebauer (eds.), *Last Hunters—First Farmers: New Perspectives on the Prehistoric Transition to Agriculture* (Santa Fe, 1995), 67.

[76] D. Brothwell and P. Brothwell, *Food in Antiquity: A Survey of the Diets of Ancient Peoples* (Baltimore, 1969), 96.

[77] T. Braun, 'Barley Cakes and Emmer Bread', in J. Wilkins, D. Harvey, and M. Dobson (eds.), *Food in Antiquity* (Exeter, 1995), 34.

[78] D. Zohary and M. Hopf, *Domestication of Plants in the Old World: The Origin and Spread of Cultivated Plants in West Asia, Europe, and the Nile Valley* (2nd edn., Oxford, 1993), 16–18. When compared with many other grains, for instance, the stickiness of the dough and ability to rise when leavened are properties which make this cereal superior to other varieties as far as palatability is concerned.

[79] Ibid., 25.

[80] White, 'Cereals, Bread and Milling', 39.

second type is free-threshing, or naked wheat, which has thinner glumes and pales; threshing separates the grain, with no further processing required before milling. Both of these classes of wheat were used in the Near East in Late Antiquity and varieties of einkorn, emmer, durum, and bread wheat were globally available. Another cereal possibly available to the Byzantine inhabitants was rivet wheat, *Triticum parvicoccum*. Evidence for this type occurs archaeologically, but no link has been drawn between grain finds (all earlier than the Byzantine period) and cereals named in farming texts.[81] Future work must link the study of plant genetics, archaeology, and textual studies, in order to create a sure framework for the discussion of the varieties of ancient grains. Until then, our identifications are open to revision.

Einkorn

Einkorn wheat (*Triticum monococcum*; τίφη), a hulled diploid wheat (having two homologous sets of chromosomes), was one of the earliest cultivated grains. It formed a component in the first agriculture of the Neolithic. Although it continued to be produced alongside it, einkorn was of less importance than emmer wheat from the very beginning of farming in the Near East.[82] Despite its Near Eastern origins, einkorn does not appear to have made much of an impact on Egyptian farming. In parts of the Near East and Europe, einkorn persisted throughout the medieval period. The remarkable longevity of einkorn cultivation is underscored by its continued presence as a crop in modern Europe.[83]

Einkorn plants are generally small, usually under 70 cm high. Their advantage of being able to grow on poor soils, where other wheat types struggle, is offset by their generally lower yields. Einkorn flour is fine in texture and yellow to grey in colour. Due to a lack of starch, it does not produce well-leavened bread but loaves that lack volume, colour, elasticity and flavour when compared with other modern and primitive cultivars.[84] Einkorn wheat is mentioned in a valuable late antique text for the study of grain: the fourth-century *Medical Compilations* of the physician Oribasius, who, following the second-century Galen, lists the properties of various grains and ranks

[81] M. Kislev, '*Triticum parvicoccum* sp. nov., the Oldest Naked Wheat', *Israel Journal of Botany* 28 (1979–80), 95–107.

[82] J. R. Harlan, 'The Early History of Wheat: Earliest Traces to the Sack of Rome', in L. T. Evans and W. J. Peacock (eds.), *Wheat Science—Today and Tomorrow* (Cambridge, 1981), 4.

[83] Zohary and Hopf, *Domestication of Plants*, 32.

[84] J. A. Le Clerc, L. H. Bailey, and H. L. Wessling, 'Milling and Baking Tests of Einkorn, Emmer, Spelt and Polish Wheat', *Journal of the American Society of Agronomy* 10 (1918), 217.

emmer and einkorn in third place behind naked wheats and barley. The grouping of emmer and einkorn together under the same heading (1.13) is significant. Oribasius explicitly stated (1.13.1) that these two grains were, for bread-making purposes, considered inferior to free-threshing wheat. Both emmer and einkorn must have been common in the late antique landscape, since they are treated as a group and compared to one another: emmer was considered the better of the two; but when einkorn was of extremely good quality, it was considered to surpass poor-quality emmer (1.13.2). Since it deteriorated quickly einkorn bread was consumed immediately (1.13.2). Based on this evidence from Oribasius and despite the predominance of other grain forms, there is some justification to conclude that einkorn wheat remained on the agricultural scene of the late antique East. Its overall hardiness and long history of successful cultivation were key to its survival as a major secondary crop. Like other hulled wheats, and especially due to its poor bread-making qualities, einkorn was often consumed crushed with its hulls, a mixture called groats.[85]

Emmer

Emmer wheat (*Triticum turgidum*, Greek ὀλύρα/ζειά, Latin *far/adoreum*) was, after its domestication in the Near East, the dominant wheat of prehistory throughout the Neolithic and into the Bronze Age when it was used for bread-making and brewing. In its hulled manifestation, it formed one of the core crops of the Neolithic spread of agriculture throughout the Mediterranean basin and Europe.[86] In the Early Roman period, emmer formed the primary grain crop, as attested by Pliny (*H.N.* 18.81), who praised its hardiness and ability to survive winter.[87] Emmer was, according to Pliny (18.83), also able to survive in under-cultivated or dry, hot places. This ability to yield under difficult conditions meant emmer was a hardy wheat, one that could thrive throughout the whole of the Mediterranean.

The texture of emmer ranged from soft to hard. Since emmer was hulled and generally bearded, it was more disease- and insect-resistant than the naked wheats with which it competed. On the threshing floor and in storage, the husk protected the grain from the damp and from pests. Modern emmer tests high for sugar content, and ancient types may have been high in energy.[88] In antiquity, emmer was consumed as porridge, although it could be made into bread. As noted above, Oribasius (1.13.2) commended bread made from good

[85] N. Jasny, *The Wheats of Classical Antiquity* (Baltimore, 1944), 125.
[86] Zohary and Hopf, *Domestication of Plants*, 44.
[87] White, 'Cereals, Bread and Milling', 39.
[88] Le Clerc, Bailey, and Wessling, 'Milling and Baking Tests', 217.

quality emmer. Modern efforts have shown that this variety can produce a good loaf, though not as well-risen as that produced from bread wheat.[89] The ecological adaptability and the potentially high yield of emmer wheat rendered it an ideal grain for many regions in antiquity, and it continued to survive the growing popularity of naked wheats.[90] In Italy, emmer persisted as an important crop into the fourth century and later.[91] In the fifth century, Palladius (*Op. Ag.* 10.4.1) recommended it be sown in September along with bread wheat.

In the late antique East, emmer also played a role in the agricultural economy. According to Oribasius, emmer and einkorn formed the second tier of wheats considered suitable for human consumption. The *sitos Alexandrinos* of the *Geoponica* (3.8) was possibly a variety of emmer (though it may have been sorghum, as discussed below), since the hulls of the grain had to be removed by pounding before the gruel *tragos* could be prepared. The fact that this hulled wheat was specifically considered Alexandrian is intriguing. We cannot know, however, if it was named after the city because of its reputation as a major grain supplier, or whether the metropolis was the source of the seed.

Durum

Durum wheat (*Triticum durum*)—today also called 'macaroni wheat' from one of its principal modern uses, or 'hard wheat' (Lat. *durum*)—is the Greek *semidalis* (ϲεμίδαληϲ), one of the most important grain crops of antiquity. Several types of free-threshing grain appeared in the Near East from the Neolithic Period onward, but there has been some doubt as to whether or not durum wheat was amongst these early types.[92] Botanical research continues to piece together the history of ancient free-threshing wheat. It is currently believed that tetraploid durum-like, naked wheats developed before hexaploid varieties, such as the bread wheat described below.[93] Present evidence suggests that the latter probably developed only outside the Fertile Crescent, whilst tetraploid wheat species were native to the Near East.

It is now highly likely that, as a member of the emmer family of grains, durum was one of the early cultivars of naked wheat grown in the Near East. Botanical remains from the Balikh Valley in Syria show that a hard wheat species, probably durum, was the major wheat after the Bronze Age.[94] Findings

89 Braun, 'Barley Cakes', 35; Le Clerc, Bailey, and Wessling, 'Milling and Baking Tests', 218.

90 P. Reynolds, *Iron Age Farm: The Butser Experiment* (London, 1979), 60–4.

91 Spurr, *Arable Cultivation*, 12–13.

92 Jasny, *Wheats*, 35–6; J. Percival, *The Wheat Plant: A Monograph* (London, 1921), 206.

93 Zohary and Hopf, *Domestication of Plants*, 46.

94 W. Van Zeist, 'Evidence for Agricultural Change in the Balikh Basin, Northern Syria', in C. Gosden and J. Hather (eds.), *The Prehistory of Food: Appetites for Change* (London, 1999), 383–4.

of durum in an Iron Age (eleventh century BC) silo at Tell Keisan confirm its production in Palestine many centuries before Late Antiquity.[95] Ease of threshing, and its climatic adaptability, allowed durum wheat to become a fixture of the region's agriculture. As early as the Classical period, durum had become the dominant naked wheat subspecies in some regions of the eastern Mediterranean, notably Anatolia. By the Ptolemaic period durum had replaced emmer as the dominant grain in Egypt. Archaeobotanical finds from Mons Claudianus show that this grain was one of the principal foods imported from the Nile Valley to the early Roman mining colony.[96] By Late Antiquity, it had spread through North Africa, Italy and Sicily. In light of this, Spurr argued that the *triticum* of the Latin sources should signify durum wheat, not bread wheat.[97] Yet archaeology has not confirmed the dominance of durum wheat in classical or late antique Italy, and its ascendance there cannot be based solely on its antiquity and climatic adaptability. At present we cannot be certain what species was the dominant grain in Italy or the Mediterranean regions.

Both Dioscorides in the first century (*Mat. Med.* 2.85) and Oribasius (1.2.3) in the fourth discuss the qualities of durum. The latter describes the best-quality *semidalis* as yellow in colour. Oribasius adds that durum wheat is dense, and heavier than other wheats. He considered this grain to be extremely nutritious. Semidalis possesses a high gluten content which gives it the special properties required to produce couscous or pasta, its main uses today. The plant's ability to tolerate drought is advantageous in the Mediterranean where winter rainfall is irregular. Durum thrives in hot climates and possesses little resistance to cold temperatures. The relatively low water content in the seed allows this grain to be stored for long periods of time without significant deterioration.[98]

A sample of archaeological and textual data suggests that durum was firmly entrenched within late antique agriculture. Besides its inclusion in the *Geoponica*, the work of the sixth-century Alexander of Tralles also described it,[99] while archaeobotanical evidence from the Byzantine period at Karanis in Egypt indicates that it was important there. The archaeobotanical assemblage from Bostra in Syria indeed demonstrates that durum wheat has been present from 1800 BC to modern times and that cultivation of durum was common

[95] Kislev, '*Triticum parvicoccum*', 361–3.

[96] M. Van Der Veen and S. Hamilton-Dyer, 'A Life of Luxury in the Desert? The Food and Fodder Supply to Mons Claudianus', *JRA* 11/101–116 (1998), 106.

[97] Spurr, *Arable Cultivation*, 17.

[98] Watson, *Agricultural Innovation*, 20.

[99] Alexander of Tralles, *Alexander von Tralles*, ed. Puschmann, 1.365; 2.61, 125, 209, 213, 217, 219, 383, 457, 473, 491.

in the Hauran during Late Antiquity.[100] It is tempting to see echoes of this cultivation in a variety called 'Haurani' wheat, which was amongst the most prized macaroni wheats of the early modern period.[101]

Bread Wheat

The Greeks called bread wheat (*Triticum aestivum*), known in Latin agricultural writings as *siligo*, *σίλιγνιον*. Bread wheat encompasses a number of varieties. Together these today account for 90 per cent of the world's total current wheat production. Hexaploid wheats, of which bread wheat is one, developed across the ages during cereal farming. Thus, these varieties were relative latecomers to the Near Eastern agricultural scene. However, by Late Antiquity, naked bread wheat already had a long history of cultivation, stretching back to *c.*4,000 BC and spreading across the Mediterranean basin over the millennia.[102] Bread wheat, as cultivated in Late Antiquity in the Mediterranean, was free-threshing, although a hulled variety of *Triticum aestivum*, the hexaploid wheat known as spelt, also existed in the Roman period. Spelt was once important in temperate Europe; it possibly entered production in the cooler areas of the Mediterranean basin in the Roman period. Diocletian's Price Edict (1.7), where it is called *spelta* (Greek *πιστικίου*), attests to its commonality. Like most ancient grains, we know far too little to draw firm conclusions about its distribution and presence in the cereal inventory, but Egypt may have been one producer of spelt.[103]

Due to its adaptability and unique gluten content (which enhanced its baking properties), *Triticum aestivum* produced the finest loaf of all ancient wheat types.[104] By the Early Roman era, bread wheat had apparently become a favoured grain and probably remained so throughout the Byzantine period, especially in the coastal zones where rainfall was abundant. It is probably the 'white wheat' noted in Talmudic sources.[105] Excavations in the harbour of Caesarea Maritima have yielded finds of *Triticum durum* or *aestivum*.[106] Bread wheat was not, however, limited to such well-favoured coastal areas. With

100 D. Samuel, 'Plant Remains from the Northwest Tell at Busra', *Berytus* 34 (1986), 92.

101 N. F. Miller and K. L. Gleason, *The Archaeology of Garden and Field* (Philadelphia, 1994), 655.

102 Zohary and Hopf, *Domestication of Plants*, 53.

103 Most, however, assume that spelt was not an important Mediterranean crop in Antiquity. Harlan, 'Early History of Wheat', 11.

104 Percival, *Wheat Plant*, 265.

105 R. Brooks, *Pe'ah* (Chicago, 1990), 125.

106 J. Giorgi, 'Plant Remains from the Sediments of the Inner Harbour (Area I14)', in K. G. Holum, A. Raban, and J. Patrich (eds.), *Caesarea Papers 2: Herod's Temple, the Provincial Governor's Praetorium and Granaries, the Later Harbor, a Gold Coin Hoard, and Other Studies* (Providence, RI, 1999), 354.

some care, it could be grown over a much greater area. In Anatolia, farmers presently grow both durum and bread wheat in arid conditions. In this case, bread wheat is grown without irrigation, while durum receives occasional irrigation which increases its yield over dry-farmed *Triticum aestivum.*[107] In light of this, we must reassess the view that bread wheat remained 'higher-status' and, due to its rainfall requirements, confined to the northern fringes of the Mediterranean world.[108] Late antique farming was protean in nature and utilized an impressive array of crops along with the ingenuity and skill of ancient farmers. It seems, then, that bread wheat shared the landscape with a whole range of cereals and was sometimes the dominant crop.

In the marginal landscape of Oriens, where hot, dry conditions prevail, one would expect barley to predominate, due to its drought-tolerant qualities. However, the Nessana excavations and papyri show us that bread wheat flourished even in marginal areas where it faced greater environmental challenges.[109] Botanical remains, recovered during excavation, yielded *Triticum aestivum* grains, indicating that this variety really was the wheat referred to in the texts.[110] Bread wheat in the archaeobotanical remains at Androna further supports the notion that cultivation and use of this grain was well-established, not only on the coast, but wherever farmers produced grain.[111] The attestation of bread wheat at these two sites has important bearing on the history of late antique agriculture in the Near East, and probably indicates that Byzantine farmers willingly invested in hydraulic infrastructure and labour intensification to produce crops that they preferred for cultural reasons, rather than growing only the most ecologically suitable plants.

Barley

Barley (*Triticum hordeum*), *κριθή*, Latin *hordeum*, one of the first crops of Neolithic agriculture, was a principal grain in the late antique Mediterranean and remains so in the present.[112] Barley is, today, divided into two major classifications: two-rowed and six-rowed, each of which comprises numerous varieties. Whilst both two-rowed and six-rowed barley appear in the archaeobotanical record, specific barley varieties within the major groupings cannot yet be easily differentiated archaeologically. Both two-rowed and six-rowed barley are predominantly hulled and thus needed to be pounded or roasted to

[107] Samuel, 'Plant Remains', 92.
[108] P. Garnsey, *Food and Society in Classical Antiquity* (Cambridge, 1999), 120.
[109] P.Ness. III.40.1, 5, 8, 11–23 and *passim*; the word used is simply *cîτoc*.
[110] P. Mayerson, 'Wheat in the Roman World: An Addendum', *CQ* 34/1 (1984), 244.
[111] Dr Mark Robinson, personal communication, April, 2001.
[112] Zohary and Hopf, *Domestication of Plants*, 54.

remove the sheath that invests the kernel.[113] Naked barley has been recovered in sites of greater antiquity, such as the second-century finds from Nahal Yattir, but, though it probably continued to be grown, specific late antique data are lacking.[114]

Written sources are more informative. Oribasius (1.11.1) attests to cultivation of barley and its use as a staple. Since classical antiquity, the dietary role of barley had changed. Whereas in Classical Greece barley production exceeded that of wheat, the Romans considered it inferior.[115] Nevertheless, wheat and barley persisted as companion crops throughout Late Antiquity, as they had done for the previous millennia.[116] What differed was the ratio of cultivation, with wheat dominating by the Early Roman period. Here Tannaitic evidence from the Mishnah (composed in Palestine in AD 200–250) is once more useful; the Mishnah mentions barley as a crop forty-seven times, whilst wheat is mentioned eighty-eight times, and it seems that this reflects realities in the countryside throughout much of the eastern Mediterranean.[117]

The Romans incorrectly perceived barley as nutritionally deficient, and therefore consigned it to poor ground. This notion of 'sympathies', ubiquitous in medical literature, is reflected here in common thinking: farmers linked poor crops with weak earth, a view repeated in Oribasius.[118] In Jewish practice, the grain allowance for a wife to be maintained by an absent husband would be double the measure if given in barley rather than in wheat.[119] The use of barley as animal food was standard throughout Antiquity and continues throughout the world today.[120]

Barley's resistance to disease and drought, its low water requirements during germination, its high nutritive qualities, and its tolerance of heat and cold allowed the crop to flourish in every region of the late antique world, even to altitudes surpassing 1,500 m.[121] The *Geoponica* (1.12) states that barley could be planted in soil of lesser quality than wheat, and on estates it was probably consigned to marginal areas. In Garnsey's study of the climate conditions of Attica before modern farming practices, he determined that in this locale, the wheat crop would fail one year in four, barley one year in twenty.[122] Thus, in classical Attica barley was more dependable and

113 Braun, 'Barley Cakes', 27.

114 Kislev, '*Triticum parvicoccum*', 185.

115 Braun, 'Barley Cakes', 25–34; Garnsey, *Food and Society*, 119.

116 Garnsey, *Food and Society*, 62–4.

117 Safrai, *Economy*, 108.

118 M. Grant, 'Oribasios and Medical Text', in J. Wilkins, D. Harvey, and M. Dobson (eds.), *Food in Antiquity* (Exeter, 1995), 373.

119 M Ket. 5.8–9.

120 T. Coello, *Unit Sizes in the Late Roman Army* (Oxford, 1996), 38.

121 Braun, 'Barley Cakes', 25; Garnsey, *Food and Society*, 119.

122 Garnsey, *Food and Society*, 10.

favoured than wheat. Barley often produced a crop in conditions in which wheat would not, and its superior yields to wheat in many circumstances make barley, in strictly practical terms, appear wheat's superior in dryland agriculture.

At least partially underlying this Roman bias against barley was the perceived nutritional superiority of wheat. Oribasius (1.10–12) disparaged the nutritive qualities of barley, though he did note (1.1.7) that the best barley appeared white after winnowing and had solid, weighty grain. Barley failed to produce a good bread loaf and was therefore consumed primarily as groats (*χόνδρος*), ground into coarse flour (*ἄλφιτα*) and made into barley cakes (1.12.1–3), or used in brewing.[123] Barley bread was consumed mainly by the poor. However, in the second century Athenaeus (3.114D) mentioned that a good sour bread, *κυλλάς*, was made from barley and enjoyed throughout Egypt. The fourth-century collection of Ps.-Apicius (4.4.1–2) preserves two recipes for barley soup and a gourmet recipe for pork cooked with barley and figs (7.10).

Types of barley known in late antique sources include the two-rowed Galatian barley, *hordeo Galaticus*, favoured for its white colour, dense grain, and rich flour. Galatian barley was a spring-harvested crop; Palladius (2.4; 3.7) recommended its sowing in winter (January or February) at the high rate of 8 *modii* to the iugum (210 kg/ha). Horse barley, *cantherium*, mentioned in the first century by Columella (2.9.14), was believed more nutritious than bad wheat.[124]

Written and material evidence shows, then, that barley was an important food plant grown throughout Oriens and the Mediterranean in Late Antiquity. Barley certainly was present as a companion crop to wheat on the coast, as archaeobotanical finds from the seventh-century ecclesiastical estate at Shelomi (Phoenice) indicate.[125] Its importance increased in the interior, drier areas of Oriens. The *Geoponica* (1.12–13) recommended sowing barley on dry ground of average quality, and Rabbinic sources show that barley was a staple crop in southern Palestine in dry regions such as the Edomite plateau.[126] The Nessana papyri provide evidence of barley cultivation in the Negev highlands around the ancient villages of Palaestina Tertia and of the hinterland of Gaza.[127] Barley was recovered at Byzantine En Boqeq in the Arava Valley and

[123] Braun, 'Barley Cakes', 27; Foxhall and Forbes, 'ΣΙΤΟΜΕΤΡΙΑ: the Role of Grain as a Staple Food in Classical Antiquity', *Chiron* 12 (1982), 53.

[124] Braun, 'Barley Cakes', 34.

[125] Dauphin, 'A Seventh Century Measuring Rod from the Ecclesiastical Farm at Shelomi in Western Galilee [Israel]', 516.

[126] Safrai, *Economy*, 108–9.

[127] C. A. M. Glucker, *The City of Gaza in the Roman and Byzantine Periods* (Oxford, 1987), 95; Mayerson, *Ancient Agricultural Regime*, 227–9.

at Nahal Yattir in Judaea.[128] Botanical finds from excavation in Nessana demonstrate that during Late Antiquity the population of Palestine continued to grow barley along with the more desirable bread wheat.[129]

Millet

Two types of millet (*Panicum miliaceum*) are known from ancient sources. Common broomcorn millet was the *milium* of the Romans and probably the μελινή / κέγχρος of the Greeks. It is today a crop of some importance in world agriculture, with the main regions of cultivation being eastern and central Asia, India, and the Middle East. Foxtail or Italian millet (*Setaria italica*; *Panicum*; ἔλυμος) is still widely cultivated throughout central and East Asia, where spectacular yields have been achieved.[130] The small size of these grains (also true for sorghum) means that archaeological recovery can be more problematic than that for large grains, such as wheat and barley.[131]

Millet possesses a host of positive characteristics which were probably recognized in the Mediterranean in Antiquity as they were elsewhere, especially in China and the Far East.[132] The grain flourishes in a wide variety of climates, and, of great importance for our area of study, it is the most drought-resistant of all cereals.[133] Able to germinate at moisture levels below other grains, the grain can tolerate dry, light soil and flourish in marginal lands, though in these areas it is helped considerably by irrigation.[134] Millet produces consistently with as little as 300 mm of annual rainfall, which is the prevailing situation throughout much of Oriens.[135] The grain is easily harvested and, due

[128] N. Lipschitz and Y. Waisel, 'The Dendroarchaeological Remains at En Boqeq', in M. Gichon (ed.), *En Boqeq: Ausgraben in einer Oase am Toten Meer* (Mainz, 1993), 396.

[129] M. Kislev, 'Contenu d'un silo à blé de l'époque du fer ancien', in J. Friend and J.-B. Humbert (eds.), *Tell Keisan 1971–1976: une cité phénicienne en Galilée* (Paris, 1980), 195; Mayerson, *Ancient Agricultural Regime*, 261.

[130] K. O. Rachie, *The Millets: Importance, Utilization and Outlook* (Begumpet, 1975), 5; Zohary and Hopf, *Domestication of Plants*, 5. This is probably the millet mentioned by Procopius (*Bell. Pers.* 1.12.18) as growing in Lazica.

[131] Extra-fine sieving is required for the archaeobotanical recovery of millet that presently is generally lacking from Near Eastern archaeology in the Late Roman–Early Byzantine period. Millets are thus almost certainly under-represented in the archaeobotanical record.

[132] Y. Youtai, 'Millet and Sorghum—The Ancients of Crops', in S. Wittwer et al. (eds.), *Feeding a Billion: Frontiers of Chinese Agriculture* (East Lansing, 1987), 201.

[133] Spurr, *Arable Cultivation*, 203; Youtai, 'Millet and Sorghum'. This drought resistance is primarily due to the fact that the grain has a low moisture requirement and short growth cycle. Millet retains the ability to flourish in temperate situations and high-moisture soils: the crop is also popular throughout temperate North America and Europe in damp climates.

[134] K. R. Chopra, *Technical Guideline for Sorghum and Millet Seed Production* (Rome, 1982), 5.

[135] D. A. V. Dendy, 'Sorghum and the Millets: Production and Importance', in D. A. V. Dendy (ed.), *Sorghum and Millets: Chemistry and Technology* (St Paul, 1995), 20.

to its hard outer shell, is readily dried and stored; sun-dried grain can survive in storage for up to twenty years without losing food value. The nutritional value of millet is high: its average protein content of 9 per cent is greater than that of rice, whilst its 3 per cent oil and fat content exceeds both rice and wheat.[136] Perhaps most importantly, millet is a fast-maturing crop, growing and ripening in only 60–75 days. Its use as an emergency crop is therefore quite rational: by the time the cultivator knew that the winter cereal crop would fail, there was only limited time before the summer heat would scorch young cereals, and millet made an ideal catch-crop.[137]

The plant is therefore well adapted to the climatic and dietary needs of the late antique Mediterranean. Although believed marginal by scholars such as Jasny, millet was probably quite common in Mediterranean agriculture.[138] Spurr argued for a revised view of millet within Italian agriculture of the early empire and it seems that millet was common throughout the late antique Levant as well.[139] The Jerusalem Talmud (edited around AD 400) (TJ Sheb. 2.7) groups millet along with rice and sorghum as common crops. In the seventh century, Isidore of Seville (*Etym.* 17.3.13) described the grain as a staple in many regions.

Medical writers frequently mention millet: Dioscorides described millet as less nourishing than other grains, but useful for various physical problems, a sentiment echoed in authors such as Oribasius (1.15) and in the sixth-century Aetius of Amida (1.145). Alexander of Tralles prescribed millet for a range of ailments, including intestinal complaints (*Frag.* 20) and cholic (2.343). The *Geoponica* cites a diverse array of uses for millet, ranging from brewing beer (7.34.1) and feeding domestic fowl (14.23.4; 14.24.4) to packing for storing grapes (4.15.9). The range of uses and the number of mentions implies millet was ubiquitous, an expected component of estate agriculture, and an important part of the diet in rural areas. The ability of the plant to grow well across the Levantine landscape and its high yield no doubt contributed to its spread and wide use.

[136] Youtai, 'Millet and Sorghum', 203. Obviously, the ancients would not have had the scientific data regarding the nutritional values and would have depended on their own perceptions, perhaps most explicitly seen in the writings of ancient physicians cited throughout this text.

[137] That millet was definitely used in this way is shown below, Chapter 7.

[138] Jasny, *Wheats*, 14.

[139] M. S. Spurr, 'The Cultivation of Millet in Roman Italy', *PBSR* 51 (1983). For dishes made from millet and Italian millet, especially as mentioned in medical texts, see M. Grant, *Dieting for an Emperor: A Translation of Books 1 and 4 of Oribasius' Medical Compilations with an Introduction and Commentary* (Leiden, 1997), 132–3. Oribasius (1.15) has backhanded praise for millet and Italian millet, favouring the former. These grains were, he says, used for bread when there was a shortage of other grains. However, they apparently were a frequent food among the rural population in the form of polenta or porridge.

Although millet was an important secondary crop, many varieties cannot germinate much below 10 °C. Consequently, it seems that millet was spring-sown, which would make it somewhat of an anomaly within the Mediterranean cereal regime, the importance of which will be explored later.

Rice

Although it was a latecomer to Mediterranean agriculture, Asiatic rice (*Oryza sativa*) was relatively common during antiquity and not the product of the Islamic 'agricultural revolution'.[140] Its spread to the Levant was gradual, but by the Hellenistic period rice was known as far west as Parthia.[141] Diodorus (19.13) knew rice, *ὄρυζα*, as an Indian crop important within the intensive agricultural year in the sub-continent. From the second century BC author Aristobulus, Strabo (15.13–18) (first century AD) knew that rice was grown commonly in Bactria, Babylon, Susis, and also in Lower Syria. During the first centuries AD the crop reached Asia Minor from Mesopotamia or the Caucasus. In the second century Galen, and later his follower Oribasius (fourth century) prescribed the grain as a medicament.

In Mesopotamia, the *Nabatean Agriculture* attests to a cultural bias against rice on unspecified religious grounds, which perhaps led to its under-exploitation.[142] Despite this, rice was a crop that was grown throughout the Sasanian empire, where it was already entrenched by the time the Muslims arrived in Iraq.[143] Its normal cultivation in Mesopotamia is known from the Babylonian Talmud, where it appears as an important food for the Jewish communities.[144]

In all likelihood, Jewish farmers introduced rice into Palestine during the Early Roman period, when it is first mentioned in late first- and early second-century sources.[145] The crop of Palestine was a fine, large-kernel rice.[146] Farmers grew *oryza* (TJ Sheb. 2.1) around Paneas-Caesarea Phillipi (Banias) and the Chrysopolis region north of Lake Tiberias, as well as in Caesarea. The

[140] Watson, *Agricultural Innovation*, 14–19. Though the early history of rice is not fully understood, present evidence suggests a South-east Asian origin. Zohary and Hopf, *Domestication of Plants*, 84.

[141] Zohary and Hopf, *Domestication of Plants*, 85.

[142] M. El Faïz, *L'agronomie de la Mésopotamie antique: analyse du 'Livre de l'agriculture nabatéenne' de Qûtâmä* (Leiden, 1995), 155.

[143] Newman, *Agricultural Life of the Jews*, 94. It is tempting to see Mesopotamia as the intermediary in the transmission of the plant, but the rice seeds recovered from Berenike may imply a more direct exchange from India to the Red Sea coast and thence up the Nile Valley.

[144] Ibid., 90–1.

[145] Safrai, *Economy*, 117.

[146] Zohary and Hopf, *Domestication of Plants*, 85.

latter produced a red variety also grown around Antioch.[147] Probably around the same time as its introduction in Palestine, *oryza* was grown in Asia Minor. Lowland Cilicia, with its numerous rivers and broad coastal plain, made ideal rice country. This would explain its probable inclusion as an item on the Anazarbus tariff inscription.[148] Certainly in the fourth century, rice was already regularly farmed in Asia Minor, since it is listed as a trade item at Ephesus in the mid-fourth-century *Expositio totius mundi* (§47). By this date, *oryza* was a common article of exchange, regulated with other daily items in Diocletian's Price Edict (1.23), with a value about twice that of wheat.

Dietary preferences curtailed the movement of rice within the Mediterranean. From Oribasius (1.16) we learn that rice was considered inferior to other cereal crops. It is questionable how deep this prejudice was; the paradigmatic epicure of the second century, Athenaeus (3.110e), did not disparage rice, but neither did he discuss it much. Both Ps.-Apicius (2.2.8–9) and Anthimus (70) preserve recipes that use the grain. The notice in Anthimus (fifth century) is especially interesting, since he wrote in Gaul and may indicate the plant's spread or regular trade in that region by Late Antiquity.

In addition to the aforementioned notices in Galen and Oribasius, there is further evidence that the plant was widely used. Aretaeus of Cappadocia (first century AD) recommended rice for the treatment of pleurisy.[149] In the first century, Dioscorides (2.95) included rice in his famous pharmacological work, noting that the grain was useful for stopping the bowels.

Rice was thus a component of Mediterranean agriculture long before the Muslim conquest. Despite acknowledging that rice was cultivated in Palestine during the centuries immediately preceding the Arab conquests, Watson considerably underestimated the range and significance of its early exploitation, and also exaggerated its later importance. Rice never became a primary cereal crop in Byzantium and the western Mediterranean, even after the Muslim conquests; nor did it make much progress (i.e. significantly expand its role in the diet or supplant wheat or barley) in the lands at the heart of the supposed Islamic 'agricultural revolution'. This was the case despite the fact

[147] Antioch-on-the-Orontes is almost certainly the city referred to here.

[148] G. Dagron and D. Feissel, *Inscriptions de Cilicie* (Paris, 1987), 182 n. 92, notes that the '– zion" is clearly present. He sees the reading Ôruzion (var. of Ôruza / Ôruzon) as possible but rejects it based on his understanding of rice as an 'Indian' crop: 'Le riz, est bien connu, mais comme une spécialité indienne qu'on ne s'attend pas à rencontrer ici.' Dagron's understandable reservations can be dealt with easily by noting the evidence from the *Expositio* which shows rice under cultivation, as far west as Ephesus, by Late Antiquity. The abundant references in the late literature suggest that the grain had been a commonplace on the Roman scene for some time. Concerning Anazarbus, see M. Gough, 'Anazarbus', *AnatSt* 2 (1954), 87, and Hild and Hellenkemper, *Kilikien und Isaurien.*

[149] Aretaeus of Cappadocia, *Extant Works*, ed. and trans. F. Adams, pp. 411–12.

that the Tigris–Euphrates corridor, the centre of 'Abbasid agriculture, was totally dependent on hydraulic agriculture adaptable to rice production.[150] In Andalusia, likewise, where Islamic water-lifting technology gave rise to the intensively watered *huertas*, wheat and barley continued to dominate, just as they did in Mesopotamia.[151]

In addition to the possible dietary prejudice just noted, there were ecological factors that hampered the spread of the grain. Rice is an extremely thirsty plant, requiring massive water supplies throughout its life cycle. The Mediterranean region is ill-adapted to provide such a volume of water, and artificial irrigation was not spent on a plant that the Greeks and Romans felt was inferior. In addition, though wheat and barley under irrigation could produce multiple crops each year and could tolerate the saline conditions common to hydraulic agriculture, in these same conditions rice was restricted to a single harvest.[152]

The role of rice in Late Antiquity was therefore that of a minor and generally familiar grain of local dietary importance, and probably a significant medicinal grain. Given the plant's environmental demands, it appeared where one would expect: in river valleys and warm areas with abundant water. The Orontes and Jordan valleys, as well as the marshy areas around Ephesus were areas of riziculture, but apart from those few regions where its production was simple, demand for the grain was not high enough to command the labour and capital outlay needed for its production.

Oats

According to Pliny (*H.N.* 18.149), oats (*Avena sativa*; *βρόμος*; Lat. *avena*) were a food of barbarians. Despite their ability to flourish on poor soils and wild oats' common occurrence amongst Levantine flora, we cannot be certain of the extent of the cultivation of domestic varieties in Oriens. The medical writer Aetius of Amida (1.73) thought the grain was binding for the belly and prescribed it as a medicament for the intestines. *βρόμος* features in the dietary work of Oribasius (1.14), where the doctor declared oats were fit only for beasts. Though they may have formed part of the diet in some regions, especially amongst the poor, it seems that oats were generally reserved for animal provender or famine food. In 582, when the Constantinopolitan food shortage grew extreme, they were a food of last resort,[153] just as in 687, the

150 McC. Adams, *Land Behind Baghdad*, 100–1; McC. Adams, *Heartland of Cities*, 217.

151 E. García Sánchez, 'Agriculture in Muslim Spain', in S. K. Jayyusi (ed.), *The Legacy of Muslim Spain* (New York, 1992), 994.

152 M. El Faïz, 'Salinité et histoire de la Mésopotamie pré-islamique', *JESHO* 33/1 (1990), 108–12.

153 Teall, 'Grain Supply', 1–2. Michael the Syrian, *Chronicle*, ed. and trans. Chabot, II.305.

famine in Syria forced people to turn to oats.[154] Other human uses are known: oats were used in brewing a kind of beer (*Geop.* 7.34.1), and they were apparently a common grain exchanged in the marketplace. Diocletian's Price Edict set the price of oats at 30 denarii per *modius castrensis*, less than a third that of wheat and half that of barley.[155]

Crops on the Land: Subsistence and the Agrarian Economy

Wadi Kurnub

If viewed as a wheel, subsistence would have demand at the hub, with radiating spokes representing technique and crops. At the rim of the wheel would be productive capacity or supply. How much the late antique farmer could ask of his or her land varied considerably from place to place and year to year. Some insight into the lot of the ancient farmer may be gained by gauging the potential capacity of a given area of land. Yield may be assessed against labour requirements, the need to feed the family, demands of the landlord, and the burden of taxes. What are reasonable assessments of the carrying capacity of the ancient landscape? Since landscapes are unequal—and factors affecting annual yield myriad—precise assessments, particular to a given landscape, cannot be considered here. Instead, my approach somewhat flattens the difference among various plots of lands and inter-annual fluctuations, and works from the simple premise that there was such a thing as an 'average' piece of land, agricultural year, and harvest. Such calculations naturally cannot account for climatic variability, and count on relative stability, maintenance of fertility of the earth, and the absence of catastrophic natural events. Some variables may be accounted for by lowering the yield within the basic models, but the calculations presented here offer only one of many possible outcomes of calculating late antique subsistence needs and grain yields.

Though account records are non-existent in former Oriens, remnants of farms do survive. The example chosen here is a runoff farm in the Wadi Kurnub in Palaestina Tertia published by Evenari.[156] In addition to its physical

[154] Michael the Syrian, *Chron.* II.474–5.

[155] Safrai, *Economy*, 117. The *Geoponica* also mentions their use as sheep feed (18.2.6) and as fishing bait (20.8.1). The grain was common enough in the minds of fourth-century Cappadocians for Basil to use it by analogy, comparing it with wheat, Basil, *Hexameron*, ed. Giet, 5.3.31.

[156] The fact that this entity is a farm dependent on floodwater management should not deter us from considering it as a representative case. Whilst the installations required to make these lands fruitful were in many ways exceptional, their tilled surface areas are in keeping with other farms known from the archaeological record. I have argued, these farms were comparable to the

situation within Oriens, two factors commend this choice. First, this farm lies within the Negev, the general zone of exploitation whence comes the textual evidence detailing yields from Nessana. Second, the farm is of a moderate size, neither the smallest known example in Oriens nor by any means a large estate. Its area is estimated to cover 40–48 iugera, very near to the 50-iugera farm Constantine alloted to discharged veterans (*CTh* 7.20.3). This land area is a useful benchmark in judging contemporary views of subsistence requirements: veterans would have received an allotment that was neither extravagant by the standards of the day, nor too poor to serve their needs and those of their families. Given this, we may classify the Wadi Kurnub holding as a medium-sized farm.

Cereals were probably one of the major crops produced at the Wadi Kurnub Farm, but it is doubtful that it was the only crop. However, this discussion assumes that there was a cereal monoculture in place—purely for the sake of contextualizing subsistence needs and transposing those needs onto a real farmed landscape. Table 3.1 outlines the cereal production capacity of the Kurnub Farm, assuming a two-field rotation system with half the land lying fallow each season. Sowing rates, and hence harvest rates, are expressed in modii. The volume and weight of the modius is somewhat problematic.[157] Columella's sowing rate of 4 modii of grain sown per iugerum, the *modius Italicus*, needs to be halved to account for the use of the double-size imperial modius.

Since a family of six, posited as reasonable for an average family size in Late Antiquity, required 94 modii of grain per annum for subsistence, the Kurnub

coastal zone in the amount of water that they actually delivered to the crops. This is not an example of a richly irrigated landscape, but rather a landscape on the extreme margins of cultivation, made productive, and I would argue only ordinarily so, through the supporting hydraulic infrastructure. For the sake of argument, it is assumed that the Wadi Kurnub farm was a discrete agricultural unit, owned and worked by a family of six identical in their demands to that used (at the start of this chapter) to calculate cereal needs. The chronological phases of 'Stage II' and 'Stage III' of the farm belong to the Roman–Byzantine periods and form the area studied here. Evenari, Shanan, and Tadmor, *The Negev*, 110–13, fig. 74a.

157 R. Duncan-Jones, 'The size of the modius castrensis', *ZPE*, 21 (1976a) argued against the conventional view that the modius *castrensis* was equal to two modii *Italici*. Pliny (18.66) clearly states that a modius is equal to twenty Roman pounds (*librae*) of wheat. *Geoponica* (2.32.1) states that a *modios* of grain weighs forty pounds (*litrai*). These figures correspond with those proposed by Schilbach: Schilbach, *Byzantinische Metrologie*, 96; E. Schilbach and A. P. Kazhdan, 'Modios', in A. P. Kazhdan (ed.), *ODB* (Oxford, 1991), 1388. He places the fourth-century modius at twenty pounds and the *basilikos modios*, the old *modius castrensis* (or imperial modius), at forty pounds. Therefore, unless specified, the following discussion uses the *basilikos modios*, as the *Geoponica* reflects agrarian custom in at least some areas of Oriens. Of course the units in use varied greatly from place to place, and we are only able to get an approximate idea about sowing and yield from the meagre evidence to hand. The calculations of the present study provide an order of magnitude, not absolute figures.

Table 3.1. Yields proposed for Wadi Kurnub farm by regime type

Rotation	Area	Wheat modii sown	Wheat modii harvested
Biennial fallow	20 iugera	40	280
Annual cropping	40 iugera	80	560

Farm would have exceeded the fundamental demands by nearly three and a half fold. Considering that 40 modii of wheat were required for the annual sowing, and assuming 20 per cent loss in storage and processing, we are left with a reasonably conservative estimated surplus of 122 modii of grain.

Having approximated yields, it is possible to address the question of the grain trade on a local scale. What was the value of the surplus crop? Various grain prices are known from Oriens during Late Antiquity. In AD 363, in a time of shortage, the emperor Julian sold grain at Antioch for 15 modii per solidus; and in Egypt, 15 modii to the solidus is recorded as a famine price.[158] Pseudo-Joshua the Stylite (§26) reports that, during a time of plenty, grain at Edessa sold for 30 modii for the solidus, but during the great famine of AD 500, shrank to 4 modii for a solidus. Morrison's data from the late fifth–early sixth century provides a price range of 30–40 modii to the solidus.[159] Based on this evidence, it seems reasonable to conclude that in the sixth century, grain generally sold for between 15 and 30 modii per solidus. The 122 modii surplus at Kurnub, therefore, would have yielded grain with a cash value of 4–8 solidi per year.

Though in practice such an area probably would have had additional hands, the Wadi Kurnub farm was small enough to be worked by one adult. He or she would have been fully engaged for about six weeks in ploughing and sowing the land during the autumn and winter in preparation for the flooding of the farm. Working a farm the size of Kurnub would have required at least a pair of oxen, whose maintenance required significant resources, perhaps equivalent to a solidus per year.[160] Another constant burden on the land

[158] Jones, *Later Roman Empire*, 446.

[159] C. Morrisson, *Monnaie et finances à Byzance: analyses, techniques* (Aldershot, 1994), 252; Ps.-Joshua, *Chron.* 26 (AD 494/5).

[160] A high estimate, based on *Geop.* 17.12, which states oxen are to be fattened on 4 *cotylae* of barley per day. As stressed above, these yield calculations are of the crudest nature and speculative. No effort has been made to account for fallowing and crop rotation, nor land fertilization. Such activities often cannot be assessed but undoubtedly had substantial bearing on yields. Nor do the calculations consider the amount of grain required to feed the farmer's flocks or transport. A more complex model, which considers such elements, would probably add little to the discussion. The purpose here is to flesh out, in a very approximate way, the data provided by written sources and archaeology, in order to frame and stimulate discussion of subsistence and trade in Byzantine Oriens.

was taxation, which can only be very roughly estimated, whilst rent, although not a universal onus, would have been a factor in a large number of cases. Before attempting to create a quantified framework of these two critical factors, we need to examine an example of a small agricultural entity, perhaps more typical of the smallholder or tenant farmer within late antique Oriens.

Michael's Farm

The so-called 'Michael's Farm' (Figure 3.1) is a small runoff farm located near the village of Sobata (Shivta) in the southern Negev, about 50 km south of Be'er Sheva. The farm comprises a series of terraced fields with a cultivable area of approximately 0.70 hectares (2.8 iugera), far smaller than the 40–48 iugera of the Wadi Kurnub farm. In the following discussion it is assumed, due to the intensive nature of floodwater agriculture and the fact that each plot receives fertilizer in the form of washed soil, that no portion of the farm lay fallow during our hypothetical agricultural year.

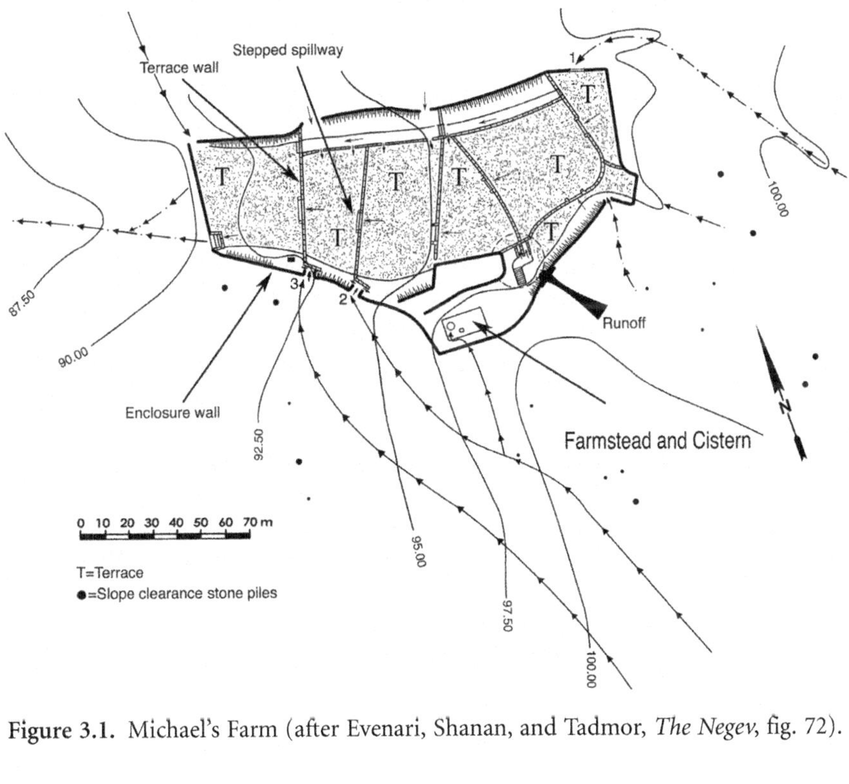

Figure 3.1. Michael's Farm (after Evenari, Shanan, and Tadmor, *The Negev*, fig. 72).

At a sowing rate of two imperial modii per iugerum, Michael's Farm required 5.6 modii of wheat, which, at a yield–sown ratio of 8:1 would return approximately 45 imperial modii. Subtracting grain (5.6 modii) for the following year's sowing, and deducting 20 per cent for losses incurred during milling and processing, we arrive at a net production of about 31½ imperial modii of wheat flour. This quantity is sufficient to provide for less than six months' grain for the hypothetical family of six used in this study. Rent, if the plot were not owned by the cultivator, and taxes would have taken an additional cut of the produce. If these rough-and-ready calculations are anything close to reality, the inhabitants at Michael's Farm must have supplemented their income with wage-labour elsewhere.

Rents and Taxation

The sources in Oriens only infrequently mention rental terms and rates. We know next to nothing of the quantity of cash or produce estate holders, such as Libanius in fourth-century Antioch, received from their lands, nor do we know much of specific rates charged. Certainly it appears that there were, at any given time and place, various types of tenancy in use; many involved payment in kind of crops. In Palestine, Safrai notes that sharecropping was a common practice: the renters gave a portion, probably a sizeable one, of the harvest to the rentiers. Based on figures from Ptolemaic royal estates, Jones argued that, in Egypt, 50 per cent of the gross product of the land was paid as rent.[161] It is likely that such high rents, well in excess of even the 30 per cent demanded on imperial estates in Roman Africa, included the cultivator's tax burden within the rent payment.[162]

The value of applying these figures (that Jones called 'very high') wholesale is dubious.[163] References to customary rates have survived, and they represent at least two broad scenarios under which the land was held by the tenant. Rabbinic evidence of the sixth century suggests that under sharecropping conditions, the worker (*'aris*) took one-third or one-half of the field's produce (TB Giṭ. 74b; TB B.M. 110a; 103b; TB B.B. 46b). In the latter instance, the owner provided the seed, and the tenant provided expenses and tools for working the plot. This is similar to the Egyptian situation noted by Rowlandson where the landlord took 66 per cent of the produce and was responsible

[161] A. H. M. Jones and P. A. Brunt, *The Roman Economy: Studies in Ancient Economic and Administrative History* (Oxford, 1974), 83.

[162] D. P. Kehoe, *The Economics of Agriculture on Roman Imperial Estates in North Africa* (Göttingen, 1988), 107.

[163] Jones and Brunt, *Roman Economy*, 83.

for the land tax.[164] However, according to the 'Farmer's Law' of the seventh or eighth century (*Nomos Georgikos* §10): 'A sharecropper's portion is nine bundles, the grantor's one: he who divides outside these limits is accursed.' This probably reflects a different arrangement in which the tenant provided the seed, utensils, and labour, and was responsible for the taxes.

Ideally we would have precise figures for each of these, but a thorough study of late antique taxation is lacking.[165] Without such documents, any valuation of tax rates is precarious and therefore provisional. In a passage dealing with the collection of imperial tax, Jones specified rental rates of half the gross produce of the land, with a third of the total crop required in taxes.[166] Approaching the problem differently, Duncan-Jones argued that the tax iugum, the standard unit of the Roman land tax, possessed a mean value of 14 iugera of land. A mean value is used, due to the fact that census methods were reasonably sophisticated, assessing arable land, vineyards, and olives at different rates according to the differing productivity and market potential of each. Duncan-Jones's model is useful, based as it is on a deeper data set and careful analysis.

According to Duncan-Jones, a mean land taxation rate in the Late Roman period was 1 solidus per 10.5 iugera of land. In the fourth century, Epiphanius of Salamis stated, according to Duncan-Jones's interpretation of the text, that a tax iugum was 12.5 iugera for first-class arable land.[167] However, the fifth-century legal compilation known as the *Syro-Roman Lawbook* (see Table 3.2) offers a picture of the links between land area and assessment.

Jones dismissed the evidence from the *Syro-Roman Lawbook*, stating that Syria appears to have been exceptional.[168] In fact, there is good reason to accept the information it provides. The law book was a contemporary compilation and preserves the historical accounts of Diocletian's assessment and tax reforms.[169] Likewise, it is geographically specific, pertaining to a region in

[164] J. Rowlandson, *Landowners and Tenants in Roman Egypt: The Social Relations of Agriculture in the Oxyrhynchite Nome* (Oxford, 1996), 248.

[165] There have been several studies on various aspects of Roman taxation, among them: A. Cerati, *Caractère annonaire et assiette de l'impôt foncier au Bas-Empire* (Paris, 1975); A. Chastagnol and I. Tantillo, *Aspects de l'antiquité tardive* (Rome, 1994); A. Déléage, *La capitation du Bas-Empire* (Mâcon, 1945); Jones, 'Census Records'; A. Piganiol, *L'impôt de capitation sous le Bas-Empire romain* (Chambéry, 1916); S. L. Wallace, *Taxation in Egypt from Augustus to Diocletian* (Princeton and London, 1938) but we still lack a detailed up-to-date global study.

[166] Jones, *Later Roman Empire*, 820; Jones and Brunt, *Roman Economy*, 184. This discussion deals only with land taxes. Poll tax probably would have also been paid.

[167] R. Duncan-Jones, *The Economy of the Roman Empire: Quantitative Studies* (London, 1974), 201; Epiphanius, *Epiphanius' Treatise on weights and measures; the Syriac version*, trans. J. E. Dean (Chicago, 1935), 59.

[168] Jones, *Later Roman Empire*, 820.

[169] *Syr.Rom.Law.*, ed. Vööbus, §157.

Table 3.2. Iugera and tax iugum according to the *Syro-Roman Lawbook*

Land type	Amount per tax iugum
First-class arable	20 iugera
Second-class arable	40 iugera
Third-class arable	60 iugera
Vineyard	5 iugera
Olives ('old')	225 trees
Olives ('mountain')	450 trees

the centre of Oriens. It is thus possible to utilize the *Syro-Roman Lawbook* as a basis from which to obtain some notion of the labour costs, rent, and tax burdens of the Kurnub Farm and Michael's Farm, assuming, conservatively, that the total area of each farm was classified as 'First-Class Land' and thus subject to a high tax rate of 1 solidus per 20 *iugera* of land.

Expressed in coin, the approximate values are as shown in Table 3.3.[170] In kind, the values in Table 3.4 are obtained.

For the sake of comparison, it is useful to look at the prices of a few goods and services. In the sixth–seventh centuries, a mason received 5–9 folles per day (there were about 180–210 copper folles to each gold solidus during the reign of Justinian). According to Procopius in the first half of the sixth century, a prostitute earned 3 folles per day; later, in AD 709, a carpenter is recorded as earning 1⅓ solidi per month. A woollen mantle at the beginning of the fifth century cost 50 folles; a pair of leather shoes cost 150 folles; an ass cost one solidus; and a horse cost 3 solidi.[171]

The simplified taxation examples noted in Tables 3.3 and 3.4 illustrate an important fundamental: that the minimum land area needed for subsistence under a low-intensity grain monoculture was about 40 iugera. Although the farm in the Wadi Kurnub would have produced a surplus in normal years, this yield would have offered little potential for long-term capital improvement. The estimated land area required for subsistence thus exceeds considerably

Table 3.3. Calculated values of grain yields

Farm	Gross yield	Rent	Tax	Net yield
Wadi Kurnub	10.6 solidi	~ 1 solidus	2 solidi	7 solidi
Michael's Farm	4.5	~ 1/3 solidus	1/12 solidus	1 solidus

[170] Some figures are slightly rounded for simplification; grain valued at 30 modii per solidus.

[171] Morrisson, *Monnaie et finances*, III.248.

Table 3.4. Calculated yields of farms in modii

Farm	Gross yield[1]	Rent	Tax[2]	Net yield[3]
Wadi Kurnub	320	32 modii	60 modii	206
Michael's Farm	45	4.5 modii	1.25 modii	31

[1] Gross yield.
[2] Here converted from coin to an in-kind payment.
[3] Including 20 per cent loss from storage and processing; figures are rounded. This of course does not include the food requirements of the family.

the minimum land requirements some scholars have estimated, which range from 9.7 to 27.7 iugera.[172]

Whilst the Kurnub Farm could have been viable under a low-intensity cereal monoculture, Michael's Farm was not. As it stands in the archaeological record Michael's Farm may be incomplete, or it could represent either a component of a larger holding, or one whose inhabitants were intended to supplement their income by engagement in other activities. Michael's Farm might also have sustained itself by alternative forms of agriculture. This possibility, described below, finds strong support in the textual and archaeological record. Michael's Farm is small by any standard, but far from unique in the late antique countryside. A significant number of people held and worked small plots that, under a cereal monoculture, were insufficient to sustain them. The fourth-century land register from Hermopolis in Egypt shows that holdings less than one aroura in size (approximately one iugerum) were common.[173] At Qirqbize in North Syria, the area of land available per family was only about 7.2 iugera,[174] while archaeological evidence from the sixth-century village of Sumaqa in northern Israel suggests that 133 families possessed a cultivated landscape of 1,303 iugera, an average of 9.8 iugera per family, sufficient to provide subsistence cereal crops under only the most promising conditions.[175] Yet Sumaqa, like so many villages throughout the

172 Dar and Applebaum, *Landscape and Pattern*, 74; Oakman, *Jesus and the Economic Questions*, 64; White, *Roman Farming*, 336, provides data from the Roman–Byzantine site of Buraq in Samaria', where an average area a small-farmer worked for cereals was about 0.67 ha—a little over 2 iugera.

173 Although the area of the mean holding was much larger. Bowman, 'Landholding in the Hermopolite Nome', 155, tbl. II.

174 Approximation based on personal prospection and Tchalenko, *Villages*. The figure is based on an estimated 217 people in the village, 6 per family, and 65 ha arable surface.

175 Dar, *Sumaqa*, 137–9, estimates that the total land area of Sumaqa was 5,430 dunams, of which 60 per cent was arable (= 3,250 dunams/1,303 iugera). With Dar's estimated population of 800, amounting to 133 hypothetical families, with the cereal demands noted above, the

late antique Levant, was an apparently flourishing community; excavations have revealed an abundance of imported ceramics indicative of regional and long-distance trade.[176] It is this paradox of scarcity of land, combined with evidence of economic prosperity, which is fully examined in Chapter 8, where full appraisal is made of the possible alternative patterns of exploitation and subsistence.

demand for grain would amount to an aggregate 16,625 modii per annum. On a biennial rotation this represents the production of 2,078 iugera. Assuming an 8-fold yield and no fallow, the landscape of Sumaqa could have provided 156 modii of grain for each family, an untenable figure if we consider elements like rent or taxation, the need to generate cash to acquire goods like iron tools not produced locally, and seasonal fluctuations in yield.

[176] Dar, ibid., 138–9; S. Kingsley, 'The Sumaqa Pottery Assemblage: Classification and Quantification', in S. Dar (ed.), *Sumaqa: A Roman and Byzantine Jewish Village on Mount Carmel, Israel* (Oxford, 1999).

4

The Vine

WINE IN THE LATE ANTIQUE WORLD

Wine played a vital role in the civilizations of Antiquity. From the symposia of the educated elite of classical Athens to the tables of the poor rustic, wine was a universal beverage for the ancient world as well as the universal substance in the sacraments of the church. All communities throughout the Mediterranean world produced wine of some sort; from the Jewish villagers in the Golan and their Arabic speaking neighbours—whose poets sang the praises of the fruit of the vine—to the Frankish noble in far off Gaul, the drink was the same. The story of wine in the late antique Levant is the story of shared social tastes and culture, of traders and farmers, the interplay of communities, technological application, and dynamism in agricultural technique.

As a dietary staple, wine offered one of the few means by which pre-industrial peoples could obtain vital calories and nutrients, and thus was constitutive in the diet. Alcohol, with its intoxicating effects and its medicinal potency, was a pleasant and useful by-product that made wine even more desirable. These attributes—especially the capacity of wine to remain stable and usable for long periods of time, even for years—explains the significance of wine in the Mediterranean food triad alongside grain and olive oil.

The drinkers of Late Antiquity imbibed the fermented juice of the European grape vine, *Vitis vinifera*, a plant that probably originated in the Caucasus. The vine had been exploited for millennia, and its spread was aided by its ecological adaptability.[1] In Late Antiquity the marvellously pliant vine flourished in some of the most inhospitable landscapes from the Negev to Nubia, where arid conditions allowed production only under irrigation.[2] Thus, across the late antique East, wine production for subsistence purposes was practised in nearly every area. Throughout Oriens, archaeological and textual evidence demonstrates the abundance of vineyards in the Byzantine period, the primary function of which was for making wine, not as a source

[1] P. E. McGovern et al., 'The Beginnings of Winemaking and Viniculture in the Ancient Near East', *Expedition* 39/1 (1997).

[2] W. Y. Adams, 'The Vintage of Nubia', *Kush* 14 (1966), 262–83.

for table grapes. The vineyards of the coastal regions formed the heart of the Early Byzantine wine trade, and their produce was enjoyed throughout the empire and beyond.

André Tchernia concludes that the *per diem* consumption of an adult Roman male was one to two sextarii (about a litre) in wine or soured wine (*posca*).[3] In order to estimate the wine intake of the late antique populace, we return to the family of six used as a basis for grain subsistence calculations in Chapter 3, which comprised three adults, two female and one male. Assuming the two youngest children drank no wine, that the father and son daily consumed one sextarius each, and the women half of that, the family's total annual consumption would be 1,095 sextarii (598 litres). On a broader scale, at this level of consumption the 150,000 intramural inhabitants of Antioch required about 15 million litres of wine per year. Urban centres, home to the majority of the late antique elite of the Mediterranean and its fringe, also formed the loci of demand for imported wine. This demand was much more than regional in its scale. With the growth of Constantinople and other eastern cities, and with the influence of eastern merchants in Merovingian Gaul and elsewhere in the post-Roman West, the eastern provinces increasingly became the vineyard of the Mediterranean.

THE LATE ANTIQUE VINEYARD: MANAGEMENT AND PRODUCTION

A thorough foundation to our knowledge of the agrarian life and economy of Late Antiquity is not possible without examining wine production. The farm lay at the core of late antique society, and the vineyard lay at the heart of the farm. It is therefore unsurprising that a culture that produced technical farming manuals also exhibits an intense interest in furthering the art of winemaking. As an indication of its importance, viticulture comprises five of the twenty chapters in the *Geoponica*. The vintage was a focus of the agricultural year. Care of the vineyard formed one of the most labour-intensive elements of ancient life. In the case of the small-scale farmer, the task of maintaining a vineyard meant arduous labour that might have been devoted elsewhere. For the estate owner, concerned with the preservation and creation of wealth, tending a vineyard demanded careful planning, concerted mobilization of labour, ingenuity, and skill. Viticulture in antiquity, as today, was

[3] A. Tchernia, *Le vin de l'Italie romaine: essai d'histoire économique d'après les amphores* (Paris, 1986), 26; cf. Patlagean, *Pauvreté économique*, 53.

complicated and uncertain, and it is thus no surprise that the handbooks offer advice on the minutiae of vine production and winemaking. Nothing could be left to chance, yet, as with most agricultural endeavours, success was never guaranteed.

Among estate owners we find considerable concern not only for quality but also for the quantity and speed of wine production, considerations that strongly suggest production for the market. Certainly, the depth of preparation and range of detail presented in the handbooks are impressive. Their approaches to vineyard organization are measured and articulated within a framework of rational exploitation, predicated on a desire for profit. The *Geoponica* presents a considerable diversity of practice, alongside debate and experimentation. It reveals farmers who were forward-looking and adaptive in their pursuit of the best possible yield and quality. Beyond this, the farming literature also illustrates a deep understanding of the production of high quality wine as well as the effects of the environment on the plant and its product. Through this window on viticultural technique, we see even more of the vital components of the agrarian economy of Late Antiquity: intensification, specialization, and a highly developed network of interlocking techniques and technologies.

Farmers made judgements in planning vineyards based on a number of considerations, including land characteristics, the kinds of grapes grown, the climate and quality of soil, how the vines would be supported, and how they would be tended. Choosing the correct variety of grape to suit the soil was the cardinal concern of the grower (*Geop.* 5.1–2). Attributes of numerous grape varieties were known and carefully assessed, and stock of the best kinds were widely available. We see them traded in Cilicia, where the sixth-century 'Anazarbus Tariff' records the taxes placed on plant cuttings brought into the city.[4] Grafting stock was transported over considerable distances (*Geop.* 10.75.22), and many farmers preserved their vine shoots for such journeys by inserting them into squills or bulbs (5.8.7).[5] The fact that farmers interested themselves in acquiring and selling different plant varieties is significant, as it demonstrates the desire to vary and adapt new cultivars to different locations in order to enrich local production. More importantly, the exchange of wine cultivars attests to the existence of markets where diverse wines were recognized and expected. This spread of famous varieties was a consequence of tastes shared by wine drinkers throughout the ancient Mediterranean world,

[4] Dagron and Feissel, *Inscriptions de Cilicie*, 179.

[5] It is important to stress that, as in Columella's day, there was apparently a vigorous exchange in plant varieties of all kinds, especially those durable enough to move across some distance.

one that continued to flourish in Late Antiquity.[6] One can see this vocabulary fully articulated in the works of Athenaeus (1.32d; 1.33d), where we see appreciation for colour, bouquet, taste, and the subtlety imparted to wines through ageing. All of these components are emblematic of a shared wine culture within an elite whose tastes were foundational, both in driving the long-distance exchange of wines and in creating an environment that influenced the viticulture of Late Antiquity toward adaptation, intensification, experimentation, and profit.[7]

Geoponica 5.15 supplies essential evidence for crop selection in the context of vineyard management. Since wines from each type of grape had such peculiar qualities, they were not to be mixed with one another: to do so risked ruining the vintage by imparting the characteristics of bad grapes to the whole batch.[8] Since wine purity was an issue, each plot of grapes was to be homogenous, but the fact that different grapes matured at varied times was also a major consideration. Unripe fruit should not be harvested with ripened grapes, since the former would lower wine quality. Although individual plots were homogenous, whole vineyards were not: Palladius (3.9.11) instructed that one should choose four or five excellent grape varieties for planting for the whole of the estate. In a given year, natural conditions might prove detrimental to one type of grape, but was unlikely to damage all at once (*Geop.* 5.16). Strategies of this kind mitigated risk and are typical of ancient agriculture; they reveal a primary interest in ensuring a safe return, rather than a reach for spectacular results. This conservatism, however, was not the result or the cause of an overall lack of specialization, market awareness, or desire to maximize yields.[9] On the contrary, in light of the capricious Mediterranean climate, strategies of this kind are amongst the more rational. Why risk the

[6] Naturally this trade in plants and the diffusion of desirable vines was long established: vine branches were found in the Hellenistic wreck at Pozzino (120–80 BC). A. J. Parker, *Ancient Shipwrecks of the Mediterranean and the Roman Provinces* (Oxford, 1992), 340.

[7] Colour and bouquet (*ἀνθοϲμίαϲ*) were elemental, I believe, to appreciation of wine and the growth of specialization. Understanding the innate qualities of vinosity (*οἰνοειδήϲ*) allowed recognition of the distinct character of wines of different regions and age. Recognition of these subtleties is the cornerstone of high quality wine production and key to ageing the vintage. The classifications of colours: black/dark (*μέλαϲ*), white (*λεῦκοϲ*), and yellow (*κιρρόϲ*) were accompanied by assessments tied to Hippocratic sympathetic thought: wines were hot (*θερμόϲ*) or cold (*ψῦχοϲ*), and to the taste dry (*ϲτυπτικόϲ*) and sweet (*γλεῦκοϲ*). In character, they were complex (*ϲύνθετοϲ*), rough (*τραχύϲ*), mild or delicate (*τρυφερόϲ*): M. Lambert-Gócs, *The Wines of Greece* (London, 1990), 272–91.

[8] This should be distinguished, however, from later blending of wines that was done once proper fermentation was complete. No doubt this rule of not mixing grapes was often ignored.

[9] Contra Duncan-Jones, *Quantitative Studies*; cf. D. P. Kehoe, 'Allocation of Risk and Investment on the Estates of Pliny the Younger', *Chiron* 18 (1988); D. P. Kehoe, 'The Sale of the Vintage in Pliny, Ep. 8.2', *Annals of the American Philological Association* 51 (1988); D. P. Kehoe, 'Approaches to Economic Problems in the Letters of Pliny the Younger: The Question of Risk in Agriculture', *ANRW* II/33.1 (1989).

entire crop for the sake of substantial but intermittent profits? Heterogeneous planting ensured a base return, and over the years could promise much more. Mixed vineyards also allowed the labour of the vintage to be spread over a longer period as grapes matured at different times, an element that Palladius lists amongst the benefits of such a planting scheme (3.9.11–12.).[10] One should thus view diversified planting as part of a scheme to maintain a constant supply and surplus, often within a specialized environment, and not as a stagnant agricultural process aimed only at self-sufficiency and survival.

Once the choices of suitable land and vines were made, a vineyard was laid out and prepared. The plot of land was measured by fractions of a *iugerum* (Palladius, *Op. Ag.* 2.9) and cordoned off from the surrounding landscape with hedges, palisades (*Nomos Georgikos* §51; §58), or ditches (*Nomos Georgikos* §50). This was primarily to protect the crop from the neighbourhood animals. Palladius envisaged a square plot 180 ft on a side, divided further into 324 sections of 100 ft^2.[11] This was an area slightly larger than one iugerum; the extra room provided for the borders and unproductive areas of the vineyard.[12]

Two major types of vineyard are evident from the sources: the *arbustum*, where vines were supported by trees, and the 'low' vineyard, where vines were propped on stakes or grew on the ground. The arbustum, the form of cultivation most suited to intensive farming systems, is especially congruous with the intensified farming of Late Antiquity. Palladius (3.10) devoted a section to the arbustum, a testament to its importance in the western Mediterranean. Similarly, the sixth-century mosaics in the Church of Deir el Adas (Figure 4.1) at Bostra show vines growing on trees. The writings of Julian of Ascalon (51.1) demonstrate that this method of planting was undertaken even within towns, where gardeners optimized limited space. The arbustum also appears in the ninth-century compilation, the *Nabatean Agriculture*, indicating the practice was common throughout Mesopotamian vineyards.[13] Vines could also be propped with reeds or pales or grown on the ground. The latter

[10] This strategy has also been used to argue for Late Roman manpower shortages, E. Frézouls, 'La vie rurale au Bas-Empire d'après l'oeuvre de Palladius', *Ktèma* 5 (1980), but is better viewed as a common-sense and economical use of available human resources.

[11] 32,400 Roman ft^2 or about 2,800 m^2.

[12] The much-discussed 'model vineyard' of Columella 3.3 was 7 *iugera* in extent. It is therefore interesting that Palladius thinks in units of 1-*iugerum* plots that in aggregate make up the vineyard. It is difficult to know if there is a conceptual difference between the two agronomists regarding their arrangement of space, or whether Late Roman farmers differed from their predecessors in their land divisions and management schemes.

[13] Salinger, *Kitab al-Filaha an-Nabatiya*, 93–5.

Figure 4.1. Deir el-Adas Mosaic (author's photo).

method, favoured since wood was often in short supply, provided high returns, but yielded low quality wine.

After surveying, division, and fencing, the vignerons cleared the earth of debris, plants, and tree roots (Palladius, *Op. Ag.* 2.10.1; 2.13.2).[14] Palladius (2.10.2) set out the dimensions of furrows; these were 2 ½ or 3 ft wide and equally deep, and ran the length of the field. Ideally vines were planted to a depth of 4 ft, or just over 1 m. This planting depth provided access to sufficient nourishment and protected the roots from the scorching sun (*Geop.* 5.12.1). Two and a half to 3 ft were left between furrows if these were worked by hand, 6 ft if ploughed by a yoke of oxen. A clear distinction was made between the heavy-producing vineyard, where vines were set close

[14] Digging for the vines then began. Palladius mentions the furrow (*sulcus*) and trench (*scrobis*) which were excavated in preparation to receive the young vines. R. Billiard, *La vigne dans l'antiquité* (Marseille, 1997), 323; White, *Agricultural Implements*, 36–50.

together, and one in which the intervening space was wider for ploughing and sowing with crops.

Like olives, vines were never planted from seed, since ancient cultivators recognized the fact that cloning (using stock) was the only means of ensuring reliable reproduction.[15] Cuttings were taken from the best ten-year-old vines in peak condition, especially those that had produced shoots with many buds (*Geop.* 4.3.2; 5.8). The cuttings were sometimes sealed for months in large in-ground ceramic jars (*pithoi*). More often they were kept in nurseries before they were planted (*Geop.* 5.8.6) in the early spring. Stock grown in nurseries was given time to develop and acclimatize, and, once transplanted, paid off in a higher rate of survival and faster production (*Geop.* 5.14.1).

Two highly effectual means of vine reproduction were layering and grafting. In layering (*Geop.* 5.18), farmers simply bent a living shoot from the parent plant to the ground without breaking it. They then buried the end of the shoot. When the newly planted shoot rooted, it was severed from the parent plant and cultivated on its own. Grafting was a far more sophisticated method, and more important. This technique enabled rapid reproduction of superior grapes and the improvement of aged, diseased, and otherwise unproductive vine stock. By Late Antiquity, vine grafting was centuries old, highly developed, and still experimental. Palladius (3.17) and the *Geoponica* (4.12–13) provide detailed insights into its techniques. Farmers grafted in the spring when shoots with nodules or 'eyes' were selected and affixed to recipient plants. The results could be dramatic, as a modern example illustrates: during the nineteenth century, grafting of immune North American rootstock to European plants saved large numbers of European vines from destruction by the *Phylloxera vitifoliae* aphid, which had previously devastated continental vineyards.

In preparation of a good bed in which the grapes could flourish (*Geop.* 5.19.1–3), vignerons dug and ploughed plots throughout the year to clear unwanted vegetation and remove roots and stones. Farmers recognized that frequent digging made fertile, moisture-bearing topsoil available to the root systems of the vine (5.19.3). This constant, annual digging strengthened the vines by eradicating shallow, spreading roots in favour of those deeper roots able to reach moisture (*Geop.* 5.25). Trenches were often left open through the winter in order to collect rainfall (*Geop.* 5.26).

Generally the use of manure was frowned upon. Though its use made vines grow quickly, in the end it yielded poor quality wine (*Geop.* 5.26.8–10).[16]

[15] Vines, like the olive, are hermaphroditic and thus reproduce vegetatively from cuttings, suckers, grafts or layers. A. J. Winkler, *General Viticulture* (Berkeley, 1962), 151; Zohary and Hopf, *Domestication of Plants*, 137.

[16] In the early Imperial period, Columella (2.15.5) warned against the use of animal manure in the vineyard, as it lowered wine quality: a stance with which Palladius (9.2) and modern French producers agree. Billiard, *La vigne dans l'antiquité*, 331.

When manure was used, ox and pigeon droppings were considered most valuable. In reality, manuring vineyards was probably quite common, since vineyard fertility had to be maintained; thus in Egypt, vineyard leases included a dovecote from which dung was obtained.[17] Composts of decaying vegetable matter were also prepared (*Geop.* 2.22), and these were considered safer for vines than animal waste. The concern expressed whether one should use manure or not demonstrates a clear understanding that wine ultimately was influenced by numerous environmental factors. It further shows that there were different means and objectives of cultivation, depending on whether one strove for quantity or quality. Finally, the search for alternatives to animal manure shows an effort to remedy known problems and to increase quality without sacrificing the health or production of the vines.

In order to produce to their capacity, vines required assiduous pruning, which necessitated a degree of skill; plants could be damaged by inexperienced vignerons (5.28), to the harm of future crops. The arbusta and ground-growing grapes were pruned in January (*Geop.* 3.1), while critical pruning to remove the buds of mature vines (three-year-old vines) was undertaken in March, probably in preparation for their maturation. The trimming of young plants was not done until the following month (3.2.; 3.4) and similar work continued into June (3.6). The task began again immediately following the vintage, when the vines in warmer climates were prepared for the winter and the next growing season by pruning and digging. In order to produce earlier shoots for the following spring—and thus earlier production—autumn cuttings were made.

Longus' (probably of the second or third century) pastoral novel, *Daphnis and Chloe*, portrays the village as a hive of activity during the grape harvest, a time of communal interaction and festivity. Preparations for harvest began in August, with the vintage commencing in September and October (2.1). During the preparations, also detailed in *Geoponica* 6.1 and 6.10, new amphorae and large ceramic storage jars (*dolia*, *pithoi*) were made, pitch was prepared and applied to jars, and the presses were cleaned and fumigated, sometimes with frankincense (*Geop.* 6.11.6).

When the harvest began, the vineyard workers cut the grape bunches from the vines using a vinedresser's knife (*falx*). The bunches were placed in baskets, then transferred to carts or asses which, as even today, carried the grapes from field to press, a scene beautifully depicted in the church of Saints Lot and Procopius (AD 557) near Nebo in the province of Arabia. The pickers dumped the panniers into receptacles placed beside or built into the presses, as in the large installations at Rehovot and Emmaus where a central

[17] Rowlandson, *Landowners and Tenants*, 228.

working surface equipped with screw-press is surrounded by storage compartments.[18] Once deposited there, the weight of the grapes caused juice to seep from the berries. This must (πρόδρομος) was prized in antiquity as being of especially high quality; in the early empire Pliny (14.85) noted the *prodromos* as *prototropum*, and Columella (12.41) recorded his high esteem for *mustum lixvium*.

After the grapes had been gathered, they were trodden in the press, which might be as simple as a wooden vat, or as complicated as those of the late antique commercial complexes (including beam or screw-presses) in Palestine, where the output was measured in thousands of litres.[19] Following treading on the treading floor, grape stalks and skins were often pressed in a lever or screw-press. The fruit was piled into a stack under the press and squeezed to extract the remainder of the must. The time and method of such pressing varied. Whatever the specifics, the basic order and aims of the operation were clear: the treading produced first quality must; pressing or second treading, that of second quality; and the third yielded inferior wine.[20]

Farmers also had the capacity to add value to their crops by producing 'wines of type', production of which depended on the variety of grape grown. In this case, they processed wines differently ('wines of process') to produce speciality wines worth more than common wine. Finally, there is good evidence that vintages from superior wine producing regions, 'wines of origin', could also fetch higher prices. I will explore each of these below.

[18] I. Roll and E. Ayalon, 'Two Large Wine Presses in the Red Soil Regions of Israel', *PEQ* 130 (1981), 111–25.

[19] *Geoponica* 6.11 discusses the dress of workers (they are to be clothed to keep sweat from mingling with the wine) and mandates that these were to have clean feet and not to go in and out of the presses frequently. In addition to this useful advice, this chapter preserves the only detailed account of the vintage from Graeco-Roman antiquity, of which we possess late antique figural representations, for example from Scythopolis (Bet She'an) and Sta Constanza. 'They who tread, also must pick them [leaves, debris], if anything has escaped those who look after the baskets; since leaves pressed with grapes make the wine more rough and likely to spoil; and great damage occurs from dry or sour grapes. Those assigned to the task [of treading] must immediately press with their feet the grapes thrown into the treading floor, and having trodden all the grapes, then collect all the kernels (i.e. the refuse) so that the greatest part of the must may flow into the vat; and when they have trodden them the second time, let them be removed.'

[20] R. Frankel, *Wine and Oil Production in Antiquity in Israel and Other Mediterranean Countries* (Sheffield, 1999), 42 describes these processes. Columella (1.54.3) stated that pressing followed the first treading A second operation, when the grape stack was cut open and run through the press again, was called *circumsicium*. The resultant liquid, *mustum tortivum* (Columella 12.36), was low quality. *Geoponica* 6.11 is slightly at variance with the Roman authorities: here two treadings to extract the must are described; only then were the stalks, skins and seeds put into the press, where they were warmed, probably by the addition of hot water, and pressed to extract *deuterios* (Lat. *lora*) or after-wine, which was of poor quality and frequently used as a drink for labourers workers.

Wines of Type

In the Mediterranean agrarian landscape, the gallimaufry of wine varieties is an expected by-product of geographical diversity and isolation and the creation of interlocking cultural and ecological concerns. It also indicates a selective process—conscious and unconscious—of those vines which were best: best in producing consistent crops, best in rendering wines, and best in producing pleasing vintages. Several vine cultivars were especially well known and produced famous wines throughout antiquity. The wines made from them correspond to our notions of 'wines of type'.[21]

Aminean

By Late Antiquity, probably the most famous wines of type were made from the *Aminean* grape that had for centuries been cultivated around the Mediterranean. Palladius (3.9.4) recognized only two major types: the greater and smaller.[22] Falernian, a dark, strong wine that aged extraordinarily well, was pressed from the Aminean grape. According to the late-second- and early-third-century author Athenaeus (1.26), Falernian was ready to drink only after ten years (an astonishingly long time by the standards of the day) and considered good for consumption for fifteen to twenty years. Indeed, so well did this wine age that Athenaeus (1.25) stated it was not ready to be consumed in peak form until it had reached twenty-five years of age! Though the first-century Pliny's statement (14.63)—that Campania's Falernian vineyards had declined in his day due to the large demand from the consumers of a burgeoning capital—may have some merit, both Aminean and the specific Falernian variety were mentioned in Diocletian's Price Edict (2.4–7) of AD 301 and were clearly commonly produced.[23] Pacatus' *Panegyric*, delivered in

[21] Lambert-Gócs, *Wines of Greece*, 256.

[22] Pliny's account depends at least partially on the agricultural writings of the philosopher Democritus (5th c. BC). Pliny (*H.N.* 14.4.21–2) notes five kinds of Aminean grape: *germana minor, germana maior, gemella minor, gemella maior, and lanata.* Each variety had specific attributes that demonstrate successful adaptation of an esteemed plant to regions in which it originally could not thrive. The *germana minor* shed its blossom easily and produced a smaller grape; it was able to withstand inclement, wet weather, whilst the *germana maior* could not. The *gemella* sub-types produced grape clusters in pairs, and the wine was said by Pliny to possess a coarse flavour but strong character, *asperrimus sapor sed vires praecipuae.* The smaller *gemella* variety was raised only in the *arbustum.* The 'woolly' grape, the *lanata* was possibly so-called due to its white colour, although an intriguing possibility is that Pliny attempts to describe a type of Aminean grape akin to the Riesling or Sémillon, where the berry is attacked by the fungus *Botrytis cinerea,* the 'noble rot' that imparts the sweet flavour especially emblematic of these modern wines.

[23] Caelius Aurelianus, *On Acute Diseases and On Chronic Diseases*, ed. and trans. Drabkin, 212.

AD 389 to Theodosius I, makes it clear that some 'effeminate' elites favoured Falernian, whilst the fifth-century Macrobius (*Sat.* 3.16.11) stated that Falernian was the drink of decadent imbibers of the past.[24] In the fifth century, Caelius Aurelianus recommended treatment of numbness with Falernian wine, attesting its continued medical use, if not its consumption as a beverage. It is thus difficult to know what to make of the statements of Pacatus and Macrobius, but it seems possible that by the fifth century Falernian was out of fashion amongst some members of wealthy society, but still produced and of some repute. The silence of the sixth-century Cassiodorus, who made no mention of the wine whilst provisioning the cellars of Theoderic from the vineyards of Italy, is notable.[25]

Planters most valued the Aminean grape for its yield, and the texts make more numerous mentions of it than any other variety. Because it offered not only consistently plentiful (sometimes spectacular yields) and high quality wines (*Geop.* 5.17.9), Aminean apparently was the most widely cultivated type of Late Antiquity. This vine was also adaptable (5.17.2); it produced excellent results in a range of soils and aspects, especially in sloping and dry fields. Growers further (4.1.14) valued the ability of the Aminean grape to grow on trees (arbustum), where it was adapted especially for intercultivation. The Aminean grape needed skilled care: vinedressers were apparently prone to under-prune its stock in the early stages of its growth, a mistake that, while it assured an abundant harvest for the first two years, led thereafter to a decline in the quantity of grapes (*Geop.* 5.17.12). This, alongside the claim (*Geop.* 5.17.9) that every *plethron* (or iugerum) of Aminean vine land produced 300 amphorae, demonstrates that eastern vintners prized the Aminean for its consistently high yields.

Mersites

The Mersites (μερσίτης) grape is mentioned briefly in the *Geoponica* (5.2.10), where exceptional praise is lavished upon it. Said to grow in Bithynia, where it produced a wine known as δενδρογάληνος,[26] the resulting wine was called by different names in Paphlagonia and the Pontus and was a celebrated type throughout north-west Asia Minor. The fact that the *Geoponica* states that the Mersites is 'very sweet' (ἡδίστη) and sought by bees for nourishment, allows

24 Pacatus, *Panegyric*, trans. Nixon and Rodgers, 464.

25 *Ep.* = *PL* 69.857–8. Presumably if the Falernian vintages of Aminean grapes were to be had, he would have procured it, given the wine's longstanding reputation. Lacking any detail, however, we can only speculate.

26 The name no doubt implying that it was predominantly produced in the arbustum.

us to identify it with some confidence as a type of Muscat grape.[27] These varieties today produce sweet or medium-sweet wines.

Leukothrakia

Leukothrakia (λευκοθρακία) rendered sweet wine (*Geop.* 5.17). It was a white grape variety common in Bithynia and characterized by clusters of oblong fruit of beautiful appearance (*Geop.* 5.17.4). As Leukothrakia probably means 'White Thracian', it is likely the same cultivar as that prized in the first century AD. Pliny (14.54) and Athenaeus note that the famous Mendean wine of antiquity came from the Thracian Chalcidike. The traffic of this wine stretches back at least to the Classical period; amphorae in the Porticello shipwreck, dated 420–400 BC, carried Mendean and amphorae for the polis of Byzantion.[28] It thus appears that this Mendean variety, originally produced in the Chalcidike peninsula, had, by the time the authors of the *Geoponica* were writing in the fourth through sixth centuries, spread to nearby Bithynia (5.2). Its exceptional longevity is impressive, but is matched by present-day varieties such as our Pinot Noir, which is known to have been cultivated at least from the thirteenth century, and is probably much more ancient.

Athenaeus (1.29e) praised 'Mendes wine' for its strength: it could be mixed with three parts water and remain palatable. In the East, Mendean continued to be produced through the seventh century, as the medical writer Paul of Nicaea (probably seventh century) mentioned it as a medicament (65.23).

Bolene

The Bolene (βωληνή), described only in *Geoponica* 5.17.5, was a vine that fruited early in the season; it was therefore better adapted to colder climates than some other grapes. The Bolene was known to produce thick white grapes in large clusters. Characteristically it was somewhat adaptable and produced good quality wine. When compared with the Aminean, however, it fared poorly. According to the *Geoponica* (5.17.5), Aminean bore as well as the Bolene and produced wine of superior flavour. The Bolene produced three

[27] This is probably the *Apiana*, the 'bee grape' of Pliny (14.4.24) and the Roman agronomists, so-called due to the fondness that insects held for the fruit. Today Muscat grapes produce generally sweet white wines best when drunk young. Pliny noted that the Apiana was sweet in its early years, but gradually grew coarse. From this we might surmise that, similarly to modern Muscat, the Mersites/Apiana were producers of medium-sweet to sweet white wines which were drunk young, within the first year of their life. Pliny ranked these varieties third behind Aminean and Nomentane grapes.

[28] Parker, *Ancient Shipwrecks*, 332.

shoots from each bud, as opposed to two for many vines, and thus tended to overbear. It therefore required assiduous pruning. When properly cared for, it was said to grow especially large and live to a great age. Like the Aminean, the Bolene seems to have been especially suited for cultivation in the arbustum.

Other Varieties

The *Drosallis* (δρόσαλλις) variety, according to *Geoponica* (5.17.3), produced sweet wine and was frequently cultivated alongside Aminean grapes. The *Geoponica* (5.2.4) also mentions three other white grape types that grew well on infertile soils and were noted for their inherent richness: the *psithios* (ψίθιος), *Corcyraean* and *Chloris*. The psithios and black psithios was used to produce raisin wine (*passum*). *Geoponica* (6.23) describes this process whereby grapes were first left in the sun for five days, after which they were soaked in must and boiled-down with seawater (adding seawater to wine was a common practice in the Aegean; Pliny, *H.N.* 14.73–4). Finally, this mixture was trodden in the winepress. The *Nabatean Agriculture*, a ninth-century compilation, discusses two famous varieties from the region of Oriens: the Syrian grape and an unnamed type from the Jazira, said to be a wine grape cultivated on trees, bearing oblong bunches with red fruit (but producing white wine), and very intoxicating.[29] To this list we must make additional mention of Macrobius' hodgepodge of grapes (*Sat.* 3.20.7).[30]

Vinification and Wines of Process

By Late Antiquity, vinification in the Mediterranean was an art that had reached a high degree of refinement, at least amongst the class who read such handbooks as the *Geoponica*. Nevertheless, as is true in any age, a great deal of substandard wine was produced and spoilage was common. Souring and soured wine, *posca* (οξύκρατον), mixed with water was a major beverage amongst the poor.[31] The vagaries of the Mediterranean and desert climates often hampered fermentation. Fermentation is a chaotic and uncertain

[29] Salinger, *Kitab al-Filaha an-Nabatiya*, 64.

[30] Some are probably table grapes; others are simply named after the region in which they achieved fame. Many are familiar: the *Aminean*, *Apiana*, the *Mareotic* grape, and the *Psithian* are all mentioned. Others, such as the *Maronian*, the donkey grape, the *Atrusca*, *Albicera* and 'cow's udder grape' (βούμαχτος)—a table grape according to Pliny—are less familiar. The list in the *Saturnalia* merely serves to underscore the fact that we will never be able to catalogue fully even the most famous grape types, much less the thousands of local varieties which existed throughout the empire.

[31] J. André, *L'alimentation et la cuisine à Rome* (2nd edn., Paris, 1981), 172–3; Tchernia, *Le vin de l'Italie romaine*, 11–13.

process even in the best of circumstances, and one which science has only recently managed to tame. A certain amount of success in fermentation and cellaring was achieved, however. The *Geoponica* contains lengthy discussions on the storage, treatment, and improvement of wines. Late antique authors make frequent mention of old wine, *palaios* (παλαιόϲ) (Oribasius 41.3.2), more than a year old, and new wine, *neos* (νέοϲ).[32] A distinction clearly existed between young wines meant for immediate drinking and 'vintage' wines, and with it a host of enological techniques to produce aged wines with special character to suit particular tastes. This is demonstrated by the existence of 'wines of process', as distinct from 'wines of type' and 'wines of origin'. Wines of type relied mainly on grape variety and, to a lesser degree, on place of production (the latter being paramount for wines of origin) to render their character; whereas wines of process depended on the intervention of the vintner in various production phases.

High-quality, ancient wines of process may be further divided into three sorts: facsimile wines, mixed origin wines, and sophisticated wines.[33] Facsimile wines were imitations of famous wines of origin. Mixed origin wines were blends of two or more wines of origin. Wines of process are so-called because they are dependent upon specific techniques to render their character. That made from sun-dried grapes, described above, is illustrative of one of the many wines made by grapes or grape juice utilized in different stages of development or treated in specific ways before pressing. Sophisticated types, a subdivision of wines of process, involved the addition of an ingredient or ingredients to enhance the end product. The fact that such wines were in demand demonstrates that specific wines (crudely analogous perhaps to modern labels) had achieved fame and widespread popularity, with inevitable attempts to copy and undersell, or to meet demands that outstripped supply.

Many sophisticated wines and wines of process were made by infusing fruit and herbal compounds into the *dolia* at any stage from initial pressing through fermentation, or even after vinification was complete. Others, such as the *theriaca* wine mentioned throughout the agricultural handbooks, were speciality wines, whose properties were sometimes supposedly enhanced by complex treatment of the vines themselves.[34] There were hundreds of

[32] The notion of A. D. F. Brown, 'Black Wine', *CR* 12 (1962), 192–5 that ancient wine was generally of one colour and quality is based on the incorrect assumption that an appreciation and snobbery surrounding wines did not exist, and that the technology and knowledge to produce aged and distinct wines was likewise absent.

[33] Lambert-Gócs, *Wines of Greece*, 258–9.

[34] For instance, a wine which was said to be a cure for venomous snake bites was produced by pouring an antidote into a cleft vine upon planting. The practice of producing antivenin from

medicinal wine concoctions, since the liquid is an ideal medium in which to place drugs or ingredients thought to soothe or cure. Whilst these wines are not varieties from specific grapes, they represent an entirely different dimension and focus of use in which the vine grower might operate.

The must was often left in the press for several days before the next pressing occurred. When the ageing must was removed, it was placed into either *pithoi* or amphorae. Amphorae, it appears, were the final destination of the *neos* wine, often used to pay workers, and frequently subject to spoilage.[35] That new wine was racked into amphorae whilst fermentation continued, is known from the finds of jars with boreholes (in the shoulders and necks of the containers), which allowed gases to escape and prevented the jars from bursting. Amphorae of late antique date with boreholes of this kind have been found at a number of sites, including Berenike in Egypt and throughout Palestine.[36]

Geoponica 7.12 gives numerous instructions about good practices of elevage. Some ingredients meant to assist in the ageing process were known only through experience, without an understanding of the mechanism behind their effectiveness. Some of these today are known scientifically to be effective. For instance, the addition of gypsum corrected the lack of acidity which apparently plagued many ancient wines. Gypsum increases the tartaric acids in the must, which fosters the growth of 'good' yeasts, and today is still added to sherry. In the same way, the *thalassan* wines of the Aegean and Cyprus contained sea water that added colour and raised acidity levels.[37] Resinating wines, which rendered the taste still familiar to us from *retsina*, acted to preserve the vintage (*Geop.* 7.12).

The porosity of ceramic *pithoi* and amphorae, which allowed limited evaporation, proved quite helpful during fermentation. This slow and limited oxidation in turn enabled the yeasts to render the beverage more alcoholic. The porosity of clay jars allowed them to 'breathe' whilst remaining airtight;

wine, however ineffective the method appears, was common throughout the Greek East and Mesopotamia: the Arabic *Nabatean Agriculture* mentions the *theriaca* vine (*diryāq*), the consumption of whose wine, juice, or vinegar saves people from snakebites. *Geoponica* Book 8 lists twenty-three wine varieties, mostly for medicinal purposes, such as apple wines (8.17), absinthe wines (8.21) and pomegranate-infused wine used to treat dysentery (8.20).

35 D. Rathbone, *Economic Rationalism and Rural Society in Third-Century A.D. Egypt: The Heroninos Archive and the Appianus Estate* (Cambridge, 1991), 258–9.

36 S. Kingsley, 'Specialized Production and Long-distance Trade in Byzantine Palestine', D.Phil. thesis (Oxford, 1999), 152; R. Tomber, 'Pottery from the Sediments of the Inner Harbour', in K. G. Holum, A. Raban, and J. Patrich (eds.), *Caesarea Papers*, II: *Herod's Temple, the Provincial Governor's Praetorium and Granaries, the Later Harbor, a Gold Coin Hoard, and Other Studies* (Providence, RI, 1999), pl. 5.12.

37 H. W. Allen, *A History of Wine: Great Vintage Wines from the Homeric Age to the Present Day* (London, 1961), 79–80.

this resulted in a more acidic wine of better and more interesting character. This action is somewhat comparable to the cork on a modern wine bottle, which permits limited oxidization insufficient to spoil the content. The process and role of the ceramic jar in winemaking was understood at least by the time of Plutarch (*Mor.* 3.8.656), who stated, '. . . and ageing increases its force, the water being separated out, [so that] the wine becomes less in measure, more powerful in strength.'[38] Wines properly racked could mature for up to twenty years according to Athenaeus and other writers, such as the pre-Islamic Arab poet Muraqquish the Younger:

Now wine of the white grape, fragrant of musk [when the jar is broached],
and set on the strainer to clear, and ladled from cup to cup.
A captive it dwelt in the jar for twenty revolving years, above it
a seal of clay, exposed to the wind and sun.
Imprisoned by Jews who brought it from Golan to lands afar and offered
for sale by a vintner who knew well to follow again.[39]

After fermentation, the production of facsimile and sophisticated wines took place. The *Geoponica* provides recipes for facsimile Aminean wine (8.22), Thasian (8.23) and Coan (8.24), an indication that these wines still enjoyed a good reputation in Late Antiquity. Infusions of fruit, herbs, spices, and unguents delivered sophisticated wines. Often ingredients were added to create a complex mixture, giving the wine a certain utility (most often medicinal) or flavour. Binding leaves or fruit in cloth and infusing the wine for a period of time was a common way to produce such beverages. *Geoponica* 8.2–22 discusses twenty specialist wines, though recipes are not always provided.

Wines of Origin

The wines featured in the textual sources are those which enjoyed a wide reputation throughout the world of Late Antiquity. Such wines, the *vin d'appellation* of antiquity, were traded far and wide and formed part of a common Mediterranean culinary currency which did not die with the fall of

[38] Lambert-Gócs, *Wines of Greece*, 263. *Pithoi*, still used in winemaking in Georgia today, brought another positive element to vinification: as they were buried in the earth, the constant temperature afforded the wine promoted steady and successful fermentation. The increasing use of barrels in the West from the first century AD onwards contributed to the death of vintage wines, the rendering of which was impossible in early barrels due to their lack of airtightness. Only with the return of the corked bottle could methods of ageing once again rival those achieved in amphorae. T. Unwin, *Wine and the Vine* (London, 1996), 55.

[39] C. J. Lyall, *Translations of Ancient Arabian Poetry: Chiefly pre-Islamic, with an Introduction and Notes* (New York, 1930), 187.

the West. From their mention in the written sources of the fourth through seventh centuries, we gain some insight into the regions best noted for quality wine production in Oriens.

Palestine and Phoenice

The fourth-century *Expositio totius mundi* (§29) declares that in southern Palestine, Ascalon and Gaza were eminent cities flourishing in trade, having all things in abundance, and that they exported a fine wine to Syria and Egypt.[40] Palestinian wine production and distribution in the East is well confirmed by statements in the fourth century by Jerome (*Vita Hil.* 17), who commented on the vineyard-laced landscape of Palestine as well as its frequent monastic settlements. The church, a major landowner in the region, was indeed almost certainly a major wine producer, as witnessed by ecclesiastical press remains at Horvat Hesheq (near Ptolemais, modern Akko), Horvat Bet Loya, near Eleutheropolis (Bet Govrin), and elsewhere in Palestine.[41] Corbo argues that the winepresses found in the monasteries of the Judaean Desert represent surplus production and monastic wine sales.[42]

According to the sixth-century poet Corippus (*In Laud. Iust.* 3.87–8), Palestinian wines were light, sweet, white wines. These qualities perhaps explain their use in medicine, since these characteristics were considered to enhance palatability. That Palestinian wine was a valued import as far away as the western Mediterranean is apparent from the writings of Sidonius Apollinaris in the late fifth century (*Carm.* 17.15–16), and of Gregory of Tours in the late sixth century (*Hist. Franc.* 7.29), both in Gaul, and of the Spaniard Isidore of Seville as late as the early seventh century (*Etym.* 20.7).

Wine production in Palestine for export has been discussed by Safrai and by Mayerson, who, in a series of articles, outlined the importance of Gaza–Ascalon wine, primarily from the written sources.[43] Sean Kingsley, whose quantified studies of winery production sites documented 899 winepresses of the Roman–Byzantine period in Palaestina I, II, and III, has treated the material evidence in depth.[44] Many of these presses (42 per cent) were of

[40] 'Ascalon et Gaza, civitates eminentes et in negotio bullientes et abundantes omnibus, mittunt omni [negotio] Syriae et Aegypto vinum optimum'.

[41] M. Aviam, 'Horvat Hesheq: A Church in Upper Galilee', in Y. Tsafrir (ed.), *Ancient Churches Revealed* (Jerusalem, 1993), 56; U. Dahari, 'Remote Monasteries in Southern Sinai and their Economic Base', in Y. Tsafrir (ed.), ibid., 348–50; J. Patrich and Y. Tsafrir, 'A Byzantine Church Complex at Horvat Beit Loya', in Y. Tsafrir (ed.), ibid., 265–76.

[42] V. Corbo, *Gli scavi di Kh. Siyar el-Ghanam (Campo dei pastori) e i monasteri di dintorni* (Jerusalem, 1955), 4–5; Hirschfeld, *Judean Desert Monasteries*, 106–8.

[43] P. Mayerson, 'The Wine and Vineyards of Gaza in the Byzantine Period', *BASOR* 257 (1985); Safrai, *Economy*, 126–35.

[44] Kingsley, 'Specialized Production', 68.

large capacity, their collection vats capable of holding between 4,000 and 59,000 litres of wine, the equivalent of 174–2,565 amphorae per press.[45]

Palestinian wine was amongst the most important export items of Oriens, and the remarkable circulation of Palestinian wine throughout the Mediterranean and beyond is now amply attested by the numerous finds of 'Carthage Late Roman 4' (LR4) and 'Carthage Late Roman 5' (LR5) type amphorae, both of reliable Palestinian origin.[46] Due to the nature of amphora studies, still at a comparatively early stage, it is likely that some eastern containers are under-represented in the current archaeological record, whilst more recognized types may be over-represented. The coming decades of archaeological research may change our presumptions of agricultural commodity movements and patterns of exchange. This is, however, unlikely to alter significantly the view that Palestinian wine was exported in large quantities. In view of the fact that many of the findspots are from areas which were outside of imperial territory, this exchange is difficult to explain using primitivist models of taxation and government redistribution.[47] The high demand for Palestinian wines may have been enhanced by their origin in the Holy Land, and was certainly helped by the well-attested dominance of eastern shippers and middlemen in all late antique long-distance trade. It also provides an interesting model for our understanding of other oriental products, including wine from other regions within the diocese, as well as oil. Unfortunately, the current state of archaeological research and publication makes comparative study difficult.

Palestine continued to produce well-known wines in the seventh century. Alexander of Tralles, in the sixth century, prescribed Ascalon wine for the treatment of liver disorders and gout, and wine from Gaza for fevers.[48] It is also clear that the production of fine wines in Palestine did not immediately end with the Islamic conquests. In the Umayyad period (AD 661–750), Arab writers praised the wines of Capitolias (Beit Ras) and Gadara (Umm Qays), both in the region of the former province of Palaestina II.[49]

[45] Ibid., 72.

[46] S. Kingsley, 'The Economic Impact of the Palestinian Wine Trade in Late Antiquity', in S. A. Kingsley and M. Decker (eds.), *Economy and Exchange in the East Mediterranean during Late Antiquity* (Oxford, 2001); F. Pacetti, 'Appunti su alcuni tipi di anfore orientali della prima età bizantine centri di produzione, contenuti, cronologia e distribuzione', in L. Quilici and S. Quilici Gigli (eds.), *Agricoltura e commerci nell'Italia antica* (Rome, 1995).

[47] Cassius Felix, *de Medicina*, ed. Fraisse, 32.12. Cassianus wrote in AD 447. See also P. Mayerson, 'The Use of Ascalon Wine in the Medical Writers of the Fourth to the Seventh Centuries', *IEJ* 43 (1993), 170.

[48] Alexander of Tralles, ed. Puschmann, 2.411; 2.539;1.419.

[49] R. Schick, *The Christian Communities of Palestine from Byzantine to Islamic Rule: A Historical and Archaeological Study* (Princeton, 1995), 261.

The two provinces of Phoenice, to the north of Palestine, are studies in topographical and climatic contrasts. Phoenice Maritima, the narrow, coastal strip of the modern Lebanon coast, is well watered and fertile, whilst Phoenice Libanensis, which stretched from the Lebanon Mountains to the Palmyrene, contained large areas of semi-arid landscape. In antiquity, both regions are recorded to have produced excellent wines, with evidence suggesting that these enjoyed a reputation and circulation overseas. The *Expositio totius mundi* (§23) describes the fertile coast of Phoenicia that abounded in grain, wine, and oil.

In the region of southern Phoenicia, now in Israel, stretching from Ptolemais (Akko) to Paneas (Banias), Frankel recorded eighty-two Byzantine winepresses.[50] As this sampling represents only a slender fraction of the countryside of southern Phoenice, the number of presses recovered is impressive, equivalent to nearly 10 per cent of Kingsley's 899 winepresses catalogued for all of Palestine.[51] Sixty-four of these presses (78 per cent) are of the direct screw type, equipment which was infrequently utilized before the Byzantine period. Although earlier silence about them does not rule out earlier production and traffic, the wines of Sarepta (Sarafand) in southern Phoenicia appear to have risen to prominence in the Late Roman period, when they were exported to Gaul, where Sidonius Apollinaris recorded them in the fifth century (*Carm.* 17.15–16). This is not the only mention of wines from that city: during the sixth century, Alexander of Tralles frequently alluded to the use of this wine in treating various maladies, including fevers and stomach problems.[52] Alexander also favoured the wines of Tyre, citing their use in six instances, though his prescriptions often suggest that several of these regional wines were interchangeable. Specific wines recommended for specific ailments were common. The wine of Scythopolis (Bet She'an) was used to treat dysentery, diabetes, fevers, and kidney ailments.[53] Similarly, the wines of Berytus, which continued to be traded, were frequently prescribed by Alexander of Tralles to treat general bodily complaints.[54]

Pre-Islamic poetry, like that of Muraqquish quoted above, praised the wine of the Golan, an area that underwent significant agricultural development during Late Antiquity. Similarly, the Muʿallaqat extolled the wine of Heliopolis and Qasirin, the latter presumably the village of Qatzrin in the Golan. In the high basin of the Biqʿa, the hot, dry climate was ideal for grapes, and the abundant waters stored within the surrounding mountains allowed for

50 Frankel, *Wine and Oil Production*, catalogue nos. 1424–2228.

51 Kingsley, 'Specialized Production', 68, tbl. 3; Appendix 1.

52 Alexander of Tralles, ed. Puschmann, 1.335, 483; 2.21, 27, 85, 95, 217, 325, 407.

53 Ibid., 1.335, 483; 2.421, 483, 495.

54 Ibid., 2.407.

irrigation throughout the region. The wines of Heliopolis (Ba'albek), imported into Palestine, were probably similar to the Berytus types, given that both cities had similar terrain and climates.[55]

Arabia

Provincia Arabia, with Bostra as its capital, possessed a number of important wine producing centres known primarily through Arabic sources. Pre- and early Islamic Arabic writers noted the vintages of Sarkhad and Der'a.[56] Gerasa (Jerash) and the city of Areopolis (Ma'ab), on the Kerak plateau also provide evidence for wine production during Late Antiquity—the former from a Greek inscription recording a vineyard, the latter from its mention by the Islamic geographers.[57] Recent excavations at the village of Sa'ad, about 17 km from Gerasa, have uncovered a sizeable winery with two screw-press installations and a basin capacity of more than 4,000 litres.[58]

The territory around Bostra has yielded a large and fine basalt screw-press. The *Jordanian Department of Antiquities Database* records sixty-one wineries in Transjordan where Late Roman or Byzantine period pottery or structures have been found, primarily during survey work conducted in the territory of ancient Palaestina Secunda and Arabia.[59] The highest concentration of presses occurs in the region of Mount Nebo, around Madaba, where the Madaba Plains Project recorded eight Byzantine period presses, and where the remains of three winepresses were excavated in the Monastery of Moses on Mount Nebo.[60] The famous mosaics depicting vintage scenes, found in the village of Nebo in the church of Saints Lot and Procopius (AD 557), emphasize the place of the grape in the local economy. Additionally, Capitolias/Beit Ras, near modern Irbid, was at the centre of the wine trade between the central Arabian Peninsula and Roman–Byzantine Arabia, as will be discussed in Chapter 8.

[55] A. Neubauer, *La Géographie du Talmud* (Paris, 1868), 298.

[56] Schick, *Christian Communities*, 447.

[57] al-Muqaddasi, *The Best Divisions for Knowledge of the Regions*, trans. Collins, p. 161; F. M. Heichelheim, 'Roman Syria', in T. Frank and T. R. S. Broughton (eds.), *ESAR* III (Baltimore, 1938), 140.

[58] J. C. Rose and D. E. Burker (eds.), *Sa'ad: A Late Roman/Byzantine Site in North Jordan* (Irbid, Jordan, 2004), 9–19.

[59] F. Villeneuve, 'Les villages antiques du Hauran', *Le Monde de la Bible* 28 (1983), 123–4. The Jordanian data may be retrieved from the website of the National Information System of Jordan (no stable URL). Accessed 14 July 2001.

[60] S. J. Saller and H. R. Schneider, *The Memorial of Moses on Mount Nebo* (Jerusalem, 1941), 194–5.

North Syria, Cilicia, and Cyprus

The written sources provide little information on the wine production of Syria and Mesopotamia from the Early Roman period to the Arab Conquests. In the Early Roman period, Martial (4.46) mentioned Syrian wine; most famous was the vintage of Laodicea (Latakia). Laodicea was a major wine exporter throughout the Hellenistic and Early Roman periods.[61] During the Early Roman period, wine from the same city was exported as far as the Red Sea coast, according to the first-century *Periplus Maris Erythraei* (§6), a trade which probably remained constant throughout Late Antiquity. Laodicea's continued export during Late Antiquity is likely, as its produce was mentioned by Alexander of Tralles in the sixth century (2.483). Unlike Ascalon wine, but like some from Gaza, Laodicean vintages are mentioned less frequently by the medical writers, possibly because the wines were, by ancient standards, highly alcoholic. In Merovingian Gaul of the 580s, when the murderer Claudius and the court official Eberulf dined together, Claudius called for stronger wine, and Eberulf's servants were sent 'one after the other to get stronger wines, those of Laodicea and those of Gaza'.[62] Such wine may have contained as much as 16–20 per cent alcohol by volume.[63]

Libanius (*Or.* 11), writing in Antioch in the late fourth century, praised the wine of his native city, and elsewhere mentioned that the fields in the city territory were rich in wine (*Or.* 31.20). An inscribed fifth- or sixth-century lintel mentions the bounty of grain, wine, and oil of the village of Kaper Pera (al-Bara) near Antioch: 'You have given joy to my heart. By our harvests of wheat, wine, and oil, we have been filled with peace.'[64] Lying as it did on the fringe of the Syrian steppe, Kaper Pera could have supplied communities of pastoralists, as was the case with the villages further east, such as Androna (al-Andarin), whose vintage was praised by the sixth-century pre-Islamic poet 'Amr ibn Kulthum among the Mu'allaqat.[65] It seems unlikely that production in these regions would have expanded in the Muslim period, but it was maintained into the medieval period.[66]

[61] Strabo (16.2.9) described the vineyards which, in his day, ascended the terraced hillsides behind the city.

[62] Mayerson, 'The Use of Ascalon Wine', 173; Gregory of Tours, *Hist. Franc.*, ed. B. Krusch, 7.29; W. Levison, *MGH* 1.348.

[63] Lambert-Gócs, *Wines of Greece*, 263.

[64] *IGLS* 4.1462.

[65] A. J. Arberry, *The Seven Odes: The First Chapter in Arabic Literature* (London and New York, 1957), 204; H. F. Lutz, *Viticulture and Brewing in the Ancient Orient* (Leipzig and New York, 1922), 24.

[66] Later Islamic geographers mention the wine of the regions around Emesa (Homs) and M'a'aret en Noman at the base of the Jebel Zawiye, as well as on the Jebel Hass, where the basalt landscape around Anasartha (Khanasir) and Zebed exhibits today the extensive terrace systems noticed by Musil. See P. Heine, *Weinstudien: Untersuchungen zu Anbau, Produktion und Konsum des Weins im arabisch-islamischen Mittelalter* (Wiesbaden, 1982) and Musil, *Palmyrena, A Topographical Itinerary*, 232.

Textual evidence provides a sketch of viticultural production and trade in this northern area; but comparatively little detailed archaeological work has been undertaken in Syria, and none in Mesopotamia, that focuses on the agriculture of the Roman–Byzantine period. Evidence in the form of the standing remains of presses does exist in abundance in the Massif of Belus, the Limestone Massif, where hundreds of presses are present amongst the village remains. However, virtually none of these installations are identified in the literature as winepresses, nor have their chronologies, or absolute dates, been determined. Although Tchalenko's notion of a predominantly olive-based economy in this region has been rejected by Georges Tate and Olivier Callot, in favour of a view that the Limestone Massif was part of a more typical Mediterranean mixed agricultural regime, there has been no systematic attempt to look at the evidence of these presses for wine production. Whilst Callot's work, in particular, sheds light on the issue through his publication of selected North Syrian presses, he fails to ask the fundamental question: If there was no olive monoculture, and the limestone hill villages were engaged in mixed farming, where are the winepresses?

Frankel argues that the roller-crushers (Figure 4.2) identified by Callot as olive mills were, in fact, wineries. He bases this reinterpretation of the North Syrian roller installation on material evidence from the Carmel in northern Israel. Obvious technical differences between known oileries and wineries, such as the presence of open, relatively flat, working surfaces and bell-shaped collection vats in wine production installations, also support the re-identification. Most roller installations are certainly different in layout and dimension from oil-presses where conventional mills are present, such as the subterranean twin oil-press at Kafr Nabo. Generally smaller in scale, they often seem to bear little relation to the oil-presses they presumably served. The village of Qirqbize illustrates this: several roller installations, presumed to operate as mills for lever and screw-presses, in fact lie on the opposite side of the village from the presses. Some of the roller installations have press niches themselves, but they universally lack the press beds found in oil installations—another argument that suggests they were wineries.

There are several, very large, rock-cut installations that lack any trace of press counterweights and rollers; these are most likely wineries. Installations such as this are found in the southern confines of Qirqbize; but the best example of this is in the village of Behyo, where Tchalenko identified 37 'oil' installations (27 roller installations and 10 presses).[67] In fact, most of these

[67] O. Callot, *Huileries antiques de Syrie du Nord* (Paris, 1984), 120; Tchalenko, *Villages* I.40–2. The available publications, and my own survey work conducted over two summers (1998–9), indicate that the installations with open working surfaces, lacking press beds and with or without

Figure 4.2. Roller crusher (from De Vogüé, *Syrie Centrale*, pl. 113).

presses are small, simple installations, and most lack any sign of a counterweight, milling or pressing apparatus and must therefore have been wineries. A thorough survey and classification of the North Syrian wineries and oileries is required before a confident assessment can be made, but, on balance, the evidence suggests that many, perhaps even the majority, of North Syrian presses were for wine. In the analysis which follows, the published archaeological data, combined with impressions from personal fieldwork in the region, help to form an estimate as to the capabilities of the Massif of Belus to produce wine in Late Antiquity. This is based on the belief that approximately 60 per cent, or perhaps 6,000 of the 10,000 estimated presses in all, are winepresses.[68] These figures, and those that follow, are intended merely to provide a rudimentary reference against which future work may be tested. The value of such estimates lies in what they offer in illuminating the potential for a given time and place.

Published Syrian press plans and dimensions are rare. Callot's corpus, however, provides a usable snapshot of the press forms of the region and, therefore, a fair cross-section of the Jebel villages' press sizes.[69] Based on twenty-one of Callot's planned installations, interpreted in the present study as wineries, the average treading-floor area of North Syrian installations is approximately 3.6×3.3 m, with the average capacity of ten published collecting vats being about 4,800 litres. This is the equivalent of about 240 LR1A

rollers and press niches, outnumber substantially those installations positively identified as oileries from examples elsewhere.

[68] Tate, *Campagnes*, 243, notes 245 presses from his limited survey work, and overall press figures have been extrapolated from this number. Given the small size of Tate's sample, the overall numbers should be treated as preliminary rough estimates only.

[69] Callot, *Huileries*.

amphorae, the storage jars of the region.[70] Comparison of the small Syrian sample with Kingsley's corpus for Palestine is encouraging: 186 Palestinian presses possessed average treading areas of 3.4×3.6 m and collecting vats which held an average of around 2,400 litres.[71] The modest sample of North Syrian press vats possess substantially more (nearly twice) the liquid storage capacity of their Palestinian equivalents.

Based on Jewish written sources, Dar suggests a seven-week vintage period.[72] Since the entire cycle of treading, pressing and removing of new wine lasted around seven days, it was therefore possible to produce seven full vats of wine each season. In North Syria, therefore, each press could produce approximately 33,579 litres, the equivalent of 1,679 LR1 amphorae, across the seven-week harvest.

How does this calculation accord with evidence from the village landscape? A good test case is the village of Qirqbize, a typical village in the Limestone Massif. Qirqbize possesses seven presses, five of which are probably winepresses. These presses could produce a combined 168,795 litres of wine. The village population was about 215 people.[73] Taking the 1–2 *sextarii* consumption per adult male as cited above, the total annual demand for wine in the village would amount to 42,815–85,538 litres. Since women and children in the village probably consumed less, these figures are on the high side; however, they are instructive, since they show that at most, 25–50 per cent of potential village wine production was required for subsistence. This left a possible annual market surplus of 83,257–125,890 litres, or 10,407–15,736 LR1 amphorae.

At a regional level, the combined intramural populations of Antioch and Apamea would have required as many as 30 million litres of wine annually.[74] However, the estimated winepresses of the cities' *territoria*, some 6,000 in number, at full capacity *could* have produced as many as 200 million litres, a vast surplus. Based on Columella's vineyard yield figures, this represents the harvest of some 4,771 ha (47.7 km^2)—9,541 ha (95.4 km^2) of land, an area less than one-tenth of the combined *territoria* of the two cities.

Turning to Cilicia, the fourth-century *Expositio totius mundi* (§39) mentions wine as a regular trade item of the region; but the archaeological evidence for Cilicia, and for Cyprus, is limited. The apparent paucity of physical evidence is due in part to incorrect identification of some press

[70] At roughly 20 litres, an approximate volume based on a range of LR1 capacities.

[71] Kingsley, 'Specialized Production', 72.

[72] Dar and Applebaum, *Landscape and Pattern*, 153–4.

[73] Based on 4.25 people per room (216.75): Tate, *Campagnes*, 183. For ease of calculations I have rounded down.

[74] Considering that the combined population of Antioch and Apamea was about 300,000 and dividing this into 50,000 of our hypothetical families noted above.

remains as oil-presses. This problem is not limited only to Cilicia, where, for example, we see the wine-treading floors at Gökören and Çet Tepe identified as oil-mills by the *Tabula Imperii Byzantini* (TIB) survey, but extends to the wider region in general.[75] Written sources partly remedy this dearth of archaeological material. Galen (*De San. Tuend.* 5.4.15–16), writing in the second century, was a proponent of Cilician wine and mentioned it on several occasions as a medicament, the Cilician Abates grape getting specific mention. The use of Cilician wine continued through the sixth century. Alexander of Tralles (2.376, 18; 2.388, 13) prescribed it, and his testimony is further enhanced by the epigraphic evidence from Corycus, where fifteen late antique funerary inscriptions name people involved in the wine trade, either *oinemporoi* or *oinegoi*, representing about 4 per cent of the workforce mentioned in the corpus of inscriptions.[76] Even more persuasive in the epigraphic evidence is the sixth-century tax inscription known as the 'Abydos Tariff', a document clearly indicative of long-distance movement of wine and oil from the eastern Mediterranean to Constantinople. The inscription outlines the taxes paid by wine merchants as 6 folles and 2 sextarii for an unspecified amount of wine. Cilician wine carriers, however, paid only 3 folles tax.[77] The Cilicians' special status as wine and oil shippers arguably attests a major role for these merchants in the hinterlands of their home cities.

Late antique Cyprus was 'distinctly prosperous', in part due to its long tradition of wine production and export within the Mediterranean, that went as far back as the Hellenistic age.[78] The island continued to be a known producer in the Roman period, noted by Strabo (14.6.5) and Pliny (14.74). Later references are scattered: in the fourth century, Synesius of Cyrene praised the wine of Cyprus in the context of comparing Cyrenaican products with the best the rest of the empire had to offer, including Attic honey.[79] Cypriot wine is also noted in other sources as having a light character. The abundance of wine on the island is expressed by the anecdote from the *Vita* of Spyridon: when the fourth-century saint, Spyridon of Tremithus, was kind

[75] H. Hellenkemper and F. Hild, *Neue Forschungen in Kilikien* (Vienna, 1986), pl. 120; 32. For a fuller picture of the Cilician wine trade and the problems in identifying it, M. Decker, 'The Wine Trade of Cilicia in Late Antiquity', *Aram* 15 (2005).

[76] J. Keil and A. Wilhelm, *Denkmäler aus dem Rauhen Kilikien* (Manchester, 1931), nos. 207, 271, 282, 357, 363, 444, 467, 471, 552, 574, 599, 652, 680, 682, 709; F. Trombley, 'Corycus in Cilicia Trachis: The Economy of a Small Coastal City in Late Antiquity (saec. V–VI)—A précis', *The Ancient History Bulletin* 1/1 (1987), 19.

[77] G. Dagron and D. Feissel, 'Inscriptions inédites du Musée d'Antioche', *TM* 9 (1985), 451–5; J. Durliat and A. Guillou, 'Le tarif d'Abydos', *BCH* 108 (1984), 581–98.

[78] D. Michaelides, 'The Economy of Cyprus During the Hellenistic and Roman Periods', in V. Karagheorghis and D. Michaelides (eds.), *The Development of the Cypriot Economy* (Nicosia, 1996), 142.

[79] Synesius, *Ep.* 148; *Greek Anthology*, ed. and trans. Paton, 9.487.

enough to pray for rain, the heavens opened in response to his call, and the resultant showers produced abundance in the vineyards around his home city.[80]

The major link connecting Cyprus with the pan-Mediterranean wine trade of Late Antiquity is archaeological. Publications of late antique winepresses are infrequent, with examples from Ohmodhos and Lania providing only slight remedy; but ceramic evidence is abundant.[81] The finding of kilns producing LR1 amphorae on several sites greatly strengthens the case for the island's place as a major wine producer and exporter.[82] Cyprus was apparently one of the principal producers and trans-shippers of LR1 containers. Some of the prosperity enjoyed by Cyprus may be directly ascribed to its role as a producer and entrepôt in the wine and oil trade.[83]

Conclusion

Far from the chaotic image of irrationality, combined with stolid caution, which the harshest critics paint of the handbooks, our picture of the late antique vineyard is one of considerable dynamism. Wine growers were vividly concerned with producing, for the sake of pride, pleasure, and profit (the only quantifiable motivator), the greatest possible quality and quantity of wine. To achieve this, producers followed clearly developed precepts, but also debated correct method and faced many uncertainties. The grapes themselves were scrutinized; their strengths and weaknesses addressed with a sophisticated appreciation of ecology yoked to the dominant tastes of the day—all playing against the backdrop of the market. We see further the advantages and

[80] Theodorus, *Vita Spyridon* = *La légende de S. Spyridon, évêque de Trimithonte*, ed. van den Ven, 2.

[81] LR1 was prominent, for example, in the inland sites explored by The Sydney Cyprus Survey Project, which found late antique material particularly abundant in the archaeological record: pottery from the period AD 300–700 represented 7 per cent of the survey assemblage. M. Given, A. B. Knapp, and D. Coleman, *The Sydney Cyprus Survey Project: Social Approaches to Regional Archaeological Survey* (Los Angeles, 2003), 279–82.

[82] J. Y. Empereur and M. Picon, 'Les régions de production d'amphores impériales en Méditerranée orientale', in M. Lenoir, D. Manacorda, and C. Panella (eds.), *Amphores romaines et histoire économique: dix ans de recherche: actes du colloque* (Rome, 1989); V. Karageorghis, 'Chronique de fouilles à Chypre', *BCH* 113 (1989); S. Manning et al., 'Late Roman Type 1a Amphora Production at the Late Roman Site of Zygi-Petrini, Cyprus', *RDAC* (2000); A. Papageorghiou, 'Chronique de fouilles et découvertes archéologiques à Chypre', *BCH* 114 (1990).

[83] S. Hadjisavvas, 'Olive Oil Production in Ancient Cyprus', *RDAC* (1988), 74, fig. 143. In 1999 numerous unprovenanced screw-presses, many of likely late antique date, were stored in the courtyard of the Temple of Aphrodite in Paleopaphos.

disadvantages of tried-and-true practice that carried growers every step of the way, from January through December, in a neatly mapped and planned avenue. The careful work regimen was established, with every task purposefully designed to do nothing more than to improve the chances of the vineyard's success. Through this regimen, we see much more than the conservatism of an owner who wished nothing more than to prevent a loss, but rather an ideal of a well-managed landscape and even, perhaps, the will to make it so.

When at work, this sort of pragmatic science transformed cultivation, limited risk, and established new paradigms of practice: new plants were generated by cutting and cloning, and through experimentation in grafting—to ensure that more plants survived, that different grape varieties could be introduced, and that the vines fruited faster, better, or heavier. Pruning, digging, fertilizing, none were new to viticulture. There were new adaptations of these operations, and, more importantly, their application was wrought within a new framework. The arbustum was an ancient method, but the attention given it offers an oblique view of a crowded, densely farmed countryside where arable land was in short supply and maximization was the order of the day. The precision with which these ways of winemaking were articulated and applied shows that there were clear ends in mind, the final component of which we see in the booming wine trade discussed in Chapter 8.

Underpinning the trade was the desire for healthy and productive vintage, that appeared in its many apparitions of white and dark, of dry and sweet, of first wine and after-wine. These suited a variety of tastes and occupied a crowded hierarchy in the wine market. Enological sophistication is a honeycomb in which all the bees buzz: without a deep appreciation of wine and the need for practical medicines, we would not have heard of *Leukothrakia*, nor would we appreciate how different were the wines of Laodicea from the wines of Gaza, nor would we be aware that one could copy Thasian wine (nor understand why one would want to), nor indeed would we know that one could improve a nursing woman's milk by infusing wine with thyme (*Geop.* 9.19). These generally held tenets are not merely peculiar or anecdotal, but part of the culture of the day, in which wine was an accepted, appreciated, and sought-after beverage that catered to the whole range of society, often for very different reasons.

Underlying this discussion is the hypothesis that a market for wines existed, and that farmers actively managed and responded to this market. Despite Julian of Ascalon's noting that vines were growing on trees in cities, the numerous urban centres of the East bought their wine, to the tune of millions of sextarii per year, to say nothing of the possibility of export overseas. The clear advantage that wine had over grains leads to important conclusions: not everyone could make their daily bread by growing it, and not everyone wanted to. For those who chose to—and choice seems to be the right word given the cheapness of plant stock—wine offered a marketable commodity from which one could render a profit even from a small plot of land, enough perhaps to buy food for the entire family and ensure survival in difficult times.

5

The 'Queen of All Trees'

The Olive in Late Antique Agriculture

THE OLIVE IN DAILY LIFE

The vine and olives were not staples in the same way that grain was; one could survive without them. Wine and olive oil were, nonetheless, indispensable cornerstones of the diet and, what we might term, cultural staples—crops without which Mediterranean life was inconceivable. In drinking wine and eating olive oil, even the humblest citizen of the empire partook of foods that imparted a taste of the *dolce vita*, civilization expressed in victuals. As signifiers of Mediterranean life, oil and wine offered two of the fundaments of *romanitas*. It is, thus, not surprising that already in Pliny's day, during the height of the paradigmatic Mediterranean empire, the vine and the olive—both symbols of Mediterranean farming—spread as fast and as far as their growth permitted.

At the edge of that world, in northern Mesopotamia, Simeon, the early eighth-century bishop of Harran, received a portion of a treasure hoard and wished to invest the gold. He descended from the uplands of Tur 'Abdin (now eastern Turkey) to the irrigated plain of Beth 'Araboye near Nisibis (Nusaybin) and planted 12,000 olive trees.[1] Saint Simeon placed a hedge of thorns and a wall of reeds around the olive plantation, and then engaged ploughmen and workers to care for it. In five years, the trees bore fruit. Under Simeon's guidance, monks assigned to the harvest did not strike the tree limbs with rods, as was commonly done in order to harvest the berries, but picked them by hand and transported them to the monastery at Qartmin, near Mardin, where the fruit was pressed. The monks used a portion of the oil squeezed from Simeon's olives for lighting the monastery. The rest was sold in the highlands of the Tur 'Abdin, where olives did not grow. The handsome profit from this olive trade ensured the saint's ability to build and provide generously for his monasteries throughout Upper Mesopotamia.[2] The saint's life illustrates a clear economic constant: olives

[1] Brock, 'The Life of Simeon of the Olives', 175.

[2] Ibid., 176–9; A. Palmer, *Monk and Mason on the Tigris Frontier: The Early History of Tur 'Abdin* (Cambridge, 1990), 163–4.

yielded oil, a valuable commodity in universal demand and which, because it could last over time, could be transported long distances. The olive was thus one of only a handful of crops that could bring wealth to its cultivators.

By Late Antiquity, the olive, the 'queen of trees' (Columella, 5.8.1), had long been a major feature of Mediterranean agriculture. In addition to the food products of table olives and edible olive oil, it had many other uses.[3] Medicine required sizeable quantities; oil frequently served as the base through which a number of active ingredients were administered, as for example as a treatment for various fevers using green oil (*omphakinos*) as suggested by the sixth-century physician Alexander of Tralles (I.323). The seventh-century writer Paul of Nicaea (4.41) prescribed an oil bath to treat fevers. Veterinary medicine also employed a good deal of oil, especially as a topical treatment for wounds and skin ailments.[4] Finally, much oil was consumed in the form of a cosmetic base at the bath house where it was used to protect and cleanse the body.

There were other uses that today we might call industrial. Although lamp oil usage is rarely mentioned in estimates of ancient oil consumption, it was an important part of daily life.[5] Amouretti states that an ancient oil lamp would have provided 250–300 hours of light per litre of oil, while O'Dea provides a lower estimate of approximately 65 hours per litre.[6] Even the poor households of the empire owned at least one lamp, thus consumption of oil for lighting purposes in Late Antiquity was significant. At three hours' usage each night, one lamp consumed annually 3.65–7 litres of oil. A city like Edessa, with perhaps 50,000 people, could have required 182,500–850,000 litres of olive oil each year to fuel its private lamps alone. Additionally, several

[3] D. C. Hopkins, 'Agriculture', in E. M. Meyers (ed.), *The Oxford Encyclopedia of Archaeology in the Near East* (Oxford, 1996), 24. In the Near East, exploitation of the olive began as early as the pre-Pottery Neolithic, according to evidence from the Carmel. By the Chalcolithic period, finds of large quantities of olive stones at Teleilat Ghassul in the Jordan Valley indicate cultivation was already underway, possibly under irrigation; see Frankel, *Wine and Oil Production*, 36. The origin of the domesticated olive is unknown, although the latter finds indicate the Jordan Valley is a strong possibility. The earliest human exploitation of the wild olive, the oleaster, focused on industrial purposes, mainly the manufacture of perfumes and unguents. The *Iliad* mentions oil as a luxury, whilst the *Odyssey* makes reference to olive cultivation. Olive oil, pleasant tasting and versatile, is the tree's principal product, but oil is only the first amongst many edible and industrial components which made the tree invaluable to ancient Mediterranean societies. On early uses of the olive in the Mediterranean, see J. L. Melena, 'Olive Oil and Other Sorts of Oil in the Mycenaean Tablets', *Minos* n.s. 18 (1983), 102. Roman–Byzantine-period uses are described by M.-C. Amouretti, *Le pain et l'huile dans la Grèce antique: de l'araire au moulin* (Paris, 1986), 177–96; Frankel, *Wine and Oil Production*, 43–5.

[4] Grattius, *Cyn.*, ed. Vollmers, 394–5.

[5] Anastasius the Persian, *Saint Anastase le Perse et l'histoire de la Palestine au début du VIIe siècle*, ed. Flusin, M318; P285. Olive oil was probably the most common oil used in lamps, though other oils, such as sesame and other vegetable oils, would have been employed as well.

[6] Amouretti, *Le pain et l'huile*, 190; W. O'Dea, *The Social History of Lighting* (London, 1958), 32.

cities had street lighting on the major thoroughfares, amongst them Alexandria, Edessa, and Gaza. The sixth-century rhetorician Choricius of Gaza described (*Marc.* 2.63) his city as having glass street lamps that burned throughout the night.[7] In 504/5, the governor of Euphratensis took 6,800 *sextarii* of oil from the holy shrines to light the porticoes of Edessa.[8]

The source of all of this oil—the domesticated olive tree (*Olea europaea*)—probably descends from the wild plant known today as oleaster (*Olea oleaster*), with which it shares many physical characteristics. The physical traits of the domestic olive plant include grey-brown bark, numerous branches, and lance-shaped leaves that are green on top and silver-grey below. Its durable rootstock often survives substantial trauma such as burning, a key to its survival and adaptation throughout the Mediterranean world. The tree typically ranges from 3 to 12 m in height. Although the trunk frequently becomes hollow with age, making its lifespan difficult to determine, the olive is renowned for its extreme longevity. In the winter months, the olive lies dormant, blooming in March or April. By June, the first fruits appear, maturing through July and August. Fruiting often varies considerably from year to year, with alternate heavy/light cropping a common pattern. Varro (1.55.6) and Columella (5.9.11–12) noted this phenomenon. In the eastern Mediterranean, the fruit ripens from September through November, and the harvest begins. Depending on the region, the picking season may last ten weeks; and if there is insufficient labour available, some olives might not be plucked until the springtime.

The Mediterranean—the olive region *par excellence*—is home today to more than 385 million olive trees.[9] In fact, the olive is adapted almost exclusively to the sub-tropical climate specific to the Mediterranean, where the hot, dry summers and cool, damp winters especially suit its requirements. Whilst the olive tree demands high levels of sunlight and warmth at key stages of its development, it also requires a sufficient period of cool weather (between 600 and 2,400 hours below 7 °C) in order to fruit.[10] It cannot tolerate temperatures below freezing (0 °C) for any length of time, however, and the 3 °C January isotherm is therefore generally regarded as the limit of olive cultivation.[11] This lack of tolerance for the cold has led to the olive being

[7] Ps.-Joshua, *Chron.* 107; D. Sly, *Philo's Alexandria* (London, 1996), 37.

[8] Segal, *Edessa: "The Blessed City"*, 124. Oil also had religious uses, particularly in anointing and purification. Olive oil also served as a lubricant in a range of functions, from greasing axles, to wool combing, to soap production (the last probably not important in Late Antiquity). Olive by-products, the watery lees or amurca, possessed a host of uses around the farm, which ranged from firewood improvement to use as a herbicide and fertilizer, to protecting grain from pests.

[9] L. Denis, 'The Present Economic Importance of Olive Production', in FAO (ed.), *Modern Olive Production* (Rome, 1977), 1.

[10] Zohary and Hopf, *Domestication of Plants*, 137.

[11] Koder, 'Climatic Change in the Fifth and Sixth Centuries?', 273.

linked in the minds of writers with the warmer coastal climate of the Mediterranean world. The fourth-century BC Theophrastus (*Caus. pl.* 6.2.4) for instance, observed that olives generally do not grow far inland. Whilst temperature is a primary factor in the olive's success, soil and water requirements are varied and flexible. The olive is a xerophyte, growing where other crops cannot succeed.[12] The tree will tolerate, and even thrive within, a broad range of soils and rainfall conditions and manages to grow in the semi-arid environments of the Libyan and Syrian deserts. The pomegranate and olive were the only fruit trees recognized as able to flourish in dry places (*Geop.* 2.8.3).

OLIVE CULTIVATION

Farmers in Late Antiquity had attained a high level of arboricultural sophistication; each technique of olive reproduction mentioned in the farming manuals is acknowledged, and many remain in use today. Some olive varieties, however, are presently grown from seed, a method never mentioned in the late antique farming literature and apparently considered unviable. Producing olives required patience; devoting land and labour to an olive grove meant making a mid- or long-term investment. Seeds take up to twenty-four months to germinate. Saplings are slow-growing and need several years after planting to yield their first crop—often up to six years. Peak production begins only when the trees mature, at twenty-five to fifty years of age. This slow growth cycle caused farmers to try numerous methods intended to speed propagation.

Planting took place in the autumn, winter, or spring, with the 15 November through 20 December considered optimal (*Geop.* 9.41–2). Cuttings made from superior individual plants were a trusted source of the parent plant's genetics, and clone shoots, taken from trees that possessed desired traits, were either set directly in the grove or placed in the intermediary of the nursery.[13] *Geoponica* 9.5 and 9.7 describe the selection process: young trees which produced a yearly crop were preferred, with shoots of one cubit (0.6 m) taken from the branches of the tree. These cuttings were smeared with cow dung and ashes and set into the ground so that only a piece the length of four

[12] F. Scaramuzzi, 'Intensive Olive-Growing', in FAO (ed.), *Modern Olive-Growing* (Rome, 1977), 61.

[13] Nurseries (*Geop.* 9.5.3) were to be similar in soil quality and aspect to those places where the trees eventually were set, so that the transplants would acclimate more easily, thus increasing their chances for survival. When it received the new plants, the nursery was tilled monthly for seven months (*Geop.* 9.5.9–10).

fingers-breadths (four *daktuloi*) remained exposed. The olive grove workers then marked the location of the plants with a reed and dug a trench flanking the cuttings to permit their irrigation by winter rains.

Once the young trees had reached sufficient size, usually in the fourth year, the plants were removed from the nursery (*Geop.* 9.5.11; Palladius, *Op. Ag.* 11.8.1). In many cases, trees were transplanted at two or three years of age; those mentioned in the Babylonian Talmud (TB B.M. 101a) were expected not to bear fruit for at least three years. The use of nurseries, where young trees could be carefully monitored and grown in close confinement, ensured greater chances of the plant's success once it was placed in the olive grove.

Olive growers also experimented with methods of reproducing plants and speeding up their development. Truncheons (larger cuttings) were taken from thicker branches, cut to a length of two cubits (1.2 m), then set at the bottom of a trench on a wide stone (*Geop.* 9.11.5), a practice of questionable efficacy. More likely to succeed were the smaller clippings that were set in trenches, as were large cuttings, called *tropaia* (*τρόπαια*), measuring four to five cubits (2.4–3 m) in length (*Geop.* 9.11.7). These cuttings yielded fruit more quickly than plants from other methods, but they also failed more frequently. Experimentation of this sort suggests an assertive approach to improving the cultivation of a tree whose development was slow, a shortcoming that estate owners were obviously trying to modify.

Olives also often form protuberances, called ovules, at the base of the tree due to circulatory problems. Ovules may produce vegetation and roots. They often are encouraged to do so in arid regions such as Libya, where they were a favoured source of planting stock.[14] *Geoponica* 9.11.8 seems to refer to use of nodules when it describes cuttings taken from the bark and placed in the earth, bark upward. Another passage mentions a propagation method favoured in Syria. There, the local inhabitants called the lowest branches—suckers which grow in clusters around the base of the olive tree—*gorphia* (or *jorfia*).[15] From these they took their cuttings for transplanting (*Geop.* 9.5.12).[16]

Layering and grafting also were used to propagate trees, as depicted in late antique mosaics from Mount Nebo (Figure 5.1). There is a single reference in the *Geoponica* (10.3.6) about the propagation of olives through layering, in which the shoots of living plants have one end buried in earth while the other

[14] N. Jacobini, M. Battaglini, and P. Preziosi, 'Propagation of the Olive Tree', in FAO (ed.), *Modern Olive Growing* (Rome, 1977), 30–1.

[15] Probably from the Arabic *jarafa* (جرف), 'to remove, or sweep away', implying that these ovules were taken as a result of pruning and clean-up.

[16] Despite the success of the planting of *gorphia* directly into the olive grove, the author of this section of the *Geoponica* (9.5.14), probably Anatolius of Berytus, recommended that cuttings first be set in nurseries where they could be irrigated.

Figure 5.1. Mt. Nebo mosaic showing grafted trees (courtesy of Ze'ev Radovan).

remains attached to the parent. Layering was probably easier in dense plantations, since only flexible (and thus generally younger and shorter) branches could be bent to earth and buried without their splitting from the parent tree that provided nourishment until roots could eventually form on the layered individual. Since the layers were taken from shorter foliage, the newly established plant would be close to the parent tree and probably resulted in denser, varied plantations such as those favoured by modern producers.[17] The

[17] Scaramuzzi, 'Intensive Olive-Growing', 78–9.

advantage that layering offered to ancient farmers was that it was probably more efficient in cloning superior individuals.

In their discussions of grafting, writers of the late antique handbooks show a keen interest in plant crosses and botanical trials. The season and methods of tree-grafting most common in Late Antiquity are described in the handbooks. The *Geoponica* in its agricultural calendar (3.4) recommended olive-grafting in April, and elsewhere (9.16.2) in May–June. Palladius (3.17) detailed three principal methods of grafting, two of which could be practised in the spring, and one in the summer. In the first of these, bark-grafting (*insitio sub cortice*), the scion was cut from the parent and placed in an incision made underneath the bark of the stock. Secondly, Palladius then described *in trunco* grafting, in which a graft was placed at the top of the trunk of the stock. The third sort of grafting was plastering (*emplastro*) in which a small measure of bark is removed carefully and the bud scion applied, using clay or another adhesive compound. According to Palladius, a worker used an auger to bore a hole into the centre of the stock trunk, and then cleared it of sawdust and debris. A budded branch, trimmed of bark and cut to fit the borehole precisely, was then inserted and covered with clay and moss. The *Geoponica* (10.37.1; 10.77.4; 10.65.2; 9.14) further describes root grafts, rind grafts, bud-grafting, and inoculation, amongst other forms. Grafts of several different kinds were inserted, for many purposes, across multiple tree varieties; the practice was clearly a thriving art considered vital to agricultural success.

Care of the Olive Grove: Digging, Manuring, Pruning, and Harvesting

Olives needed less care than vines, but there was nevertheless a year-round schedule of tasks required by the olive grove. The handbooks recommend frequent digging around the trees. This took place around the time of the harvest. According to Palladius (*Op. Ag.* 11.8.9–11), the tree roots were laid bare in October. The *Geoponica* (3.11.1; 3.15.5) ordered digging in August and December, whilst 9.9.2–3 address the issue in more detail, stating that one should dig round the trees according to their size and the quality of the earth.

Farmers recognized the need for young stock to be cared for when their underdeveloped root systems made them vulnerable. Growers therefore attempted to maintain fertility and promote growth, mainly in young stock, by manuring. Any kind of manure, excepting human excrement, was considered acceptable (*Geop.* 9.15.1). *Amurca*, a waste product of pressing, also was used as a fertilizer and applied to olive trees (*Geop.* 9.10.1—8), as were leguminous green manure and seaweed. Palladius recommended manuring every three years, whilst the *Geoponica* (9.15.2) is less specific as to time between compostings, but offered (9.9.3–4) pointed advice on which situations were better served by various manuring methods. In wet and fertile ground, little manure was required and at longer intervals. Compost retained soil moisture and allowed rapid penetration of the short bursts of winter rains. It was therefore recommended that it be applied generously and heaped up to protect the tree from the sun. It is noteworthy that early Roman and late antique cultivators did not express any concern of being short of fertilizer.[18] The problem is not overtly raised in the *Geoponica*, which presumably would have offered composting alternatives or other ideas to render soil more fertile, if the conventional animal waste were unavailable: the handbook is a treasury of alternative methods.

Pruning (*Geop.* 9.12) was also a recognized necessity; trees were trimmed during both the spring and autumn months (3.5.3.), but more commonly shortly after the harvest, since new growth was thought to produce the most olives. As is the modern practice, the handbooks recommended (*Geop.* 9.9.51) pruning of adult trees only. Likewise, Palladius (*Op. Ag.* 11.8.21) advised pruning only after trees had reached eight years of age and that then cutting should be limited to dead wood. *Geoponica* 9.9.5–10 offers more detailed advice: damage caused by pruning should be offset by manuring, and branches were to be kept short to lessen the chance of wind damage; their maximum length was reckoned at 10 cubits, or 6 m. Although this suggests that mature olives in the eastern provinces were quite large, they were on the whole apparently smaller than those of Africa, where Pliny (*H.N.* 17.93) reported North African trees of extraordinary size.[19]

At the end of the agricultural year, the olive harvest, along with the vintage, was one of the most labour-intensive tasks of the countryside. In September–October, workers began to gather unripe olives for preparation of green oil. The most careful picking was done by hand.[20] Since many harvesters had difficulty ascending trees and picking berries in the uppermost parts of the foliage, the olives often were allowed to fall and then collected, or were beaten from the tree using canes.[21] *Geoponica* 9.12 indicates that hand-picking, the most arduous and laborious method of collecting the fruit, was considered best. Certainly, we should be aware of the context of the *Geoponica*, whose intended reader was

[18] *Contra* White, *Roman Farming*, 136–7. Palladius 11.8.16 may hint at the problem, but the passage leaves us uncertain: 'Now, if you can, after a three-year hiatus, olive-groves are to be manured' (*nunc, si subpetit, intermisso triennio stercoranda sunt olivueta*).

[19] D. J. Mattingly, 'Olea Mediterranea?', *JRA* 1 (1988), 60.

[20] White, *Roman Farming*, 227.

[21] Amouretti, *Le pain et l'huile*, 73. At Anazarbus in Cilicia, prepubescent boys were preferred for the harvest, since their presence was thought to make the olives more fruitful (*Geop.* 9.2.6).

probably an estate owner with the means to hire outside labour and whose concern was the production of the highest quality and greatest amount of fruit.

A good harvester picked between 50 and 150 kg of olives per day or the fruit of 5–15 trees.[22] The bigger olive plantations, such as Simeon's 12,000 trees, would have taken 800–2,400 man-days to harvest. Since olives keep on the trees for months with only slight changes in quality, the individual farmer could have harvested as many as 750 trees over a three-month period. The harvest spread over many weeks, with the exceptional-quality green oil produced first from unripe olives, while common oil derived from those which had fully matured.[23] Since the harvest could stretch for many weeks, the labour requirements of the olive grove were less demanding and intensive than the vineyard, whose produce was at peak for only a short time. Even without access to hired labourers, farmers could have managed hundreds of trees with only the help of their immediate family members.

Once harvested, olives were sometimes left unpressed for some time, although, according to Pliny (*H.N.* 15.14), this decreased the quantity of oil. Allowing a hiatus between harvesting and processing of olives may have been more common in the West than in the East. Whilst Columella instructed that olives be processed immediately, Palladius favoured immediate processing in a section (*Op. Ag.* 12.17) devoted to oil production in the 'Greek way'.[24] Here the farmer is told to gather only as many olives as can be processed the same night, which may indicate that it was the harvesters who also worked in the pressing room. Nothing approaching the detail of Columella's description of cleaning and softening the berries before processing is found in the later handbooks. However, a certain amount of preparatory work was always required, as the fruits, especially those beaten from the trees, had to be cleaned and sifted, placed on mats, sprinkled with salt, and dried (*Geop.* 9.16). The berries were then often softened either with boiling water, light pressing (*Geop.* 9.19.6), or, for green oil, light milling.

Before pressing, olives were ground into a paste. This milling could be achieved by very simple means: treading, pounding with mortar and pestle, and crushing using a stone roller. Treading probably was done using wooden clogs in vats; rock-cut examples of these early milling installations, where treading or pounding took place, are known from Israel.[25] In Oriens, milling

[22] Amouretti, ibid., 74 (50–100 kg); D. J. Mattingly, 'Maximum Figures and Maximizing Strategies of Oil Production?', in M.-C. Amouretti and J.-P. Brun (eds.), *La production du vin et de l'huile en Méditerranée* (Athens, 1993), 49 (up to 150 kg).

[23] Mattingly, 'Maximum Figures?', 491.

[24] '*De oleo faciendo secundum Graecos et emendando*'. Apparently a section from Anatolius of Berytus which has been omitted from the *Geoponica*.

[25] On the simplest means of processing olives, see White, *Farm Equipment*, 226, and Frankel, *Wine and Oil Production*, 68.

seems to have been performed primarily using rotary crushers driven by either people or animals. Cato (66.1) and Columella (12.52.6) assert that olives could be milled without crushing the kernel, the inclusion of which imparted a bad flavour to the oil. But most scholars doubt the ability of ancient oil millers to remove the majority of seeds without damage.[26]

Following milling, the olive paste was placed in frails (baskets) (Lat. *fiscus*; Gr. *κυρτός*), often constructed from willow osiers (*Geop.* 9.19.7) and placed on the press bed for the first pressing.[27] In antiquity, several different types of presses were used for processing olives, which, for the sake of simplicity, are divided in this study into four major categories. Type 1 is the simple lever and weights press. The beam lever is fixed into the anchor fulcrum (sometimes built upright or a masonry wall), while weights are attached to the free end of the beam. The olive stack is placed in the centre. Force is exerted by lowering the beam onto the stack of olives. Type 2 (Figure 5.2) works on the same principle, with the free weight being replaced by a system of windlasses and pulleys that allow easier manipulation of the weight. Type 3 (Figure 5.3)

[26] M.-C. Amouretti, 'Des agronomes latins aux agronomes provençaux: les moulins à huile', *Provence Historique* 24 (1981), 92; Mattingly, 'Olea Mediterranea?', 156. The *Geoponica*, like Cato (66.1) and Columella, clearly states that olive kernels were not to be crushed during milling because the crushed pits impart an unpleasant flavour to the oil. Columella's statement deals with the use of the *mola* rather than the rotary crusher. If the *mola olearia* is indeed the rotary mill from Spain and North Africa, argued by Frankel, then it seems possible that the machine could be adjusted to strip the flesh off the kernel without the latter's destruction. Unlike the evidence from Columella, which hinges on the type of machine used, the *Geoponica* (9.19.5–8) stresses that great care should be exercised in milling for the making of green oil. The wheel (*τροχός*) should be turned by hand gently and lightly, so that only the skin and flesh are stripped from the kernel. Separation of the kernels, once milling was complete, appears out of the question due to the labour involved; those kernels that survived the milling process intact would certainly have been crushed during pressing. E. L. Tyree and E. Stefanoubaki, 'The Olive Pit and Roman Oil Making', *BiblArch* 59/3 (1996) noted in trials, using a pre-industrial *trapetum* with convex millstones, that 50–75 per cent of the pips were crushed. Perhaps more importantly, the same trial included taste-tests that demonstrated virtually no appreciable difference between oil produced from paste containing seeds and that free of seed components. If the kernels did somehow survive and were removed between mill and press, this would have been an arduous task, and one for which no evidence exists. Although objections raised against the possibility of milling without crushing the kernel are significant, the testimony of the primary sources should not be discounted without further investigations into the range of ancient techniques and technologies available.

[27] There are several excellent works that deal with presses and press technologies. The major starting-point is now Frankel, *Wine and Oil Production*, to which we might add R. Frankel, S. Avitsur, and E. Ayalon, *History and Technology of Olive Oil in the Holy Land* (Arlington, Va., 1994); J.-P. Brun, *L'oléiculture antique en Provence: les huileries du département du Var* (Paris, 1986); J.-P. Brun, *Archéologie du vin et de l'huile dans l'Empire romain* (Paris, 2004); J.-P. Brun, *Archéologie du vin et de l'huile de la préhistoire à l'époque hellénistique* (Paris, 2004); J.-P. Brun, *Archéologie du vin et de l'huile en Gaule romaine* (Paris, 2005); Callot, *Huileries*; and Amouretti, *Le pain et l'huile*.

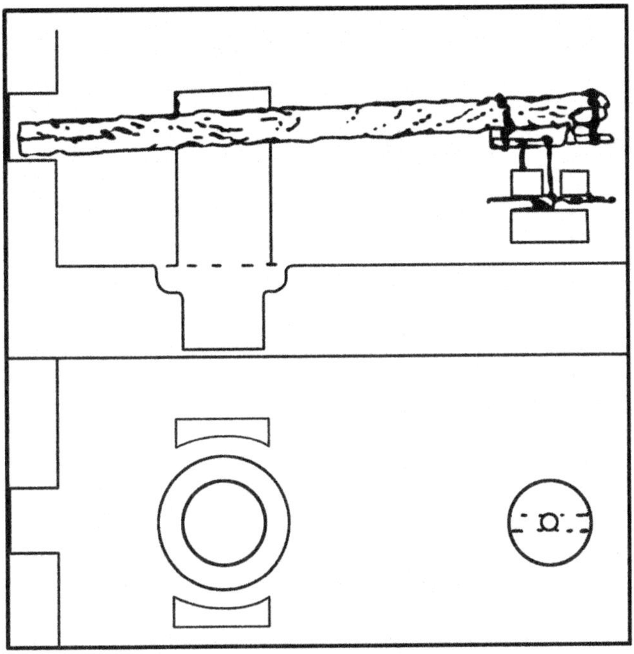

Figure 5.2. Schematic of Type 2 press (after Frankel, *Wine and Oil Production*, Appendix 2, 'T40111002').

likewise uses the lever and fulcrum, with the downward pressure on the beam achieved with a wooden screw. Turning the screw lowered the beam. Type 4 (Figure 5.4) is the direct screw-press, which eliminated the beam entirely and used a screw-driven mechanism to deliver force on the stack. Pressing often took place in the evening after the day's harvest.

The processing of a single press-load of olives was arduous, sometimes requiring as much as 24 hours per load, as the same stack might be pressed more than once. *Geoponica* 9.19 prescribes at least three pressings, though regional variation is likely. Each load required new frails, according to Columella (12.52.22). The latter envisaged production of ordinary table oil and superior oil, accomplished in perhaps only two pressings. In order to aid the extraction of oil, hot water was often applied to the olive paste. As the press squeezed the stack, and the latter decreased in height, hot water was applied to swell the mash and flush the oil from the stack into the receptacle. When the stack decreased to the point where effective pressure was no longer possible, the beam was anchored lower in its socket, and pressure again applied.

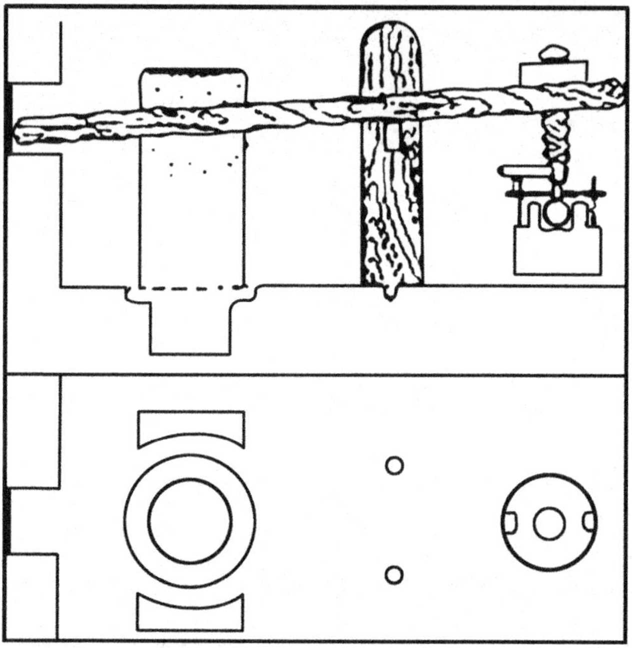

Figure 5.3. Schematic of Type 3 press (after Frankel, *Wine and Oil Production*, Appendix 2, 'T40111008').

During pressing, the mix of pulp, watery lees, and oil flowed into the collection vats where purification took place. Since lighter oil floats on the heavier water lees, Columella (12.52.11) suggested that workers skim the oil from one jar to another with ladles in order to purify it. The *Geoponica* (9.26.2) also mentions removing oil from the surface of the vats with a ladle. In North Syria, twin vats allowed decantation by gravity in many press installations. At Kafr Nabo in the Jebel Sim'an, twin vats received the oil from the press; the first of these allowed the heavy sediment to settle to the bottom. The second vat was separated from the first by a low partition wall. In a process called 'overflow decantation', the light oil in the first vat floated to the surface to the level of the partition and thence into the adjoining second vat, from which it was ladled for storage or final purification.[28] Overflow decantation seems to have been the only method of oil refinement used in late antique Oriens.

[28] Callot, *Huileries*.

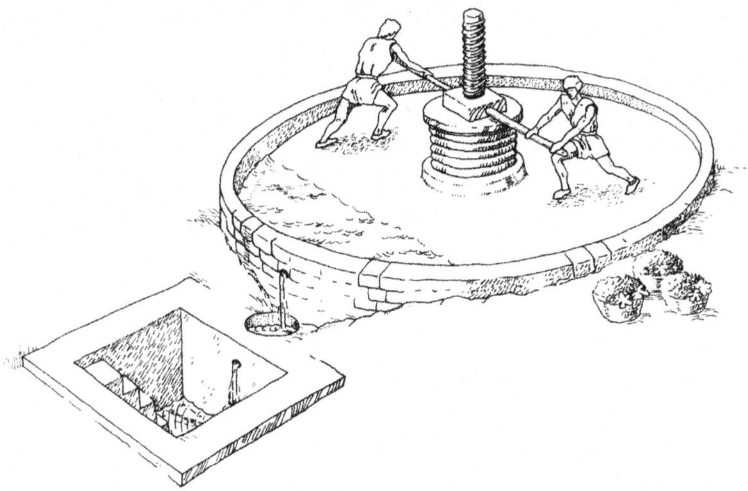

Figure 5.4. Schematic of Type 4 press, here shown as used in wine processing (from Hirschfeld, *Ramat Hanadiv Excavations*, fig. 145. Reproduced by permission of The Israel Exploration Society, Jerusalem).

Modern Scholarship and the Olive in the Roman Economy

The oil trade has been at the centre of some of the most important discoveries in the study of the ancient economy. Amongst the first to forge the link between supply and demand, Bonsor and Dressel pioneered work in Baetican oil transport amphorae and their terminus at Monte Testaccio in Rome.[29] Clark-Maxwell and Ponsich added another link through their survey work in the Guadalquivir River valley (within Roman Baetica); kiln sites and press remains formed the final strand in the chain that has allowed scholars to trace massive olive oil supply from source to terminus.[30] It is now known that huge quantities of Baetican olive oil supplied both the Rhine legions and the city of Rome. At the peak of the empire, Baetica thus acted as an extended hinterland of the metropolis. Mattingly and Hitchner synthesized the North African archaeological evidence of press components, farms, kilns, and ceramic evidence, and indisputably established

[29] J. E. Bonsor and C. L. Penney, *The Archaeological Expedition Along the Guadalquivir: 1889–1901* (New York, 1931); H. Dressel, 'Ricerche sul monte testaccio', *Annali dell'instituto di corrispondenza archeologica* (1878).

[30] W. C. Clark-Maxwell, 'The Roman Towns in the Valley of the Baetis Between Córdoba and Sevilla', *ArchJ* 56 (1899); M. Ponsich, *Implantation rurale antique sur le Bas-Guadalquivir* (Madrid, 1974).

Tripolitania and Byzacena as major oil provinces for Rome and the western empire.[31]

Beyond the production and transmission of annonary payments to Rome and its legions, Mattingly maintains that African oil was a viable commercial commodity of sufficient value to fill ships on its own during the first three centuries AD—not merely as piggyback cargo stuffed amongst African grain freighters. Shipwreck evidence upholds this view; cargoes from the high empire (first three centuries AD), comprised in the majority of African amphorae, are known throughout the western Mediterranean, away from militarized provinces. Findspots of African oil amphorae are too diffuse and frequent to suggest exclusively state distribution.[32] As a profitable product, and one of the few agricultural endeavours that the semi-arid and arid jebel hinterland could support, the olive fuelled agricultural prosperity.

In the eastern Mediterranean, Georges Tchalenko's work rendered a picture of Syrian oil production similar to that of Spain and Africa. Tchalenko's thesis is that the prosperity of farms and villages of the limestone hills in the Antiochene and Apamene lay in olive cultivation and oil exports, which ended with the Persian conquests of the early seventh century.[33] Whereas Tchalenko believed that large landowners, military veterans, and small-scale free and peasant farmers had been key actors in the play, Tate's reassessment downplayed veteran involvement and instead focused on smallholders and a few more substantial landlords.[34] Tate likewise relegated the role of the olive in socio-economic life to an ensemble participation and discounted the importance of overseas olive exports, stressing instead an economy centred on local and regional economic activity.[35]

Areas of Regional Specialization

The imperial authorities tacitly recognized the potential profit from olive groves. In the fourth century, the *Syro-Roman Lawbook* assessed that 1 iugerum

[31] R. B. Hitchner and D. J. Mattingly, 'Ancient Agriculture', *NatGeogRes* 7/1 (1991); D. J. Mattingly, 'Oil for Export? A Comparison of Libyan, Spanish and Tunisian Olive Oil Production in the Roman 'Empire', *JRA* 1 (1988); Mattingly, 'Farmers and Frontiers'; Mattingly, 'Maximum Figures?'.

[32] P. Reynolds, *Trade in the Western Mediterranean, AD 400–700: The Ceramic Evidence* (Oxford, 1995), 126–7.

[33] Tchalenko, *Villages*, 433–6.

[34] G. Tate, 'Mutabilité des économies antiques: l'exemple de la Syrie du nord (IVe–VIe siècle)', in P-L.Gatier, B. Helly, and J-P. Rey-Coquais (eds.), *Géographie Historique au Proche-Orient* (Paris, 1988), 254–5; G. Tate, 'Le Latifundium en Syrie: mythe ou realite?', in *Du Latifundium au Latifondo* (Paris, 1995), 248; G. Tate, 'The Syrian Countryside during the Roman Era', in S. E. Alcock (ed.), *The Early Roman Empire in the East* (Oxford, 1997), 67–9; Tchalenko, *Villages*, 310.

[35] Tate, 'Mutabilité des économies antiques", 249–55; Tate, *Campagnes*, 247–56.

of 225 olive trees were taxed as an equivalent to 5 iugera of vineyard and 20 iugera of arable land.[36] What is not clear is how much of this profit derived from regional specialization and the long-distance export of olive oil.

The presence of Syrian and Cilician merchants throughout the empire presumably reflects the movement of oriental products, and some evidence does link these traders to the oil trade. Regions that produced oil in quantities exceeding their needs are, upon occasion, noted in texts.[37] More crucially, abundant press remains, in certain regions, indicate sites with surplus production and potential regions of specialized oleiculture. Finally, ceramic (amphora) evidence links production sites to shipment and/or consumption points. Admittedly, the evidence is, at present, far less abundant than that from the West. Whilst amphora evidence unequivocally shows the large-scale transport of oil between Spain/North Africa and Rome, this is nothing comparable to the massive pottery dump of Monte Testaccio in Rome, discovered in the East. However, the lack of eastern evidence is certainly in part due to the lesser development of quantified ceramic studies in the region and in Istanbul itself. The nature and scale of local and long-distance eastern oil trade, therefore, remains difficult to assess. Despite these shortfalls, the data link several eastern amphorae types to the oil trade.

Olive oil for subsistence was produced throughout coastal Greater Syria, but the evidence of presses—combined with scattered textual and amphora evidence—suggests that there was a more specialized eastern 'oil belt' that lay along the coastal regions of northern Palaestina I, Phoenice, Syria I–II, and Cyprus, and continued through the fertile Cilician plain along the coast of Cilicia I–II. Oil production probably also formed a major component of the local economies just to the west of Oriens along the Lycian coast: multiple presses have been found at Aperlae, a site with an extremely rugged and truncated hinterland, whose survival probably depended on its place in the local and long-distance sea trade.[38]

Cilicia and Isauria

Cilicia and Isauria enjoyed a reputation as prime olive country; the *Geoponica* mentions both the suitability of the Cilician climate for oil cultivation (9.3.1) and the harvest around Anazarbus (9.2.6). Oil installations found in Cilicia and Isauria—while subject to cursory archaeological exploration only—nevertheless exist in sufficient number to support the other evidence for

[36] *Syr.Rom.Law.* §157.
[37] *Expositio totius mundi* §15.
[38] Dr R. Hohlfelder, personal communication, November 1999.

widespread oleiculture in Late Antiquity. Remains were abundant enough for Hild and Hellenkemper to propose that, in the sixth century, the small settlements of western Cilicia and Isauria flourished on a virtual olive monoculture—a prosperity that came to an end with the Islamic invasions. Their survey for the *Tabula Imperii Byzantini* located forty-two Late Roman sites where oil-pressing equipment was present.[39]

From Hierapolis-Castabala (north of modern Ceyhan)—a city that appears on the Byzantine city lists of both the sixth-century Hierocles and the seventh-century George of Cyprus (*Desc. orb. rom.* 825)—comes essential evidence for estate production of Cilician oil.[40] Excavation in the hinterland village of Domuztepe unearthed the remains of a 20-room fifth–sixth-century 'villa' which included a lever and screw oil-press and storage *pithoi.* On the strength of other regional finds, Rossiter and Freed propose that this installation is for oil manufacture, a view supported by the absence of the treading surfaces usually used for wine. The size of the Domuztepe press vat—5,000 litres in capacity, about 1,500 litres of which would be pure oil—is on a similar scale to small pre-industrial North African presses. The vat filled to full capacity would have far exceeded the villa's subsistence requirements, and probably is a good indicator of surplus production.[41]

The metropolis of Isauria, Seleucia (Silifke) lay near the Cilician coast and possesses hints of considerable ancient olive production. The discovery of thirteen oilery sites around the city demonstrates that focused rural survey work would likely reveal a strong oil-producing region and thus provide a picture of Seleucia's late antique economy.[42] Scattered press finds from elsewhere in Cilicia and Isauria are also suggestive, but further survey is needed before we can say anything for certain about press frequency and hence olive cultivation here in Late Antiquity.[43]

[39] Hild and Hellenkemper, *Kilikien und Isaurien*, 100,09 The *TIB* does not publish plans of presses and there are only a handful of photographs.

[40] J. J. Rossiter and J. Freed, 'Canadian-Turkish Excavations at Domuztepe, Cilicia (1989)', *EchCl* n.s. 10 (1991), 145–74, pls. I–III. See Jones, *Cities of the Eastern Roman Provinces*, 247–65 for city lists of Cilicia-Isauria.

[41] Dependants, such as those labourers living on the estate *epoikion*, may have processed their own oil in the villa press; but, in view of the vast numbers of village presses from neighbouring North Syria, this is questionable.

[42] These sites comprise 28 per cent of the total number of mapped late antique sites (thirty-three) within the approximated city *territorium*. *TIB*, 215, 265, 272, 297, 413, 453.

[43] Elsewhere, scattered finds allow only piecemeal reconstruction. In mountainous Isauria, archaeological exploration is more circumscribed than in the neighbouring provinces of Cilicia. Unsurprisingly, work has focused around the provincial metropolis of Seleucia and the city of Claudiopolis. The latter, sited in the Taurus foothills overlooking an arm of the Kalykadnos River, possessed a rolling hinterland well suited for vines and olives. In the territory of Tarsus,

Cyprus

Cypriot oil does not feature widely in late antique textual sources, the only explicit reference of the Early Roman period being that of Strabo (14.6), who called Cyprus 'blessed with oil'. The Cypriot archaeological record also requires a thorough survey of the entire countryside, but Hadjisavvas has collected and analysed a considerable amount of Cypriot material. He concludes that oleiculture was of prime importance throughout antiquity, and Michaelides stresses that oil was one of the constants in the ancient export trade of the island.[44]

Sixteen of the twenty-two dated press stones in the small catalogue of Hadjisavvas are classified as Late Roman/Early Byzantine, but there are some impressive remains and well-preserved sites that indicate surplus production on some scale.[45] Site continuity is especially evident on Cyprus, unsurprising in view of the island's long peaceful existence during the centuries of Hellenistic, Roman, and Early Byzantine rule.

The press at Kourion, near Limassol, dates to the fourth–seventh centuries, and was found in the Villa of the Achilles Mosaic, a dwelling with which it appears contemporary. This press utilized a lever and screw over a circular press bed with an inner diameter of 85 cm and an incised *canalis rotunda* and projecting rectangular spout.[46] This example is distinctive in that it clearly demonstrates the modification of press elements from one type to another. Initially a hanging beam-weight with a round perforation was used to manipulate the press arm. During Late Antiquity, this weight was modified into a mortised anchor, able to accommodate a windlass or possibly a screw

the prosperous metropolis of Cilicia I, villages such as Bingüç (where were found 'numerous' oil-press stones), Hacihamzali (where an Early Byzantine church was discovered, as well as a press), Keşbükü, Meşelik, Mopsukrenai (possibly a roadside *mansio*), Siraköy (also with an early Christian church), and Hisarkalesi (near Zephyrion (Mersin)), represent a group of Byzantine press remains clustered amongst two cities' *territoria*. *TIB*, nos. 215, 265, 272, 299, 360, 413, 453. In the region of Elaiussa-Sebaste and Corycus, eleven rural settlements with oil-presses are known. Çet Tepe, where the remains of at least four oil-presses were found, presents the best case for production on a commercial scale, *TIB*, no. 226. Village remains at Halilin Öreni and Kanytella territory represent portions of the domain of Sandaios, of unknown date but probably 1st–3rd centuries AD, perhaps continuing into Late Antiquity. Dagron and Feissel, *Inscriptions de Cilicie*, 50–1, no. 20.

44 D. Hadjisavvas, 'The Economy of the Olive', in V. Karagheorghis and D. Michaelides (eds.), *The Development of the Cypriot Economy* (Nicosia, 1996), 135; Michaelides, 'The Economy of Cyprus During the Hellenistic and Roman Periods', in ibid., 139.

45 Notably the ecclesiastical press at Kato Paphos and presses from Salamis and Dameftis. For these, see respectively S. Hadjisavvas, 'Windlass Vs. Screw: A Case-Study for the Reconstruction of an Olive Press', *RDAC* (1990), 181–5; G. Argoud, 'Un huilerie à Salamine', *Salamine de Chypre IV* (Paris, 1973), 201–19, pls.1–19.

46 S. Hadjisavvas, *Olive Oil Production in Cyprus from the Bronze Age to the Byzantine Period* (Nicosia, 1992), 41–2.

mechanism. The owner of an opulent urban estate house certainly would have been able to afford to improve his equipment in the light of technological development. That he did so in an inexpensive manner, through the simple re-cutting of the weight, offers a rare glimpse into the accommodation of new agricultural methods by a fiscally conservative farmer.

The oilery at Amathus (just east of Limassol) lay in a building complex to the west of the basilica which was excavated in the 1970s.[47] In the north-west corner of the building lies the oilery of which the core is the rectangular press room. The vat has a total fluid capacity of 3,600 litres (720–900 litres of which would have been oil when full). A crude estimate of the potential load capacity of this press is 170 kg of olives, with each pressing extracting 34 kg (37.4 litres) of oil. The vat therefore could hold the liquid from nearly twenty pressings. Over a ninety-day pressing period, assuming a daily operation with one pressing per day, the Amathus press would have produced approximately 3,100 litres of oil.

Palestine, Arabia, Phoenicia

In extolling the general fertility of the three Palestines, Arabia, and Phoenicia, ancient authors offer few specific details. Oil, however, certainly formed an important export commodity attested in ancient sources (especially the Rabbinic literature). Modern scholars agree that Palestine II, the Galilee in particular, produced oil of export quality and quantity.[48] By Late Antiquity, Galilee was indeed a long-standing oil producer, having created surpluses traded with the coastal cities during the lifetime of the first-century writer Josephus (*BJ* 2.21.2). Early Christian pilgrims commented on the fertility of the places they saw, including the olive groves that spread around Nazareth in the territory of Sepphoris.[49] The Babylonian Talmud, whose materials were collected mainly in the fourth and fifth centuries AD (TB Men. 85a), also mentions Galilean olive oil, whilst the fourth-century *Expositio totius mundi* (§31) remarked on the wealth in grain, wine, and oil of the Palestinian–Phoenician coast generally. Rabbinic sources also mention the oil produced at Capernaum on the north coast of the Sea of Galilee, Scythopolis (Bet She'an), and Jerusalem.[50]

The southern coastal provinces of the Levant (within modern Israel) also provide the best evidence for an increase in late antique productivity linked to technological innovation and development. As noted in Chapter 1, although a

[47] Described by Hadjisavvas, ibid., 49–51, and M. Loulloupis, 'Excavations at Amathus', *RDAC* (1975).

[48] Neubauer, *La Géographie du Talmud*, 180; Safrai, *Economy*, 194–5.

[49] Wilkinson, *Jerusalem Pilgrims*, 81l.

[50] Heichelheim, 'Roman Syria', 137.

Hellenistic invention, there is no concrete evidence for the introduction of the screw into oileries in Syria–Palestine before the Byzantine period. Frankel's work in Israel, however, recorded the remains of more than 150 screw-press components (either press counterweights or screw-press bases) in former Palaestina I, II, and Phoenice Maritima. These remains almost certainly belong to the Byzantine or Early Islamic period and represent about half of all late antique presses included in Frankel's catalogue.[51] Among these finds, Frankel noted considerable variation in the technology. Several major types of screw-press were utilized, showing differing approaches to technical improvements from one place to another. The range of press types and their ubiquity illustrates a trend toward technological innovation and investment in agricultural specialization and suggest that the oil-producing regions of Palaestina Prima and Secunda, along with southern Phoenice Maritima, made a significant technological shift from the awkward lever and weights press towards the more efficient (Type 4) direct-screw presses.

One of the agriculturally marginal regions of Phoenice–Palaestina shows that olive cultivation, wine production, and the adaptation of screw technology on a wide basis all coincided. In the Lower Golan, especially, the remains of direct-screw press (Type 4) bases in basalt have been found at a number of sites, and several oil installations possess multiple presses. These press remains, and traces of terracing, domestic structures, and water management systems, argue for considerable agricultural development during Late Antiquity.[52] This expansion apparently began in the Early Roman period, to which surveyors assigned 143 sites in the region.

By the Byzantine era, the number of sites increased to approximately 173, whilst Early Islamic era pottery was discovered on only about 14 sites.[53] The population decrease implied by this ceramic assemblage was probably not as drastic as the evidence suggests; the Byzantine–Early Islamic transition remains somewhat difficult to spot in the ceramic record. Jodi Magness has

[51] These conclusions are based in large part on data appearing in Frankel, *Wine and Oil Production* with additions originally compiled for my dissertation work and appearing in M. Decker, 'Agricultural Production and Trade in Oriens, 4th–7th Centuries AD', D.Phil. thesis (Oxford, 2001), Appendix 1. My dating of the presses within this study is based on technical features, settlement patterns, and survey work. The widespread introduction of screw technology throughout Syria–Palestine during the Late Roman/Early Byzantine period is probable: see Frankel, *Wine and Oil Production*, 117.

[52] Frankel, *Wine and Oil Production*, catalogue II.81–9; D. Urman, *The Golan: A Profile of a Region during the Roman and Byzantine Periods* (Oxford, 1985), 147.

[53] Ma'oz, 'Golan', 536–8. On the archaeology of the Golan in the Byzantine period, see also C. Dauphin and J. J. Schonfield, 'Settlements of the Roman and Byzantine Periods in the Golan Heights: Preliminary Report of Three Seasons of Survey (1979–1981)', *IEJ* 33/3–4 (1983).

cogently argued that the late antique pottery chronology of Palestine should be shifted later, thus calling into question any sudden demographic collapse in the seventh century.[54] Nonetheless, by the tenth century, much of the agrarian development that the Upper Golan had witnessed in Late Antiquity had vanished. The people there, as the Arab geographer Muqaddasi reported, lived on acorns and desert barley. The same source indicates that the Lower Golan was primarily a grain producer; the vine and olive had apparently once more become subsistence crops.[55] It thus appears that Golan oil and wine fabrication on a significant scale was predominantly a late antique phenomenon that in all likelihood persisted into, and perhaps throughout, the Umayyad period.

Syria

The limestone hills of the Massif of Belus (the Limestone Massif) are a marginal landscape today, but were much more productive in Late Antiquity. Although these uplands are very poor for arable farming when compared with the rich irrigated valleys of the Orontes and Afrin rivers and the coastal plain around Seleucia-Pieria and Antioch, they were ideally suited to the olive and vine. The largely featureless, broken chain of hills that comprise the massif rises to 600–800 m in height. Today, the hills receive rainfall in excess of the minimum required by cereals, and certainly enough to sustain olives and vines. There are no perennial sources of water; the area is a lunar landscape, where the fertile *terra rossa* soil is restricted to pockets trapped in decaying limestone.[56] The proximity of one village to another facilitated communications amongst them, and external contacts were made easy by the major Roman road running from the coast to Chalcis (modern Qinnasrin), seat of the *dux* and a major centre in the Byzantine period. Bordered by cities—Chalcis and Beroia to the east, Cyrrhus to the north, Antioch to the west, and Apamea to the south—there was no shortage of markets for those dwelling on the escarpment. Thus the region, at first glance seemingly peripheral, was fully integrated within a nexus of communities.

Whilst oil-press remains are known to exist throughout Syria, North Syria provides the most abundant archaeological material and forms the basis of analysis here. Tchalenko's work in the Limestone Massif led him to the

[54] Magness, *Early Islamic Settlement.*

[55] Le Strange, *Palestine Under the Moslems*, 90–1.

[56] In Late Antiquity, the region well may have supported woodland; some areas of former Oriens supported native forest cover into the nineteenth century. See S. Merrill, *East of the Jordan: A Record of Travel and Observation in the Countries of Moab, Gilead, and Bashan* (London, 1881), 292, for the case of the south of Oriens.

conclusion that the presses there were 'innumerable'; Tate's survey of forty-five villages in the Jebels Sim'an, Halaqa, Bariša and al-'Ala yielded 249 presses.[57] Many others remain buried; still others have gone unnoticed in the tumble of buildings. Nearly all villages contain at least one press, and some—such as Kafr Nabo with ten presses, and Behyo with thirty-seven installations—contain an extraordinary number of presses, relative to the small size of the village.[58] Despite its excellent state of preservation and the considerable interest scholars have shown in the region over the past half-century, the data from North Syria remain unsatisfactory, with, for instance, only one site, Dehes, having been partially excavated and published.[59] The picture that one gains from this evidence is therefore necessarily painted in only broad-brush strokes; a thorough survey of every massif village would be needed to clarify problems of press identification and number.

The development of the Syrian and Tripolitanian oileries resembled one another. The Tripolitanian jebel and the Syrian hills of the Massif of Belus belong to a broad transitional zone, a region between the fertile coast and expansive steppe. The climate in both regions is ideal for the olive, which is well-adapted to the calcareous soil and upland conditions; and, as in Syria, the Tripolitanian development of oil in the early and high empire took place in a region with a significant pastoral presence. The Limestone Massif overlooked a largely pastoral semi-desert steppe south and east of Chalcis, known as the Chalcidike, that was, in the Early Roman period, agriculturally underdeveloped. By the mid-sixth century, however, the Syrian Chalcidike was filled with settlements, with nomadic groups providing labour for the harvest. Some nomadic groups perhaps even settled amongst the farming villages.

Oileries in North Syria: the Example of Kafr Nabo

The village of Kafr Nabo lies in the hinterland of Antioch on the Jebel Sim'an, north and east of Deir Sim'an. It comprises 74 houses, remains of a fifth–sixth-century church and chapel, possibly an inn, and ten or eleven press installations.[60] Kafr Nabo is a medium-sized village. Compared with others village of its size, it does not possess an especially high concentration of

[57] Tate, *Campagnes*, 243, notes 45 presses, but in the list of presses according to jebel, the total is 49.

[58] Tchalenko, *Villages*, 360. Callot, *Huileries*, 120, has examined the technical aspects of the Limestone Massif presses. Whilst not comprehensive, his study is based on the author's extensive experience in the region and presents a representative sample of installation types.

[59] J.-P. Sodini et al., 'Déhès (Syrie du Nord) Campagnes I–III (1976–1978)', *Syria* 57 (1980).

[60] Tate, *Campagnes*, 209.

presses, making it a useful proxy in determining oil exploitation throughout the North Syrian villages.[61]

According to Georges Tate's survey work, the 74 houses in Kafr Nabo comprised a total of 170 rooms, a bit more than two rooms per house.[62] Tate further posited that each house contained multiple families, and Clive Foss estimated that 4.25 persons inhabited each room.[63] Under these conditions, Kafr Nabo would have housed a maximum of 723 people. A fair estimate for general annual olive oil consumption in antiquity is 20 litres per person.[64] Thus, in order to meets its subsistence needs, the village needed an annual supply of 14,460 litres.

Of the ten or eleven presses known to exist in Kafr Nabo, four or five were probably winepresses. The villagers of Kafr Nabo operated at least six oil-presses, of which only two examples, those belonging to the subterranean installation published by Callot, could be used for quantitative purposes.[65] For the sake of illustration, we make several assumptions regarding the output of the Kafr Nabo presses: first, that six of the ten or more total press installations likely to have been oil-presses were in operation at the same time, over the same ninety-day pressing season, and, second, that each produced about the same amount of oil. These assumptions are reasonable, given that the Limestone Massif villages experienced peak occupation in the fifth–sixth centuries, which corresponds to the growth and development of Kafr Nabo, and thus the full exploitation of the village hinterland and its presses. Regarding the functioning of the presses, it seems unlikely that these installations represent equipment designed for the contingency of a bumper crop. Given the cost of oil-presses (especially screw-presses, which required specialist skill to build and maintain), their remaining idle for years at a time, while waiting for a superabundant season, appears doubtful. It is more likely that each one of

[61] The ten presses at Kafr Nabo are not an extraordinary number. Behyo possessed 37 installations and only 29 houses, clearly an extreme case. Qirqbize, which I consider paradigmatic in terms of settlement-size and production in the Limestone Massif, had 22 houses and 7 presses, a ratio of about 3:1 houses to installations. Kafr Nabo has an even lower density of presses: 10 installations amongst 74 houses, a ratio of 1:7.

[62] It is questionable whether ground-floor rooms would have been inhabited in many Limestone Massif houses, since many seem to have been stables and storerooms.

[63] Tate, *Campagnes*, 183; Foss, 'Syria in Transition', 201.

[64] Amouretti, *Le pain et l'huile*, 181–3, 95–6, and Mattingly, 'Oil for Export?', 34, reach similar conclusions, calculating that an urban ancient Greek household required 28.5 litres (52.1 sextarii) per person annually, whilst a rural household might use 15 litres per person.

[65] Callot, *Huileries*, 121, pls. 22–3. Callot mentions twelve presses, but his village plan indicates ten presses and one probable example, with two mills, which I interpret to be serving presses close to them, and not themselves independent press facilities. My 'Press 1' is the installation labelled 'ab' in Callot's illustrations and is the larger of the two installations (see Callot plates 26, 57a, 118a and b). My 'Press 2', the smaller of the two, is 'cd' in Callot's plan.

these presses was built in response to average-yearly processing needs, and that they annually operated at least at their minimum, with the chance for production to be ramped up should the need arise. Further, in order to present a conservative set of estimates, we have selected Press 2, a smallish example capable of processing 4,000–8,000 litres per season. Press 1, capable of processing 5,200–13,675 litres per season, was a larger machine, and may in fact represent the archetype of the village installations.[66]

Press 2 could have accommodated a stack of olives with an estimated capacity of 176–353 kg of olives (44–88 kg oil) per load, or 3,960–7,920 kg (4,237–8,474 litres) of oil over the three-month season.[67] Six presses of this size, fully engaged over the three-month pressing season and producing the minimum output for each press, could have yielded an aggregate 25,422 litres. This represents oil manufacture more than one-and-a-half times that likely needed for total subsistence needs, and suggests that the inhabitants of the village were engaged in regular surplus production of olive oil.

The presses at Kafr Nabo processed the fruit of the lands of the village. Do these calculated possibilities of potential output harmonize with what the land could produce? Two boundary stones, remnants of the cadastral survey of Diocletian, help define the limits of the settlement's lands.[68] Based on these finds, Callot estimated the village territory to have covered 600 ha (2,400 iugera). Assuming for a moment an olive monoculture and maximal use of this landscape, the olive groves around Kafr Nabo could have contained 60,000 trees bearing 231,000 kg of fruit. An average load for Press 1 at Kafr Nabo is 373 kg, with one load processed per day. In order to process the hypothetical harvest of Kafr Nabo, the village would have needed seven presses of the scale of Press 1, working continuously over the ninety-day season, to produce a total oil output of 57,500 kg (61,793 litres), more than four times the amount required to support the village population. Despite the hypothetical nature of our calculations, the correlation of the presumed productive potential of the land and the probable output of the presses is suggestive of an agrarian regime in which the people of Kafr Nabo engaged in sustained production that generated surpluses.

[66] Cf. the 2,500–5,000 kg seasonal oil output of smaller North African, and 9–10,000 kg of oil estimated for the Kasserine presses. Mattingly, 'Maximum Figures and Maximizing Strategies of Oil Production?', 490–91; Mattingly, 'Oil for export?', 193, which represent large examples.

[67] Each load produced around 60–150 litres of oil. The extraction rate of oil from olives is 25 per cent. D. J. Mattingly, 'Megalithic Madness and Measurement or How Many Olives Could an Olive Press Press?', *OJA* 7 (1988), 192.

[68] Tchalenko, *Villages*, III.10, 51 nos. 8e, 8f. The survey must have been conducted in 293–305, since Constantius and Maximianus are mentioned.

Nor was Kafr Nabo unique. We have noted above in Chapter 4 that a sizeable number of presses in the Massif should be considered winepresses and linked to a Syrian wine trade that was already centuries old by the time of the arrival of the Romans in the region. This has obvious implications for the potential supply of olive oil from the jebel villages and the hinterlands of Antioch–Apamea in general, since it suggests a significant number of identified presses in the region were not in fact used to process olives. Nevertheless, Tate's survey data from forty-five villages giving a total of 249 presses are a departure point for estimates of the region's oil production capabilities.[69] There are 700 ancient villages in the massif of Belus. An estimate based on 5.5 presses per village therefore yields a total of 3,850 for the region. If this calculation is erroneous, it is in fact more likely to be too conservative, since many presses are buried or were destroyed. This is particularly true in the valleys where settlement continuity is greater.

Comparison with the evidence from the olive-rich region of Tripolitania, with which the Limestone Massif shares many similarities, is instructive. The whole of the territory of Lepcis Magna contained an estimated 1,500 presses. Based on the extrapolation from our limited data, the Antioch–Apamea region may well have exceeded this figure, with nearly 4,000 presses in the limestone hills alone. Syria under the Romans was known for its vast city territories, and while it is true that the Limestone Massif is an important and territorially significant component of the hinterland of Antioch (and to a lesser degree, Apamea), the hill country does not in fact comprise a majority of the territory of either city.[70] In the case of Antioch, the region belonging to the city probably approached 6,000–7,000 km^2.[71] Within this was the rich valley of the Orontes, the fertile coastal plain towards Seleucia and Daphne, the productive valley of the Afrin River as far as the territory of Cyrrhus, and the southern extent of the Amanus range. The Jebels Sim'an, al-Ala, Bariša, Wastani, and northern Zawiye, which belonged to Antioch, cover about 2,500 km^2, under half the entire hinterland. If the rate of 1.5 presses per km^2, which we have found in the Limestone Massif, can be extended over the whole Antiochene territory, the total number of presses could safely be doubled to around 9,000–10,000. When we add to this another approximately 5,000 km^2 of lands around Apamea, a vast sweep of river valley and plains opening to the Syrian steppe, we are able to further contextualize the North

[69] Tate, *Campagnes*, 243 notes 45 presses, but in the list of presses according to jebel, the total is 49.

[70] G. W. Bowersock, 'Social and Economic History of Syria Under the Roman Empire', in J.-M. Dentzer and W. Orthmann (eds.), *Archéologie et histoire de la Syrie* (Saarbrücken, 1992), 66.

[71] Liebeschuetz, *Antioch*, 41.

Syrian villages within the cities' *territoria*. Whilst the territory of Apamea contains a significant area of plains and river bottom lands more suitable to livestock rearing and cereal culture, the presence of the Jebels Zawiye and Nosayri would allow us to add considerably to the number of presses estimated for the territory of Antioch, to a figure of well over 10,000 installations for the region of Antioch and Apamea as a whole.

Such a figure accords reasonably well with Mattingly's Tripolitanian evidence, where he estimates the presence of one press per km^2. The combined hinterland of Antioch–Apamea was approximately 11,000–12,000 km^2, and the density of presses was thus similar. Even if only half the presses were engaged in oil production, and at the same level as the Kafr Nabo Press 2, they potentially processed 2.1–4.2 million litres of olive oil annually, a great deal of it produced in the jebel villages.

The characteristics of oil production at Kafr Nabo were repeated in thousands of villages throughout Oriens. The quantity of oil that these small villages could generate is something of a marvel when one ponders their rather bleak setting today. Whilst archaeological remains demonstrate the extent to which oleiculture was a major endeavour within the countryside, the handbooks allow insight into a carefully planned orchard landscape that was managed with some rigour. Although most of their techniques were ages old, late antique farmers seem to have added a few new methods of planting and grafting, and the farming literature again demonstrates a strong tendency to improve strains and develop new varieties through experimentation. The taste for olive oil, and its fundamental place in the diet, meant that a sizeable market was generated in the cities, countryside, and military garrisons of the empire. This basic demand was itself a motivation for the planting of olive groves on a large scale throughout the eastern provinces, especially along the Levantine coast where transport to the Mediterranean was not difficult. In addition, the fifth and sixth centuries also saw the relative decline of African and Spanish oil production, creating some space in the western market that was partly filled by eastern oil (see Chapter 8). The value of oil as a commodity, and the extent of its trade, will be explored below.

6

Invading the Desert

Irrigation and Agrarian Expansion

Intense and expansive settlement, as described in Chapter 2, made significant inroads into the margins of the Roman world. The frontiers that settlement invaded were both political and climatic. The lands dominated by Constantinople overlapped into those of its enemy, Persia, without clear boundaries. The intervening landscape, a kind of no-man's land, was marked by the pastoralists who roamed there. Yet these lands, like the borderlands of North Africa during the first and second centuries, increasingly became the objective of an industrious and land-hungry body of farmers and estate owners who possessed the capital to open new lands, and who found competition along the coastlands intense. This was a group that naturally turned inland. They were probably people 'on the make' rather than members of the established elite. A flurry of building in the countryside reflects in part the ethos of an aggressive aristocracy, a vivid picture provided to us by Chrysostom.[1] Inscriptions provide another snapshot of these new settlers: near Anasartha, we find building inscriptions commemorating the work of the brothers Leontius and Stephanus, who climbed up the imperial ladder rung by rung. In 553, they were *peribleptoi* and *lamprotatoi*, and in the following year, they had advanced into the office of the Praetorian Prefect as *megaloprepstatoi* and *trakteutai*.[2] Such men were of the same cut as their untitled peers, such as Thomas, whose buildings lie a day's march south of Anasartha at Androna. Their numbers and their works were repeated up and down the desert margin of Oriens.

This parched landscape needed villages to settle it, hands to work it, but, above all, water to make it green. By Late Antiquity, canal irrigation had been practised in the Near East for millennia; the floodwater farms that made the desert bloom in the Negev had roots in the Nabataean era. Some of the water-lifting machines employed harked back to the Hellenistic period or earlier. Given the rich tradition of Roman engineering on the one hand, and the rich technical

[1] *In Acta Apost.* 18 = *PG* 60.149ff.

[2] Mango, *Artistic Patronage*, 63–4; *IGLS* 2.316, 318, 320.

heritage of the Levant and Mesopotamia on the other, it is only natural that the farmers of Oriens availed themselves of hydraulic works that were varied in origin and design.[3] The use of such varied technologies, and especially the breadth and depth of their use in response to unique conditions, makes Late Antiquity an especially interesting period in the history of the Levant.

Irrigation is an 'intensifier', bringing water to crops by expending capital and energy in a concentrated effort to improve the yield of the land. Irrigation can either be applied to existing areas of cultivation, in order to improve yields, or can be used to bring wholly new, and dry, areas into production. I must stress that in many contexts expansion of the area under cultivation (the subject of this chapter) does not exclude the intensification of areas already cultivated (the subject of Chapter 7) or vice versa. Both depend on internal and external conditions within the landscape and the farming communities that worked it. Much of what is addressed in this chapter has a bearing on, or is influenced by, issues discussed in Chapter 7, and vice versa. Indeed, there are many instances where the same mechanisms that provide the capacity for extension allow—even predicate—an intensive approach to agriculture. A good example of the latter is terracing, a fundamental component of concentrated agriculture, for which reason it is treated in the next chapter. However, the net result of terracing is more arable land, which therefore warrants discussion below. Bearing these complexities in mind—and bearing in mind that some overlap between chapters is inevitable—is necessary to provide the latitude required for a discussion that is both wide-ranging and in-depth.

IRRIGATION AND THE FARMED LANDSCAPE OF LATE ANTIQUITY

In the hinterlands of many of the cities in Oriens shown in Map 1, the dry-farming of cereals was impractical or impossible. Many cities, such as

[3] Despite the objections of B. D. Shaw, 'Lamasba, an Ancient Irrigation Community', *AntAfr* 18 (1982), and B. D. Shaw, 'Water and Society in the Ancient Mahgrib: Technology, Property and Development', *AntAfr* 20 (1984), the balance of the evidence from Roman North Africa argues for Roman improvement of indigenous irrigation works. The approach of Roman irrigators was eclectic: they continued to add to their repertoire of water management techniques by employing technologies from outside the Italian landscape. C. Vita-Finzi, 'Roman Dams in Tripolitania', *Antiquity* 35 (1961); A. I. Wilson, 'Classical Water Technology in the Islamic World', in C. Brunn and A. Saastamoinen (eds.), *Technology, Ideology, Water: From Frontinus to the Renaissance and Beyond* (Rome, 2001). This is important to consider when one considers that the Roman-Byzantine state involved itself in at least one of the major projects described below.

Hierapolis (Mambij), Theodosiopolis (Ras el-'Ain), Barbalissus (Meskene), Callinicum (Raqqa), Zenobia (Halabiye), and Constantia (Viranşehir), lay on land at or beneath the limits of rainfall needed to ensure a steady annual cereal supply.[4] Unless they could rely on a supply of imported cereals, these centres would have required hydraulic structures to survive. It is thus no wonder that each is situated on a major perennial watercourse that supplied domestic needs and agriculture. In the environs of these cities, we find traces of dams and canals that supported the agricultural wants of city and countryside. Whilst paucity of data curtails our understanding of the extent and nature of irrigation, there are strong indications that the late antique inhabitants of Oriens actively shaped their countryside and expended no small quantity of sweat and gold to improve their lands and render new ones habitable. In their efforts, these farmers were perhaps part of a larger trend throughout Eurasia, where people responded to a general growth in population, as the Sasanians did in the irrigation development of the Diyala Basin.[5]

Broadly speaking, irrigation and water collection sometimes were conducted on a small scale where individual farmers took care to provide for their own needs (microsystems). There is evidence, however, of community systems (mesosystems), and of large works involving more than one community, such as the long canals in Roman Mesopotamia (macrosystems).[6] In the countryside, each landowner had to assure his or her own water supply. Both Palladius and the *Geoponica* describe water management on agricultural estates. According to Palladius (*Op. Ag.* 1.17), each farm should be equipped with a cistern, and he recommended (9.9) digging wells and building aqueducts or pipelines. It is interesting, but entirely unsurprising, that seasonal water shortage seems to have been a common problem. In his agricultural calendar, Palladius (10.8) addressed summer shortages and what should be done to alleviate them, then discussed wells (10.9), water testing (10.10), and rural aqueducts (10.10).

The use of irrigation varied widely in form and practice. There is no discussion in the handbooks, for example, of watering grain or other crops under extensive cultivation. The explanation for this is that rainfall in most of the Mediterranean basin was adequate to ensure a harvest of cereals under dry-farming conditions. However, the encyclopaedic nature of the *Geoponica*

[4] On the Jazira and its past land use, see in particular T. J. Wilkinson, 'The Structure and Dynamics of Dry-farming States in Upper Mesopotamia', *CurrAnthr* 35/4 (1994).

[5] McC. Adams, *Land Behind Baghdad*; R. McC. Adams, 'Historic Patterns of Mesopotamian Irrigation Agriculture', *Irrigation's Impact on Society* (Tuscon, 1974); McC. Adams, *Heartland of Cities.*

[6] The classification is that of K. W. Butzer et al., 'Irrigation Agrosystems in Eastern Spain: Roman or Islamic Origins?', *Annals of the Association of American Geographers* 75 (1985).

is also possibly to blame. The work was compiled in the tenth century, by which time the Levantine provinces, where most of its material originated, had long been lost, and while irrigated farming of cereals was standard practice in places like Egypt and Mesopotamia, it was impractical or unnecessary over most of Anatolia and the Balkans. At least one of the early Greek farming manuals that survives only in Arabic translation calls for the irrigation of grain crops, but this aspect of cultivation was perhaps cut by the compiler(s) of the *Geoponica*, who considered it unnecessary information.[7]

While we can only surmise that grain was commonly irrigated in the drier parts of the Levant, some field crops, such as lucerne, were certainly commonly grown under irrigation (Palladius, *Op. Ag.* 5.1). *Digest* 39.3.3.2 mentions the case of a farmer who annually turned his grain land into an irrigated meadow. Where rain-fed agriculture predominated, cultivation under irrigation was often restricted to the garden and orchard, where the intensively worked plots provided the greatest return on investment. This certainly is the impression given by the agronomists, as in Palladius (*Op. Ag.* 6.5, 7.4, and 8.3) for whom the garden and orchard are the focus.[8]

Analysis of individual sites, regarding environmental concerns and the historical background of water management, allows us to begin to produce a picture of irrigation in late antique Oriens.[9] However, no study has focused on Roman- and Byzantine-period canals in the East, and the few studies of other regions that exist discuss canals for transportation, not farming.[10] Due to this dearth of data, and a lack of precision within the present corpus of material, it is only possible to sketch the types of hydraulic structures and their significance within the agrarian life in dry areas. The various installations surveyed have been chosen based on their scale, context, and the extent to which they cast light onto the wider issues of subsistence and the expansion of the farmed environment.

Canals

From legal texts we learn that canals were generally unlined, although sometimes a lining of stone is mentioned (*Dig.* 43.21.10; 43.21.3.1). The fifth-century

7 Ps.-Democritus, *Kitab al-Mukhtasar*, Bibliotheque Nationale MS Arabe 2802 f.13v; on the date of this epitome of a lost treatise (thought to be Hellenistic) see M. Wellmann, 'Die *Georgika* des Demokritos', *APAW* 4 (1921).

8 In the second century, Galen (*Nat. Fac.* 3.15) had used a literary picture of garden irrigation as an extended metaphor to describe an anatomical condition.

9 Butzer et al., 'Irrigation Agrosystems in Eastern Spain', 498–500.

10 N. Smith, 'Roman Canals', *Transactions of the Newcomen Society* 49 (1977 8); C. Wikander, 'Canals', in Ö. Wikander (ed.), *Handbook of Ancient Water Technology* (Leiden, 2000).

Theodosian Code (11.42.1) exempted from other corvées those estates through which canals passed, as maintenance of the channels was already their burden. In a telling passage, the sixth-century *Digest* (43.21.4) stated that the upkeep of canals was considered more important than road maintenance. The water that flowed through such channels was regulated by a hierarchy arranged according to use: priority was given to domestic needs, followed by watering of stock and gardens, then field irrigation, and finally that for industry and mining.[11]

The Euphrates Complex

In Mesopotamia, only one example of a major Byzantine-period canal has been investigated using geomorphological tools that permit confident dating.[12] The canal systems of this and other regions are notoriously difficult to date, because they are rarely associated directly with datable archaeological evidence and because most of them consisted, in part, of older systems (since large-scale irrigation in Mesopotamia began in the Bronze Age). However, there is enough circumstantial evidence—above all, the flourishing late antique cities and rural settlements in a landscape that needed irrigation to prosper—to show that our period saw considerable continuity and even some possible extension of the region's irrigation canals.

As noted in Chapter 1, the Euphrates was not an easily managed river and thus had somewhat limited potential for irrigation. Low, rocky bluffs restrict the width of the valley of the Euphrates over much of its length. Without major damming of the kind made possible only in modern times, these hills virtually preclude the potential for long major canal branches north or south of the river. Not until the river reached the alluvium of Iraq were the waters easily exploited. In recent years, largely due to the construction of dams, the water has dropped markedly relative to the shore; but in Antiquity, the depth of the banks was still often substantial. The general lack of vast dendritic canals on the Euphrates was therefore true of any era, with mesosystems dominating. In Late Antiquity, lifting gear was employed in the form of *norias* and *saqiyas* instead of canal construction. Nevertheless, as the following survey demonstrates, the Roman–Byzantine-period inhabitants, and probably the imperial government, maintained and enhanced the irrigation network of the Euphrates complex. This is especially true of the more manageable tributaries of the Euphrates, the Balikh and Khabur Rivers.

[11] E. F. Ware, *Roman Water Law* (Littleton, Colo., 1985), 94.

[12] T. J. Wilkinson, 'The Archaeological Landscape of the Balikh Valley, Syria: 1994–95 Annual Report', http://www-oi.uchicago.edu/OI/AR/94–95/94–95_Jazira.html (updated 14 November 1997), accessed 21 May 1999.

The riparian province of Euphratensis (Map 5) included the important cities of Hierapolis (Mambij), Cyrrhus (Nabi 'Uri), Zeugma, Germanicia, Neocaesarea (Dibsi Faraj?), Sura, Barbalissus, and Zenobia.[13] The Euphrates watered all of these lands, save Cyrrhus and Germanicia. The province of Osrhoene comprised lands on the left bank of the river, including Callinicum, Circesium, and Dausara, all of which have yielded traces of probable Byzantine-period canalization.

Near Barbalissus (Meskene), a fortress city on the right (south) bank of the Euphrates, a canal began on a bend of the river at the modern village of Tannuz, flowed along the river terrace and passed within 300 m of the city walls.[14] From the city, it followed the shoulder of the low hills with traces of it found more than 80 m above the plain at the Byzantine ruins of the village of Qsubi. Traces of this 30 km-long system remain further downstream at Abū Hurayrah.[15] In the ninth century, the Islamic historian al-Baladhuri wrote that the Umayyads built this canal, called the Maslama Canal (Nahr Maslama).[16] But Barbalissus could not have survived without irrigation, since it was a settlement considerably larger than that which could have been supplied by *norias* or other lifting gear alone. Therefore, it seems that the early twentieth-century archaeologist Alois Musil was correct in his opinion that the Umayyad work at Barbalissus entailed restoration of an existing canal network and not the creation of a new one.[17] When the Umayyads seized power in the 660s, the system had perhaps decayed due to the disruption of previous decades of warfare in the Fertile Crescent.

The so-called Nahr Maslama flowed to Dibsi Faraj (Neocaesarea?), where Harper traced its line for 8 km. Harper also recorded a small mesosystem, whose main channel flowed north of the city near the city walls.[18] Another small irrigation system (or part of one of the former networks) lay along the

[13] In northern Euphratensis, the twin-city complex of Zeugma-Seleucia and Samosata yield scant evidence of irrigation in Late Antiquity, but substantial irrigation is deduced from a Roman inscription and the notice of the Islamic geographers, who stated that Samosata particularly depended on rainfall and irrigation. The evidence falls on either side of our period, but there is little reason to suppose either city would ever depend entirely on rainfall since they were situated very near the river. Around Zeugma-Seleucia, in particular, the broad river terrace was fertile and easily exploited. Undoubtedly water-lifting gear played an important role in this exploitation. That this was so is in part confirmed by *ILS* 8903, which records the construction of a water-screw by *legio III Gallica* at Aini, near Samosata in the first century AD. Le Strange, *Palestine Under the Moslems*, 539.

[14] Any hope of tracing the canal networks around Meskene with certainty is now impossible, due to the construction of the Assad Dam and the subsequent obliteration of the city and most of its hinterland.

[15] Lauffray, *Halabiyya-Zenobia I*, 53.

[16] Hitti, 'The Origins of the Islamic State', 232.

[17] A. Musil, *The Middle Euphrates* (New York, 1927), 316.

[18] R. P. Harper and T. J. Wilkinson, 'Excavations at Dibsi Faraj, Northern Syria 1972–1974: A Preliminary Note on the Site and its 'Monuments', *DOP* 29 (1975).

north city wall. Although its name belongs to the Islamic era, this system was certainly in use during Late Antiquity, when the city walls were rebuilt and a programme of renewal was undertaken, probably by Justinian.[19] Downstream of Neocaesarea lay Sura, an important city in Late Antiquity, and headquarters of the *legio XVI Flavia Firma*.[20] Justinian rebuilt the walls of the city following the Persian devastation of 540. In the Islamic period, however, the city is rarely mentioned and seems to have sunk into obscurity.[21] Jean Lauffray discovered traces of an ancient canal there, which may be associated with some confidence with the Roman–Byzantine period.[22]

Further down the right bank, another canal had its beginnings near the Qa'lat Nemrod (probably ancient Thapsacus) ruins and watered the right-bank village lands opposite Callinicum to which they belonged. The canal continued for 15 km, ending slightly below the confluence of the Balikh with the Euphrates. In the nineteenth century, Sarre and Herzfeld dated this canal to the reign of the Umayyad caliph Hisham (724–43), who is mentioned as its builder in the *Chronicle of Zuqnin* (completed in the ninth century); but the same reservations raised above can be observed regarding this system.[23] This canal was a key component in the city's food supply, irrigating an estimated area of 3,000 ha.[24] The right bank further south, opposite the confluence of the Khabur and Euphrates at Circesium (Buseira), shows only a handful of sites yielding late antique pottery; the development of extensive irrigation networks is therefore improbable.[25]

There are also traces of hydraulic engineering of the Byzantine period on the left bank of the Euphrates. Two canals were associated with ancient Dausara (Qalʿat Jabar), an important Byzantine town and Islamic fortress.[26] The old 'Semiramis' Canal that cut through the left bank of the Euphrates downstream of Zalabiya and Deir ez-Zor (ancient Auzara?) certainly functioned in Late Antiquity, since the remains of villages with fourth–sixth-century pottery have been discovered along *c.*12 km of its length that were supplied during the

[19] Ibid., 326.

[20] A. Poidebard, *La trace de Rome dans le désert de Syrie, Le limes de Trajan à la conquête arabe* (Paris, 1934), 83.

[21] Lauffray, *Halabiyya-Zenobia I*, 53; Musil, *Middle Euphrates*, 325.

[22] Due to the narrowness of the valley at this point (the upper gypsoid terrace is about 1 km away from the city), the canal ran parallel to the river, and was probably only a short mesosystem.

[23] Ps.-Dionysius of Tell Maḥre, *The Chronicle of Zuqnin, parts III and IV: A.D. 488–775*, trans. Harrak, 160–1; F. P. T. Sarre, E. Herzfeld, and M. V. Berchem, *Archäologische Reise im Euphrat- und Tigris-Gebiet* (Berlin, 1911).

[24] Lauffray, *Halabiyya-Zenobia I*, 56.

[25] B. Geyer and J.-Y. Monchambert, *La basse vallée de l'Euphrate syrien du Néolithique à l'avènement de l'islam: géographie, archéologie et histoire* (Beirut, 2003), I.275–9.

[26] Lauffray, *Halabiyya-Zenobia I*, 56.

fourth–sixth centuries by norias and other lifting gear. Recent archaeological fieldwork around the confluence of the Khabur suggests that the late antique settlement presence in this part of the Euphrates valley was limited, probably due to its proximity to the Roman–Persian frontier. Only after the Muslims unified the area was a large-scale canal, the Nahr Dawrin, built there.[27]

The Balikh–Khabur Complex

The Balikh (ancient Balissus) and the Khabur (ancient Aboras), both tributaries of the Euphrates, coursed mainly through the province of Osrhoene and met the great river on the right bank. At the joining of each watercourse lay important settlements. On the confluence of the Euphrates and Balikh was Nicephorium-Callinicum (Raqqa), whose canals that lay along the Euphrates have been described. At the Khabur–Euphrates confluence was the frontier city of Circesium. Further upstream on the Khabur lay Theodosiopolis (Ras el-ʿAin). Behind this city, in the upper catchment of the Mygdonius/Khabur watershed, lay the late antique cities of Constantia (Viranşehir) and Dara. Due to their low banks and broad alluvial terraces, the Balikh and Khabur were easily tapped for irrigation waters. A major Seleucid-era (fourth- through first-century BC) canal that flowed more than 12 km was cut parallel to the Balikh and marked a new phase of agricultural expansion that continued through the Roman and Byzantine periods.[28]

Two canal complexes in the basin of the Balikh, when the valley was dominated by the city of Callinicum, can shed light on the agrarian setting of Late Antiquity.[29] The first of these, on the upper Balikh, dubbed the Sahlan-Hammam by its investigators, carried water from the Balikh, past Tell Sahlan, and thence south toward Nicephorium (later Callinicum). This was a Hellenistic work, probably related to the Macedonian build-up of the region reflected in the first-century AD *Parthian Stations* of Isidore of Charax.[30] A sizeable canal 6 m wide, this channel probably diverted most or all of the flow of the Balikh to the new Hellenistic settlements along its banks. According to geomorphological analysis conducted by Wilkinson, the

27 S. Berthier, *Peuplement rural et aménagements hydroagricoles dans la moyenne vallée de l'Euphrate, fin VIIe–XIXe siècle: Région de Deir ez Zōr-Abu Kemāl, Syrie: Mission Mésopotamie syrienne, archéologie islamique, 1986–1989* (Damas, 2001), 161–5; Geyer and Monchambert, *La basse vallée de l'Euphrate syrien*, I.274–82.

28 T. J. Wilkinson, 'Water and Human Settlement in the Balikh Valley, Syria: Investigations from 1992–1995', *JFA* 25 (1998).

29 Bartl and Hauser, *Continuity and Change in Northern Mesopotamia.*

30 Isidore of Charax, *Parthian Stations by Isidore of Charax: An Account of the Overland Trade Route between the Levant and India in the First Century, B.C.* ed. and trans. W. H. Schoff

channel apparently went out of use around the sixth–eighth century when the Byzantine–Early Islamic canal, the Nahr al-Abbara, was built.[31]

The second canal, the Nahr al-Abbara, apparently originated at a dam near Tell Sahlan, whence the water was directed into a channel east of the river and passed close by the site of the early Islamic steppe village of Madina al-Far.[32] During Late Antiquity, the valley of the Balikh belonged to the hinterland of Callinicum, the only large late antique site within the valley. The Nahr al-Abbara was fed by a dam, and its course planned; the channel split the middle ground between the flow of the Balikh and the Wadi al-Kheder. Its elevation ensured delivery to the greatest possible area of the river terrace. The growth of Callinicum—which, according to Ammianus (23.3.7) in the fourth century, was a busy trading post and stronghold—required this canal. Callinicum remained a vital frontier military station and caravan-staging area into the sixth century, one of only three eastern cities where trade between Sasanians and Byzantines could be conducted legally (*CJ* 4.63.4).[33]

On the Khabur, where Byzantine settlement centred on Theodosiopolis in the north and Circesium in the south, several presumed Byzantine forts exist close to irrigation canals of local significance, such as that at Tell ʿAgaga (ancient ʿArban or Araban).[34] A canal, called Qattina in the modern period, terminates in the territory of ancient Theodosiopolis at ancient Themeres

(Philadelphia, 1914), 4–5; Lauffray, *Halabiyya-Zenobia I*; Wilkinson, 'Water and Human Settlement in the Balikh Valley', for the canals of the Balikh watershed. Lauffray noted Byzantine ceramics at many of the sites he visited, and his association of many of the developments in the Euphrates basin with Late Antiquity fits well with the picture rendered by Wilkinson.

[31] Wilkinson, 'Water and Human Settlement in the Balikh Valley', 67–9.

[32] K. Bartl, 'The Balīkh Valley, Northern Syria, During the Islamic Period, Remarks Concerning the Historical Topography', *Berytus* 41 (1993–4); C.-P. Haase, 'Madīnat al-Fār: the Regional Late Antique Tradition of an Early Islamic Foundation', in K. Bartl and S. Hauser (eds.), *Continuity and Change and Change in Northern Mesopotamia from the Hellenistic to the Early Islamic Period* (Berlin, 1996). I have written elsewhere about the problems regarding our understanding of the settlement of Balikh during Late Antiquity: M. Decker, 'Frontier Settlement and Economy in the Byzantine East', *DOP* 61 (2007).

[33] A third canal, which Lauffray, *Halabiyya-Zenobia I*, 58–9, believed was used in the Byzantine period, began near Ichnae and drew a shallow arc west of the main riverbed, but apparently stopped short of the Euphrates. This may have been the canal initially dug during the Old Babylonian period (second millennium BC), when residents of Tuttul (Tell Biʿa) complained that their water was being stolen by the new canal.

[34] Theodosiopolis (Ras al-ʿAin) lay at the foot of the scarp of the Mesopotamian highlands, and numerous springs fed into the Khabur and lent themselves to irrigation of the lands around the city. Poidebard, *Trace de Rome*, 137, noted the military posts of the area. Assuming that Poidebard was correct and that these forts are Roman, the purpose of their placement was probably manifold: many of the forts rested on tels and thus on sites of centuries of human habitation, guarding the fertile districts and surrounding water resources, and monitored the movement of the pastoralists of the region. The garrisons would have drawn on the canals for drinking water and for watering their animals.

(Tell Tamer). Qattina irrigated several thousand hectares along its 40 km length. Another canal on the upper stream, named Ubeyan, flowed on both banks of the river and watered the vicinity of 'Umm Gharqan and the fortified town of Thallaba. Lauffray considered the section on the right bank the more important, and estimated the irrigated area at 12,000 ha.[35] East of Thallaba, the dam-fed canal of al-Breij flowed around a Roman camp and continued eastwards to Dabausa (Tell Dibs).[36] Near ancient Thannouris (Tell Tunaynir), below the confluence of the ancient Mygdonius and Khabur, a bridge-dam with Roman foundations diverted water to fill the 3 m-wide Sfaya Canal.[37] South of Haseke, many more ancient canalizations have been discovered, testifying to prolonged and deep development of this frontier region.[38]

W. J. Van Liere and Jean Lauffray traced four of the Khabur systems, which include some of those mentioned above. They estimated that the canals carried an average of 2–5 m^2 per second.[39] From dimensions and the rate of flow, it was possible to estimate the land that was irrigated. The results are summarized in Table 6.1.

An estimated 80 per cent of the total area (i.e. 15,500 ha) was irrigated. An additional 12,000–15,000 ha were watered on the terraces of the lower Khabur. If these relics of canals do in fact reflect ancient practices, as they seem to, then the total irrigated area probably approached 30,000 ha.

The investigation of these systems sheds some light on the possible extent of irrigated landscape in northern Oriens during the Roman–Byzantine period. Until more detailed archaeological work is done, the full extent and precise chronology of these canal networks will remain uncertain. But urban sites beyond

Table 6.1. Mesopotamian canals and irrigated area

Canal	Length (km)	Hectares irrigated
Sfaya	27	3900
Nahr al-Taff	35	4750
Nahr al-Jum	32	8000
Nahr Saruat	12.5	1750
Total	106.5	18400

35 Lauffray, *Halabiyya-Zenobia I*, 61.

36 Mouterde and Poidebard, *Limes de Chalcis*, 151–2.

37 Lauffray, *Halabiyya-Zenobia I*, 60–1; Poidebard, *Trace de Rome*, 138–42.

38 These include the canal of Saba' Skur and the Nahr at-Taff. Portions of both canals lay on either bank of the Khabur. The Nahr al-Taff provided irrigation for more than 30 km around ancient Arabana, home of the *Legio III Parthica*. The Nahr Sarruat, on the right bank, continues for 4–10 km until ending south of Tell Suwar. Nahr Marqada, also on the right bank, watered the upper terrace. Nahr al-Jum/Nahr al-Hama, north of Tell Suwar on the right bank, flowed into the area around Circesium at the terminus of the Khabur.

39 J. Lauffray and W. Van Liere, 'Nouvelle prospection archéologique dans la Haute-Jézireh syrienne', *AAAS* 4–5 (1954–5), 147.

the margins of dry-farming were thoroughly dependent on irrigation and those that lay on the edges of dry-farming needed artificial watering to even out annual variations that made cropping the desert fringes particularly precarious. Since we know that the settlements of Osrhoene and Euphratensis provinces attained a significant population and prominence in the fourth to seventh centuries, it is no surprise that their territories exhibit traces of ancient canal networks. Despite our limited data, some change is perceptible. Major investment was apparently made to rehabilitate the valley of the Balikh, with the construction of a dam and the excavation of the Nahr al-Abbara. In the late sixth or early seventh century, due to dense settlement, engineers apparently attempted to augment or replace the old Hellenistic canal of Sahlan-Hamman on the Balikh. In order to create maximal irrigated land area, the canal was cut into the highest portion of the valley floor; but the long course also meant that the gradient decreased. This led to rapid siltation and the failure of the canal nexus. Despite the considerable effort expended in transforming the landscape, the ceiling of human development ultimately rested on the feeble flow of the Balikh itself.[40]

The Khabur discharged over eight times more water than the Balikh, and consequently its landscape had a much higher agricultural potential. The data from the Khabur Valley, though incomplete, produce a picture of a richly appointed agricultural territory. The investment was sizeable, involving multiple canal systems over the entire length of the river. While present-day damming of the river means that we shall never fully understand the entire river system in Late Antiquity, the old evidence of Lauffray and Van Liere is suggestive of a substantial settlement core. Under an intensive regime and optimal conditions, the aggregate 30,000 ha area irrigated, suggested above, would have produced enough grain to feed approximately 62,000 inhabitants.

Irrigation underlay the fabric of life on the margins of the steppe and was an integral part of the human-worked landscapes that developed over millennia. In these many regions, irrigation was so central to life, and so much a part of the background, that detecting change is extremely difficult or even impossible. In other areas, where massive short-term investments are clearly visible in the archaeological record, the results of late antique expansion or intensification are clearly discernible.

Creation of Artificial Oases

In the Early Roman Period (first and second centuries AD), the Biqā^{c} valley (al-Biqā^{c}), the ancient Massyas Valley of Phoenicia, was one of the most

[40] Wilkinson, 'Water and Human Settlement in the Balikh Valley', 82: a possible maximum population of the Balikh Valley was perhaps 8,000.

agriculturally undeveloped regions in the Levant. The northerly extension of the Great Rift Valley, in the heart of modern Lebanon behind the city of Beirut, the Biqāᶜ forms an upland trough around 1,000–1,400 m in elevation, roughly defined by the catchment of the Litani and Orontes rivers.[41] The Biqāᶜ is a complex mosaic of microenvironments, having substantial changes in climate and terrain throughout. In the south, for example, in the basin of the Litani, a marshy lake predominated through Early Antiquity, and the river itself was utilized in only limited fashion.[42] Other challenges included an inconsistent matrix of good agricultural soils, forest cover, and a variable climate that witnessed significant shifts as one traversed the region: the southern valley around Mashgarah today receives mean rainfall of 1,440 mm, whilst the north-east limits of the valley at Hirmil receive 260 mm, already at the limits of dry-farming. When one arrives at nearby al-Qaa, the desert is on the threshold with an annual precipitation of only 180 mm each year.

Prior to the Roman period, agricultural settlement had reached a peak in the Bronze and Early Iron Ages and had declined during the Achaemenid era. The Hellenistic period that followed is one for which the Roman sources are instructive, since it is the age of their first interventions in the Levant. During this time, settlement increased and far-reaching changes took place in the Biqāᶜ, including the seminal elements that Marfoe identified: increasing population, new urban centres, the appearance of two distinct settlement zones of north and south (with the centre of the valley less populated), and the populating of the mountain slopes of the Lebanon and Anti-Lebanon ranges. Despite the lawless image given by Strabo (16.2.18) of the Aramaic-speaking Ituraeans who controlled the territory of the Biqāᶜ at the end of the Hellenistic period, settlement of Chalcis sub-Libanum (Majdel Anjar) and Baalbek probably occurred during this time.[43]

Roman occupation left an enduring mark. In addition to the above changes, there were long-term patterns of deforestation; settlement of veteran colonies at Berytus and Baʿalbek; and drainage of large tracts of marshland in the southern Massyas by Roman engineers, who had already honed their skills on large-scale drainage and engineering projects.[44] Sometime in the Roman–

41 Marfoe, 'Between Qadesh and Kumidi', 17.

42 The marsh began to recede in the Early Bronze Age, when sites appear within its older area. The Litani had cut deep into the alluvium of the valley floor, and high banks and winter flooding challenged irrigation. Its tributaries were utilized more readily. See Marfoe, ibid., 212.

43 Marfoe, ibid., 566ff., 630–60, who notes that Chalcis sub-Libanum and Baʿalbek must have been modest centres at this time.

44 With Nero's efforts in the Pontine marshes, and the construction of the canal in the Peloponnese being notable examples reported by Cassius Dio (44.5), while there was an earlier failure at the Fucine Lake during the reign of Claudius. On Roman drainage and canal projects see Y. Calvet and B. Geyer, *Barrages antiques de Syrie* (1992); N. Glueck, *Rivers in the Desert*

Byzantine period, the Massyas witnessed profound agricultural expansion, leading to a settlement density that would not be matched until the present day.[45] In the 1930s, Weurlersse noted one of the traces of that development: a canal originating in the oasis of Lebwe in the central Massyas Valley. He traced the channel northward, over more than 24 km, to al Qaa. The latter site, as noted above, lies below the safe limits of dry-farming.[46] The canal continued beyond al Qaa and passed Ras Baʿalbek, a site with a Byzantine church first settled during the Roman period.[47]

René Mouterde and André Poidebard also traced the Lebwe Canal and found that it bifurcated at Ras Baʿalbek. The north branch, surveyed by Jacques Weulersse, was also investigated, and Mouterde and Poidebard were able to show that it continued past al Qaa and the remains at Jusiyah al-ʿAmar and Jusiyah al-Kharab (which lie 2 km apart from one another, approximately 37 km south of Homs (Emesa)). In the hinterland of Emesa (the Emesene region), Jusiyah al-ʿAmar contains the remains of a large fortified complex and continues to be inhabited. Jusiyah al-Kharab is an extensive ruin-field on the Lebanese–Syrian border. Based on a notice in the *Notitia Antiochena* (a document belonging to the sixth century, with later reworkings), Mouterde and Poidebard identified the sites of Jusiyah as those of the late antique city of Mauricopolis Gausithon, named after the emperor Maurice (580–602).[48] The site may not have been an entirely new settlement, but its name suggests that it attained urban status in the late sixth century, thus implying deeper urbanization in the northern Biqāᶜ in Late Antiquity.

After its bifurcation at Ras Baʿalbek, the second branch of the Lebwe Canal, with the Anti-Lebanon scarp at its northern limits, swept to the east into the lands south of Emesa and descended to Sadad in the Syrian steppe.[49] Mouterde and Poidebard located a series of *qanats* there that approached from the

(London, 1959); F. G. Moore, 'Three Canal Projects, Roman and Byzantine', *AJA* 54/2 (1950); White, *Roman Farming*, 149–50, 70–1; A. Wilson, 'Land Drainage', in Ö. Wikander (ed.), *Handbook of Ancient Water Technology* (Leiden, 2000), 309–11.

45 Marfoe, 'Between Qadesh and Kumidi', 666–8.

46 Ibid., 666–7; Weulersse, *L'Oronte*, 52.

47 Marfoe, 'Between Qadesh and Kumidi', 666, believes Ras Baʿalbek is the ancient Conna, whose bishop was present at the Council of Chalcedon in 451. He accepts Mouterde and Poidebard's description of the Lebwe Canal. In September 2000, I interviewed local sources who stated that there was a large canal on the slopes of the Anti-Lebanon, above the ruins of Jusiyah, near al-Qaa. Due to the politically sensitive nature of the area, I was unable to visit the canal.

48 Mouterde and Poidebard, *Limes de Chalcis*, 33, describes the two sites of Jusiyah al-ʿAmar, and Jusiyah al-Kharab ('Jusiyah the Village' and 'Jusiyah the Ruin', respectively). E. Honigmann, *Historische Topographie von Nordsyrien im Altertum* (Leipzig, 1923–4), 85, noted the presence in the *Notitia Antiochena* of Mauricopolis Gausithon.

49 Mouterde and Poidebard, *Limes de Chalcis*, 149–51.

south and supplemented the main branch of the Lebwe Canal. These qanat systems drained the Wadi Nebk and continued north of Sadad to Qnaye where they irrigated a walled enclosure encompassing 144 ha. From Qnaye, another branch apparently flowed to Ghuntur, where traces of a qanat and an open canal were detected for a further 12 km to the east, near the late antique palace complex of Qasr al-Hayr al-Gharbi.

This palace complex is associated with an impressive array of hydraulic installations that include an underground canal running 16.5 km. A bifurcation of this canal fed the early Islamic bath house and a reservoir with a capacity of 13,000 m2, followed by a secondary canal which propelled a mill further downstream. The enclosed garden area fed by this canal covers approximately 46 ha. A series of qanat-supplied reservoirs and a large enclosed garden (*paradeisos*) were fed by waters stored behind the nearby massive Harbaqa Dam.[50]

Harbaqa is a large gravity-dam, 365 m long and 20.5 m high, extending the length of the Wadi al-Barda. It could store 5 million m^3 of water, enough to irrigate 2,200 ha of wheat land. I visited the site in October 2004 and found it still exceptionally well preserved. Its two faces are constructed of large, rectilinear limestone blocks set over a core of unworked stones. Engaged piers provide extra support. The width at the dam's base is 18 m. Gates probably controlled the outlets at the base of the dam, as they did for the outlet 11.2 m from the crest, and supplied a canal that fed the gardens of the site, probably an Umayyad addition, as argued by Guenequand, who also argued that the dam was an early Islamic endeavour.[51] While the Umayyads certainly developed the agricultural potential of the region, the Palmyrenes and Romans/Byzantines had already utilized the landscape for agriculture and herding and either political entity may have organized construction of the dam. The remains of a nearby village show evidence of Roman, Byzantine, and Umayyad occupation and more work is needed before a definite chronology is established.[52]

The canal at Lebwe and nearby hydraulic structures are impressive monuments to ancient engineering skills and developmental planning, and would certainly merit a dedicated survey to confirm what various sources have stated thus far. Even if the Lebwe canal is comprised of a series of discrete canals developed over time, rather than the single-phase gargantuan project that

50 Schlumberger, 'Les fouilles de Qasr al-Heir al-Gharbi'; Schlumberger et al., *Qasr al-Heir al Gharbi*.

51 D. Genequand, 'Some Thoughts on Qasr al-Hayr al-Gharbi, its Dam, its Monastery, and the Ghassanids', *Levant* 38 (2006), 82.

52 Calvet and Geyer, *Barrages antiques de Syrie*, 86; N. Saliby, 'Les installations hydrauliques à Qasr el-Heir el-Gharbi', in B. Geyer (ed.), *Techniques et pratiques hydro-agricoles traditionnelles en domaine irrigué: approche pluridisciplinaire des modes de culture avant le motorisation en Syrie: actes du Colloque de Damas 27 juin–1er juillet 1987* (Damascus, 1990).

Mouterde and Poidebard conceived, the overall scale of these endeavours, taken as a whole, cannot be ignored. There can be little doubt that the two major branches of the Lebwe watered a vast area, since the proposed total length of the eastern branch alone exceeded 100 km. In cutting this canal, the engineers enabled, at a crude estimate, the irrigation of 12,600 ha.[53] The result was the transformation of a sizeable band of steppe land into productive agricultural tracts.

A final scrap of information, which perhaps also points to imperial involvement in the project, must be considered. About 40 km to the south of Sadad, where the Lebwe Canal struck the Syrian plain, and some 30 km south-south-east of Lebwe, is the town of Yabrud. In Late Antiquity, Yabrud was the centre of an imperial estate, the *Klima Iabrudōn* mentioned by the seventh-century compiler George of Cyprus (990).[54] It is possible that the lands of this estate incorporated the Anti-Lebanon range and part of the Biqāᶜ around Lebwe. Those lands to the east comprised part of the imperial domain of the *Klima Anatolikon* and were apparently confiscated after the conquest of Palmyra by Aurelian in 272. The Lebwe canal system may, at least, have been conceived and extended to improve the profitability of an estate that had comparatively recently fallen into imperial hands.

Among the settlements within the boundaries of the vast swathe of estate land, we find Sadad, Qnaye, and Euaria (Hawarin). Euaria was elevated to city status probably around the fourth century. In AD 451, its bishop attended the Council of Chalcedon; later, it became an important centre for the Umayyads.[55] The development of the irrigation works on this imperial estate in the fourth–seventh centuries corresponds with the growth of other settlements on the *limes* and seems to indicate efforts by imperial authorities to settle and develop the region for farming.

The Lebwe canal system and nearby irrigation installations may, at least in part, have been conceived and extended to improve the profitability of an estate that had recently fallen into imperial hands. Sometime around the fourth century, a persistent irrigation programme was centred on the Orontes river and land east of the Massyas valley, extending settlement, urbanization, and arable land use deep into the steppe. Nor was the Lebwe canal the only remaining strand of the fabric. The northern Massyas also witnessed placement of three dams at Hirmil, Said Ali, and al-Umari on the Orontes

[53] The output of the source at Lebwe is unknown. However, if it produced enough water to sustain irrigation of wheat on 500 m of each side of the canal the quantity of water needed would have been perhaps 0.7 m² per second, about one-eighth of the total flow of the Balikh. The Orontes rate of flow was about 12 m² per second. The Lebwe Canal potentially diverted about one-tenth of the flow of the river to the north and east.

[54] Jones, *Cities of the Eastern Roman Provinces*, 267.

[55] Ibid., 269.

that fed water into an extensive, sophisticated network of qanats that watered the plains to the east of the river.[56] At the same time, the marginal landscape beyond the limits of reliable dry-farming witnessed concentrated agricultural development and intensified urbanization. The elevation to cities of relatively young settlements like Mauricopolis, Danaba, and Euaria signals real transformations of demography, not simply 'paper' changes in administrative titles. They reflect the realities of sizeable urban and rural extension that pushed the margins of settled farming deep into the Syrian steppe. The likelihood that at least a portion of this work took place on imperial estate lands suggests potentially considerable intervention of the state in the agrarian economy of Oriens. The sheer volume, depth, and scale of the work had a profound impact on the Syrian countryside for the next three centuries.

The ghūta of Androna

Considerably north of Euaria—occupying the same agriculturally marginal steppe zone—the village of Androna (al-Andarin) is attested first by the third-century *Antonine Itinerary* as a *mansio*, on the route from Palmyra to Chalcis (Qinnisrin). From its apparent beginnings as a staging-post, Andarin grew into a settlement nearly a mile across and developed extensive urban features, notably circuit walls, a large *kastron*, public bath, and at least twelve churches.[57] Several seasons of excavation and survey in Androna and its vicinity have yielded evidence of marked late antique development. The data testify to the regional importance of the site, its prosperity, and its connectivity to the nexus of late antique trade.[58]

Androna lay between the cities of Salaminias (Salamiyah) and Anasartha (Khanasir) on a route that linked Emesa, via the crossroads at Anasartha, with the Euphrates frontier and the fortress cities of Barbalissus and Dibsi Faraj (Neocaesarea?). Like Salaminias and Anasartha, Andarin flourished in the steppe environment at the limits of dry-farming (on the edge of the present-day 250 mm isohyet). This was a land where traces of earlier urbanization are scant, typically exploited by pastoralists.[59] Unlike its two neighbouring cities,

[56] Marfoe, 'Between Qadesh and Kumidi', 666–7; these features and other hydraulic structures on the Orontes are discussed in Chapter 7.

[57] Mango, 'Androna, 1999', 307.

[58] M. M. Mango et al., 'Oxford Excavations at Andarin (Androna): September 1998', *AAAS* (1999); M. M. Mango et al., 'Oxford Excavations at Andarin (Androna): September 1999', *AAAS* (2000); Mango, 'Androna, 1999'; Mango, 'Androna, 2000'.

[59] Several prominent tells dot the landscape around Androna, and it is not unlikely that the steppe zone here followed a pattern similar to that elsewhere in the Levant, with a settlement peak in the Late Bronze/Early Iron Age, subsequent ruralization, and dynamic development in the Late Roman period.

however, Androna never gained civic status; an Early Byzantine inscription identifies it as a *komē* within the territory of Apamea.[60] Its location on communication routes and its physical size argue for Androna's position as a primary node in the eastern steppe zone. The density of stone buildings, numerous inscriptions, public structures and adornment, and a concentration of agricultural installations, indicate that the late antique village prospered, especially in the mid–late sixth century.[61] Further accentuating the vitality of the settlement are the trade connections that indulged a taste for imported goods, known from numerous finds of LR1 (Late Roman Amphora type 1) amphorae from the coastal regions, ARS (African Red-Slip Ware) from central North Africa, and Late Roman C ware from western Asia Minor.[62]

Natural barriers largely defined the limits of the *komē*; only directly to the east was the unbroken steppe accessible for the grazing of wandering herds. In the west, the modest hills, known as the Jebel al-'Ala, provided land for grazing and vine terracing. In the south lay the last, low fringes of the Palmyrene hills, and to the north, the jebels Hass and Šbeit. The gentle declination of these squat hills forms a large, shallow basin that Androna dominated; a large area of about 400 km^2. Numerous seasonal watercourses, upon which the settlement depended, flowed into this basin before its kinetic transformation at the hands of late antique irrigators.

Androna, whose primary limitation was a lack of water, depended on the careful exploitation of its agricultural hinterland.[63] In the immediate environs of Androna, the work of late antique hydraulic engineers manifests itself in the form of an extensive mesosystem of qanats (chains of wells that harvested water from aquifers and transported it underground over hundreds of metres or several kilometres) and reservoirs (Figure 6.1). About 1 km from the southern edge of the fortification wall, the first, roughly square reservoir, measuring about 61 m on each side and with a depth of approximately 4 m, was fed by a single drainage gallery that extended over 1 km to the south-east. The reservoir, of masonry construction, is carbon-dated to the sixth century, with the surrounding spoil heaps indicative of its initial construction and subsequent cleaning.[64]

[60] H. Salame-Sarkis, 'Syria grammata kai algamata', *Syria* 61 (1989). The reason for Androna's failure to gain city status may well have been due to the size and importance of Apamea, which undoubtedly avidly sought to maintain control over its vast territory.

[61] Notably, the bath house yielded substantial marble fragments used in beautifying the building, as well as *opus sectile* work, glass mosaic, and wall paintings. Mango, 'Androna, 2000'.

[62] Mango, 'Androna, 1999', 314.

[63] The gypsoid soils that cover substantial portions of the lower areas of its hinterland are not terribly rich. The present-day settlement, which continues to grow today, depends on non-renewable water sources tapped by the drilling of deep wells that will, in all likelihood, render the region uninhabitable within several decades.

[64] Marlia Mango, personal communication, July 2004.

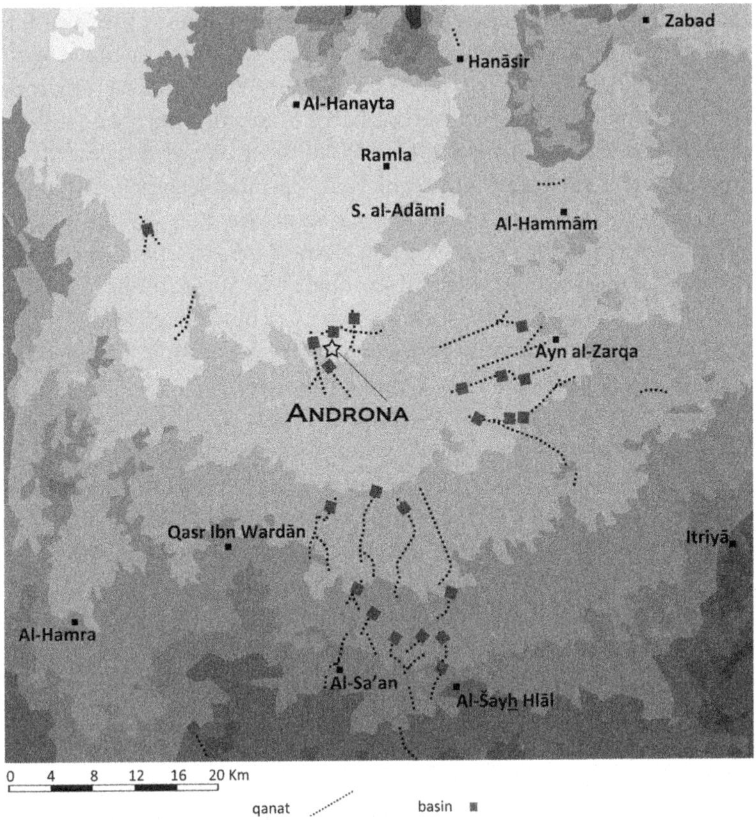

Figure 6.1. Androna and its environs (after Jaubert et al., *Land Use and Vegetation Cover*, map 13).

The inhabitants built another basin of equivalent dimensions, also constructed in stone, about 1 km to the north-west of Androna. Here a qanat line also fed the reservoir, with galleries extending southward for about 2 km and a bifurcation at about 1.5 km along its course. These qanats tapped the waters that flooded the wadis during the winter rains and subsequently exploited water that percolated from the hill catchment area. North and north-west of Androna, two more reservoirs were supplied by the flow of additional qanat lines, one of which exploited the aquifers east of the city, whilst the more easterly one abstracted water from a spring south of the foot of Jebel 'Ubaysān.[65] Ten to 15 km to the

[65] In 1998 and 1999, I visited these systems with Dr Andrew Wilson, to whom I am grateful for sharing his knowledge of Roman hydraulic systems. These systems were explored by Bernard Geyer and summarized in B. Geyer, 'Aux origines de l'archéologie aérienne', in L. Nordiguian and J.-F. Salles (eds.), *Aux origines de l'archéologie aérienne* (Beirut, 2000).

east of the village, a second aggregation of four qanats watered the eastern flank of the village lands. Each of these lines extended for 10 km or more, with the most southerly example a V-shaped network whose branches approached 15 km in length. These handled the discharge of a spring and aquiferous wadi soils. Beyond the extent of these systems, the most remarkable feature of the eastern complex is its design and regulation. The southernmost qanat system tapped the subsurface waters over an area of approximately 225 km^2. No fewer than six reservoirs punctuated their lines, giving access to irrigation and drinking water, and regulating the daily and seasonal flow. At the southern edge of the territory of Androna are four more qanat systems. These represent the outliers of a heavily watered band of cultivation, articulated by a multiplicity of smaller systems developed around reservoirs and village fields.

The inhabitants of Androna thus created a *ghūta*, an artificial oasis, within the basin surrounding the village. The vast investment in money, time, and labour required to excavate the Androna qanat complex was apparently commensurate with the expected return. Although cereals, particularly barley, were grown at Androna without irrigation, there is no doubt that other crops were irrigated. The finds of fragments of hourglass grain mills and multiple *mola olearia*-type crushing basins for olive processing indicate a highly developed agrosystem in Late Antiquity.[66] Indeed, olives grow in the area today only under irrigation, and it was probably olive groves that at least a portion of the waters supported.

Bearing in mind that Androna was never a city and was a relatively young settlement, the prodigious feat of creating an artificial oasis is all the more impressive. The inscriptions on the public works in the village, for example, bear no marks of title or office—neither ecclesiastical nor secular. The bath, for example, was built in 558–9 by Thomas as a memorial to himself and 'for the sake of the farmers'.[67] The same Thomas constructed the *kastron*, dated AD 559.[68] In neither case does he offer any titles. None of the other 38 extant inscriptions from Androna mention a single official.[69] In other monumental buildings in the environs of Androna, such as Stabl Antar and Qasr ibn Wardan, the same dearth of titulature is evident.[70]

[66] Mango, 'Androna, 1999', 315.

[67] *IGLS* 4.1685 The interpretation of *τοῖϲ γεωμόροιϲ* as farmers is that adopted both by the *IGLS* editors and by Prentice: Butler et al., *Syria*, vol. III, no. 918. In agreement with Prof. Cyril Mango, who recently read the inscription, I maintain this interpretation. However, Anderson interprets '*τοῖϲ γεωμόροιϲ*' as 'for the masters of the land' ('aux maîtres des lieux') and the bath as a place for the local elite landowners. See *SEG* XLV.1902.

[68] Butler et al., *Syria*, vol. III, no. 915; *IGLS* 4.1682.

[69] *IGLS* 4.1676–1713.

[70] Butler et al., *Syria*, vol. III, nos. 906–8; *IGLS* 4.1673–4, 1841–4; Mouterde and Poidebard, *Limes de Chalcis*, 217.

These late antique farmers created a striking landscape. The impressive buildings, now badly ruined, stand in the midst of bleak country that today, for the first time since Late Antiquity, is supporting a surely brief agricultural revival, made possible only by the exploitation of deep aquifers. The numerous churches, the baths and fortress, reservoirs and qanats symbolize a determined effort to bring Roman life, as they knew it in the eastern provinces, to the steppe. The pleasures of the bath, the comfort of the church, the security of a curtain wall and *kastron*, were placed here at enormous expense. Barely audible, amidst the din of the flourishing daily life of the village of Androna, would have been the trickling of that catena of qanats on which the entire enterprise depended.

The lack of evidence for state involvement here is striking; the work at Androna was that of private landowners. By pressing further into the desert, men like Thomas—who evidently possessed substantial wealth—and probably many others of more modest means, opened up vast tracts to cultivation. Their efforts betray a scarcity of high-quality cultivable land and concern for water. Their endeavours furthermore indicate mobile capital, brought by competing landowners who sought their fortunes where land was available, and a pool of labourers who made ready-colonizers, possibly drawn from the semi-nomads of the region. Moreover, the investments in hydraulic agriculture supported much more than bare subsistence: olives were grown for cultural reasons, and probably for sale; and wine was a famous commodity of the village, trafficked among the Arab tribesmen who passed through its territory.[71]

The Negev

The Negev comprises a triangular area of more than 17,000 km^2, defined by the present-day Israeli–Jordanian border around Eilat, through the Wadi Arava, with Beersheba (ancient Birosaba) forming the apex. In Late Antiquity, the Negev belonged to the province of Palaestina Tertia and possessed three cities: Birosaba (Beersheba), Mampsis (Mamshit), and Elusa (Halutza). Several well-known, large villages, such as Oboda (Avdat), and Nessana (Nitzana), filled out the landscape. Like the Biqāᶜ, the region around Emesa, and the Chalcidike region of Androna and Salaminias, the Negev landscape saw considerable investment in agriculture through the building of floodwater farms and, to a lesser degree, qanat building.[72] Although the Nabataeans

[71] Wine (documented in pre-Islamic Arabic poetry): Lutz, *Viticulture and Brewing in the Ancient Orient*, 24.

[72] M. Evenari, L. Shana, and N. Tadmore, 'The Ancient Desert Agriculture in the Negev IV: Chain Well Systems in the Wadi Arava', *Ktavim* 9 (1958).

occupied the Negev in the early centuries AD and undoubtedly farmed in the wadis, most extant runoff installations in the Negev belonged to the Byzantine period, with many continuing into the Umayyad period.[73] In recent surveys, hundreds of these farms and agricultural installations have been discovered, and a settlement history is slowly emerging. Unquestionably, an explosion of human activity occurred during the fourth–seventh centuries, and, far from what one would expect from such a marginal environment, agriculture was apparently widespread. This loose patchwork of tilled land supported unprecedented village populations (though modest by the standards of the coast) and the growth of impressive rural settlements, most with multiple churches and even public amenities, such as the fifth–sixth-century baths found in the central Negev at Avdat and Rehovot.[74]

In this region it was the successful management of surface-water resources that underpinned agricultural growth. The Negev farms aimed to harness rainfall, which occurs infrequently: the average annual precipitation is today only around the 200 mm mark in the north, 160–180 mm in the central Negev, and as little as 100 mm along the eastern and southern border of Provincia Arabia. As mentioned in Chapter 1, it remains uncertain whether or not the period of Late Antiquity was wetter. The Negev farms collected runoff water from the scant rains that come to the region in brief, though often powerful, bursts. The region's geomorphology is such that most of these showers do not penetrate the soil crust, but flow to low points in the landscape, often gathering in wadis and creating flash floods. Though methods varied—and there was considerable adaptation depending on the landscape—some generalization is possible. A fairly substantial barren slope area, often twenty times the size of the cultivated plot, was required to create productive agriculture.

Water harvesting was accomplished through a number of common installations, each ingeniously employed in the landscape.[75] Along the sloping floor of the wadi, farmers formed terraces by building drystone cross-dams and levelling the resultant fields. Barrages retarded the potentially devastating flow of water; contained the precious moisture so that it percolated into the fields; and retained the silt that settled in the farmed areas, thus enhancing fertility.

[73] Of the surveys undertaken by the Israeli Department of Antiquities, the relevant ones are: Avni, *Map of Har Saggi Northeast (225)*; Cohen, *Map of Sede-Boqer—East (168)*; Cohen, *Map of Sede-Boqer—West (167)*; Haiman, *Map of Har Hamran—Southeast (199)*; Lender, *Map of Har Nafhah (196) 12–01*. On the continuation of settlement into the Umayyad period: M. Haiman, 'Agriculture and Nomad–State Relations in the Negev Desert in the Byzantine and Early Islamic Periods', *BASOR* 297 (February 1995).

[74] R. Rubin, 'The Romanization of the Negev, Israel: Geographical and Cultural Changes in the Desert Frontier in Late Antiquity', *JHistGeogr* 23/3 (1997), 274.

[75] The basic techniques of floodwater farming in the Negev are discussed by Evenari, Shanan, and Tadmor, *The Negev*; Mayerson, *Ancient Agricultural Regime*.

Rainfall that flowed down the slopes of the catchment area, off the wadi floor, was captured in channels that ran parallel to one another, and the discharge was then directed into the fields or into storage cisterns. In order to enhance the runoff that reached their fields, the farmers of the Byzantine era cleared the slopes surrounding the wadis of their surface stones and piled them in mounds. This led to a massive increase in runoff (250 per cent), compared to those surfaces where the stones were not removed: it is thus clear that ancient farmers understood some of the dynamics of slope-runoff and actively sought to transform the environment to their advantage.[76] But constructing these installations was not easy: as many as 2,000 man-days of labour were needed to build an average runoff farm.[77]

Michael's Farm (Figure 3.1), near Sobata, is a runoff farm that has been used as a paradigm for small holdings in the quantified studies throughout this work.[78] It also provides a good example of many of the structures used by the Negev farmers to control water. A four-room house, situated at the southern flank of a tributary wadi, formed the core of the farm. The 'run-on' area, the flooded farmland, was part of this seasonal watercourse, cordoned off from the desert by an irregular stone wall about 700 m in length. Within the enclosed area, eight terraces were built of unhewn fieldstones, forming discrete units, today called division boxes. Each terrace field was separated from the next by a low, well-built stone wall. The regulation of water within the division boxes was realized by the use of stepped spillways that communicated with the adjacent terrace. In order to control the water, the farmer blocked or unblocked the stepped spillways with wooden beams, thus controlling the flood.

Michael's Farm also managed its runoff area with relative sophistication. Within the courtyard of the farmstead was an underground cistern supplied by waters flowing from the slopes to the south through an incised channel. Once the cistern filled to capacity, the overflow coursed into the down-slope terraces. The hill slopes of Michael's Farm were cleared of stones, which were gathered into mounds as described above. When a rainstorm occurred, the streams coursed down the hill slopes and into the terraces, either through

[76] There have been interpretations of these mounds, including hypotheses that they were used to support grapes, to increase dew and its infiltration into the soil, or that trees were planted within them. H. L. Lavee, J. Poesent, and A. Yair, 'Evidence of High Efficiency Water-Harvesting by Ancient Farmers in the Negev Desert, Israel', *JAridEnviron* 35 (1997).

[77] Y. D. Nevo, *Pagans and Herders: A Re-examination of the Negev Runoff Cultivation Systems in the Byzantine and Early Arab Periods* (Negev, Israel, 1991), 109. This represents a sizeable investment in time and effort, but not one beyond the means of a single family; three adult males would need about two years to construct a full farm.

[78] Evenari, Shanan, and Tadmor, *The Negev*, 104–10.

sluices in the boundary wall or via a series of channels that piped water to the plots from the east, north, and south.[79]

But with such labour-intensive building and careful management, what did the Negev farms produce? We have seen already the unsuitability of these microsystems to generate grain under a monoculture system, but were cropping alternatives sought? The so-called Site 59 in Nahal Ela preserves the remains of a farm complex comprised of a number of one- and two-room dwellings, relatively well-built with double-masonry walls.[80] A stone enclosure surrounds portions of the main wadi floor and a share of a tributary wadi. The total farmed area is only 6 iugera. On the eastern side of the wadi, below a stone watchbooth, lies a winepress. Site 59 thus yields clear evidence of one product that the Negev microsystems aimed to grow: wine. As in the case of the Limestone Massif, the hilly hinterland of the coastal cities provided a region with potential for the export trade; a large share of the famous wines of Gaza and Ascalon almost certainly came from inland vineyards. Fuelled by the wine trade, these small-scale vineyards pushed further into the margins; the modest plots, being easy to work once constructed, were ideally suited to intensive viticulture. The ample sun of the region ensured fine fructification, and the floodwater delivered to the plots meant that in a year with only 100 mm of precipitation, the fields received the equivalent of 500 mm of precipitation.[81]

The Negev exemplifies one of the most successful landscape transformations in the Mediterranean in any period. Indeed, despite years of study, researchers are only now beginning to understand the skill employed by Byzantine-period farmers to create an agriculturally productive region in this desolate environment. The range of small settlements, scattered across the vast and rugged terrain of the Negev, demonstrates an impressive hunger for land and thirst for wine, but it also betokens a palpable feeling of security. As elsewhere, one of the major changes engendered during the Roman–Byzantine era was the spread of small settlements and isolated farmsteads. Nowhere are the advantages of this particular settlement pattern better illustrated than in the Negev, where remote seasonal watercourses and an upland environment required frequent attention, and, thus, the immediate presence of the worker or owner. These regions would have been inconvenient at best to manage from the traditional,

[79] Lavee, Poesent, and Yair, 'Evidence of High Efficiency Water-Harvesting', 345. At least 5 mm of rainfall is required to cause runoff; this currently happens in the drier parts of the Negev five to ten times a year.

[80] Haiman, *Map of Mizpé Ramon—Southwest (200)*, 43–4: his dating of the site to the Early Arab period should be modified to 'Byzantine–Early Islamic', since the chronological framework of the transition period from Byzantine to Islamic rule is uncertain.

[81] R. Rubin, 'Urbanization, Settlement and Agriculture in the Negev Desert—The Impact of the Roman–Byzantine Empire on the Frontier', *ZDPV* 112 (1996), 281 n. 27.

nucleated rural settlements of earlier eras. But this extension of cultivation, and the scattered settlement that it entailed, were only possible under the prevailing security the region enjoyed during the Byzantine era. Without effective policing and government control, the pervasive rural settlement of Late Antiquity would have been unattainable.

In every village and isolated farm of the Negev, as in all other areas without running water, such as the Syrian Limestone Massif, the storage of water in cisterns and reservoirs was essential. The Negev is a region in which rural cisterns are extremely common, numbering in the thousands.[82] Their primary agricultural function was probably the maintenance of flocks and their owners, but, although the tanks might be of modest size, there is evidence for their use also in agriculture. Terraces equipped with cisterns appear frequently throughout Palestine and Arabia, and some regions north of the Negev provide even better proof for cistern-irrigated agriculture. The terraced fields formed natural catchment basins, the runoff from which filled the cisterns. Archaeological survey work has revealed, for instance, such agricultural installations around Madaba. In Samaria, Dar recorded terraces with cisterns established within the centre of the plots.[83] Similar installations were found, linking terraces, winepresses, and cisterns, in Lower Galilee.[84] These systems channelled runoff from the terrace plots into storage, slowed erosion, and harvested water for irrigation.

A sixth-century Nessana papyrus (P.Ness. III.32) details the sale of a half-share of a fig-vineyard (an arbustum in which the fig trees support the climbing grape vines) to a group of men.[85] The fig orchard had a cistern in its midst, of which the men obtained a quarter-share. The men purchased not only the figs, but also the means to make their acquisition viable: irrigation rights. Figs are drought-tolerant and need only periodic watering to produce fruit, and thus are ideally suited to small applications of stored water. The fig-vineyard (*sykampelynos*) did not function without the attached cistern (*lakkos*). This cistern was almost certainly equipped with a *saqiya*-driven water-lifting wheel; evidence from Egypt suggests that the term *lakkos* often denoted a well or cistern with a *saqiya*, rather than a simple water-tank.[86]

[82] Haiman, *Map of Har Hamran—Southwest (198)*, 23; Haiman, *Map of Mizpé Ramon—Southwest (200)*; Haiman, *Map of Har Hamran—Southeast (199)*, 55–6; Lender, *Map of Har Nafhah (196) 12–01*, xvi.

[83] Dar and Applebaum, *Landscape and Pattern*, 19; Herr, *Madaba Plains Project: The 1989 Season at Tell el-Umeiri and Vicinity and Subsequent Studies*, 282–3.

[84] Gal, *Map of Har Tavor (41) and Map of En Dor (45)*, 58.

[85] Kraemer, *Excavations at Nessana*, III: *The Non-literary Papyri*, 102–3.

[86] Banaji, *Agrarian Change*, 109 n. 30.

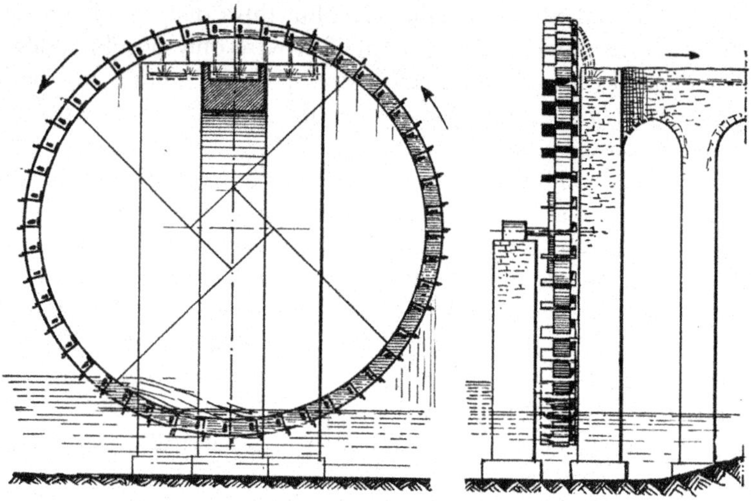

Figure 6.2. Noria schematic (from Weulersse, *L'Oronte*, fig. 28).

Noria and Saqiya

Whilst the transportation of water over distances—be it via surface canal or underground qanat—was an important facet of late antique agriculture, water-lifting machinery often played as vital a role. There were several types of water-lifting machinery commonly used in late antique Oriens, including the *shaduf*, the ancient counter-weighted beam and bucket that supplemented human effort. The *delu* was also used. The *delu* is a pulley-drawn bucket that used the linear motion of a beast: as the animal walked in a straight line away from the well, the pulley drew up the leather water bag, which dumped automatically at the mouth of the well. The beast was then led back, sinking the bag, and beginning the process once more. But it is the *noria* and *saqiya* that are of major concern for the agricultural development of the Levant in Late Antiquity, and therefore are our focus.

The *noria* is a current- or animal-driven water-lifting wheel, the most famous examples of which are those still in operation on the Orontes at Hama in Syria (Figure 6.2). These wheels can be immense and possess tremendous lifting capacities. Those that survive on the Orontes are often 20 m in diameter or more, and they have supplied houses and gardens throughout the city in Hama into the modern period.[87] Constructed of wood, with compartmented rims or pottery jars, they fill as they plunge

[87] A. Delpech, *Les norias de l'Oronte: analyse technologique d'un élément du patrimoine syrien* (Damascus, 1997), 29.

below the waterline and discharge through gravity at the top of the aqueduct. Despite a supposed Islamic origin, there is conclusive proof of their use in Byzantine Greater Syria. In the colonnade of the cardo at Apamea, a mosaic dated to AD 469 depicting a noria has been found.[88] Like many technologies with agricultural applications, the date of the invention and first use of the noria will probably be pushed further back as more evidence emerges. As for the place of origin of the noria, the middle Orontes Valley remains a strong candidate.

Both wheels and their aqueducts are found throughout the lower Orontes Valley; these are the wheels mentioned by Arab geographers who declared that they were the largest in the world and unknown elsewhere.[89] The irrigation capacity of current-driven norias is substantial. Weulersse concluded that one large waterwheel could irrigate 25–75 ha.[90] Building these wheels involved small-scale dam construction, as well as the erection of aqueducts and the wheel itself, and was a complex and time-consuming task. In antiquity, noria-building must have been a communal or estate endeavour.

Lauffray believed that irrigation of the Euphrates' lower terrace at Zenobia was through the noria, a view supported by the discovery of *qadus* (from Greek *kados*) pots at Zenobia and elsewhere along the Euphrates.[91] The Late Roman site of Dibsi Faraj (Neocaesarea?) on the Euphrates also yielded qadus jars, possibly from current-driven norias.[92] However, such pots may well derive from smaller, similar machines, driven by animals, called a *saqiya*.

The *saqiya* (Arabic 'irrigator') (Figure 6.3) was one of the most effective, enduring, and widespread irrigation machines. In its most specific sense, 'saqiya' is the cogged gearing system that translates horizontal into vertical motion. However, 'saqiya' is often used to describe the entire machine driven by such a gearing system, e.g. small, compartmented norias and pot-garlands. Pot-garlands, also called *paternosters*, employed buckets or clay qadus jars fixed at intervals on a rope that attached to the vertical wheel. As the saqiya

[88] C. Dulière, *Mosaïques des portiques de la grande colonnade: section VII, 16–17* (Brussels, 1974), 26–7, 37–8, pls. XXII, XXV, LXII–LXIII, *contra* A. M. Watson, 'The Arab Agricultural Revolution and its Diffusion', *JEH* 34 (1974).

[89] Le Strange, *Palestine Under the Moslems*, 59–60. This region always has been the area in which such wheels were most utilized. Other Syrian examples are known from the Khabur and on the Euphrates but cannot be dated securely to any period.

[90] Weulersse, *L'Oronte*, 56.

[91] Geyer and Monchambert, *La basse vallée de l'Euphrate syrien*", 279; Lauffray, *Halabiyya-Zenobia I*, 52; D. Orssaud, 'Notice sur la céramique', in J. Lauffray (ed.), *Halabiyya-Zenobia II* (Paris, 1991), 123. On the saqiya in Palestine, see E. Ayalon, 'Typology and Chronology of Water-Wheel (saqiya) Pottery Pots from Israel (Methodological Research on Roman Period Hydraulic Techniques)', *Israel Exploration Journal* 50 3/4 (2000) 216–26.

[92] R. P. Harper, 'Neocaesareia—Qasrin—Dibsi Faraj', in J. C. Margueron (ed.), *Le Moyen Euphrate: Actes du Colloque de Strasbourg 10–12 mars 1977* (Strasbourg, 1980), 339 nos. 74–7.

Figure 6.3. Saqiya (from D. Roberts, *Egypt and Nubia*, 1842).

rotated, the cogs on the vertical wheel raised and lowered the pots in continuous motion.

The saqiya made a marked impact on agriculture. Single wells that formerly watered only a small area could, with a saqiya-driven machine, supply an area many times greater. Scholars of Muslim Spain, where such wheels were widespread, have stressed the dramatic effect the application of saqiya technology had upon agriculture, especially for microscale irrigation.[93] A single saqiya installation could lift water as high as 20 m, and irrigate 2–13 ha. The machine was thus ideal for the intensive farming of vineyards, orchards, and gardens.[94]

In discussion of the so-called 'Islamic Agricultural Revolution',[95] saqiya-driven installations receive much of the credit. Watson uses the case of Africa, where he can find no pre-Islamic evidence for 'certain types of norias, particularly those driven by animals', to argue for significant Islamic technological diffusion. Similarly, Glick stated of the Islamic period of saqiya-driven noria in Spain: 'Its impact on the countryside was immense.' Caro Bajo stated that the saqiya is a machine which, 'owing to its cheapness, has permitted more than any other, the development of the smallholding or familial exploitation [of the land]'.[96]

But the common belief that the saqiya is an Islamic invention is incorrect. Recently, Lewis has argued for a date of 230 BC for its invention.[97] Certainly by the fourth century AD the saqiya enjoyed wide currency in Egypt.[98] During Late Antiquity, the saqiya-driven, compartmented wheel or pot-garland is the *mēchanē* of the papyri.[99] There is also plenty of archaeological evidence confirming its widespread use. Forty-six sites in Oriens have yielded material remains of qadus jars or saqiya installations. In addition to Androna, Beroia (Aleppo), Dibsi Faraj (Neocaesarea?), and Zenobia, finds of qadus jars and

[93] P. Guichand, 'Review of T. F. Glick, *'Irrigation and Hydraulic Technology: Medieval Spain and Its Legacy''*, *Technology and Culture* 39/3 (1998), 180.

[94] J. P. Oleson, *Greek and Roman Mechanical Water-Lifting Devices: The History of a Technology* (Toronto, 1984), 363–9. See also M. Decker, 'Water into Wine: Trade and Technology in Late Antiquity', in L. Lavan, E. Zanini, and A. Sarantis (eds.), *Technology in Transition A.D. 300–650* (Leiden, 2007), 4.

[95] Watson, *Agricultural Innovation.*

[96] T. F. Glick, *Irrigation and Society in Medieval Valencia* (Cambridge, Mass., 1970), 180. The cheapness of the machines is to be seriously doubted; they involved hundreds of parts and required diligent maintenance.

[97] M. J. T. Lewis, *Millstone and Hammer: The Origins of Water Power* (Hull, 1997), 19–32.

[98] D. R. Hill, 'Engineering', in R. Rashed (ed.), *Encyclopedia of the History of Arab Science* (London, 1996), III.771; Oleson, *Water-Lifting Devices*, 373.

[99] L. E. Tacoma, 'Replacement Parts for an Irrigation Machine of the Divine House at Oxyrhynchus P. Columbia Inv. 83, October 12 AD 549 (?)', *ZPE* 120 (1998), 128–30.

saqiya installations occur in Byzantine-dated contexts at Tass, Epiphaneia, Sepphoris, Ramat Hanadiv, Kefar Manda, Jerash, and on Cyprus.[100] At the riparian sites qadus jar finds may imply the use of current-driven norias. However, inland sites, or those without perennial surface water, such as Androna, indicate the use of pot-garlands to lift water from desert wells or qanats.

If, as the evidence suggests, the saqiya-drive and its attendant machines were a regular feature of the agricultural landscape of Oriens in Late Antiquity, this must have substantially increased farmers' capacity to irrigate their land, improving both productivity and the extent of cultivation. Especially in the heavily worked coastal orchards and vineyards, the data suggest the likelihood of a ribbon of heavily watered orchards, akin to the *huertas*, the well-known, intensively farmed plots of medieval Spain, in which cash crops were the focus. Several of the qadus jars were found closely associated with vineyards, such as that at Ramat Hanadiv. Indeed, at Akhziv, qadus jars were found in a winepress in the vineyard that they irrigated. Similarly, the saqiya installation at Zur Natan was part of a complex in which both a winepress and oil-press were major features.[101] Given the flourishing wine trade, the adoption of irrigated viticulture, which helped to ensure stable yield minimums and overall increases in production, is unsurprising.

CONCLUSION

In Late Antiquity, as in all periods, water was of paramount concern. This was true especially in the marginal zones where settlement continued to grow and outpace the natural water resources. The riparian zones, along the semi-arid Euphrates corridor and in the band of territory along the Syrian Desert and south to the Negev, were areas where precipitation was either scant, or unreliable, or both. In response to agricultural needs and the settlement of drylands, Byzantine-period farmers practised irrigation at a pitch and to an extent far greater than previously recognized.

Hydraulic management bore enormous significance for agriculture and its expansion. Throughout this period of increasing population and intensification, irrigation provided one of the few avenues, and certainly the most effective one, open to farmers to sustain crops or enhance yields. Kilometres

[100] Decker, *Agricultural Production and Trade in Oriens*, Appendix 3.

[101] E. Matthews, W. Neidinger, and E. Ayalon, 'Preliminary Report on the Excavations at Zur Natan' (unpublished excavation report), 12.

of wadi dams harvested rainfall and harnessed floods. From the Tigris to the Jordan, rivers were canalized or ladled up by machine. Aqueducts watered not only cities, but gardens. In the desert, where there was neither rain nor river to nurture the grain and olive trees, the storehouse of water underground yielded to the qanat. In southern Palestine, where cultivators had no access to perennial water supplies, hydraulic engineering extended through the use of elaborate systems to harness and preserve storm water. Agriculturists had worked the Negev floodwater farms since the Early Roman period, when the Nabataeans created a series of artificial oases in the desert. But the Byzantine period witnessed peak attainments in the numbers of farms, the extent of cultivated land and population, and the vibrancy of village life.

The widespread application of canal and qanat technology by late antique landowners and engineers allowed exploitation by sedentary agriculturists of a broad band of territory running from the Euphrates to the Hauran, on a scale unknown before and matched only today through the use of mechanized pumps. Much of this achievement was founded on the work and technology of earlier times, but the evidence strongly suggests that it was in Late Antiquity that agricultural exploitation was at its most extensive and intensive. Indeed in some areas, such as the Leja east of the Jebel Hauran, late antique farmers expanded into areas which modern agriculture has yet to reclaim. Settlement reached a peak under the later empire, based largely on the skilful application of water technology.

Although scholars of Islamic irrigation have rightly stressed saqiya-driven machinery's role in the advance of agriculture, they have ignored the evidence of the machine's heavy use in the Byzantine period. The landscape throughout Oriens, especially in the semi-arid regions, was a patchwork of ghūtas, pockets where agriculture, already intensified in many areas, was extended and developed. The desert steppe was in full retreat under a steady onslaught of water and plants.

7

Mixed Farming and Limited Specialization

Methods and Means of Intensification

Throughout this work I have noted that expanding settlement in many regions of the Levant in Late Antiquity has left a strong trace in the archaeological record. I have also stressed that agrarian expansion and intensification were closely linked to population pressure. Intensification involves increasing the input of labour or capital investment into the countryside in order to elevate the production of the land from its previous level, which in turn allows land to support a larger population.[1] One of the general prerequisites for an intensive farming system is its closed nature; new resources and space are non-existent or highly limited, encouraging instead the rise of intensive patterns of land management. The dynamic of increasing population and the supply of arable land was constantly shifting.

There were clearly social mechanisms in play. For instance, the supremacy of an estate in a given area would encroach on the acquisitive capabilities and the room to manoeuvre amongst the local smallholders and tenants. If the supremacy of an estate increased, tenants and smallholders were faced with difficulty. Emigration and intensification were likely responses. Likewise, any increase in the demands of taxation may have influenced the conversion from conventional agricultural practices—such as monocropping or cultivating an orchard space with fruit crops alone—to more intensive forms of land use. Nor can we ignore the market, a complex forum dominated by the elites and the state, but where those of middling and lower wealth also participated. The forces of increasing population, an ascendant, profit-driven, and land-hungry aristocracy, a grasping state, and a thriving market drove intensification, expansion, and specialization.

MIXED FARMING AND THE MEDITERRANEAN AGRICULTURAL REGIME

As has been demonstrated in Chapter 3, cereal monoculture was not an ideal form of exploitation in the Mediterranean, and was probably rarely employed.

[1] On theories of population pressures and the dynamics of agriculture change, the classic work is that of E. Boserup, *The Conditions of Agricultural Growth* (Chicago, 1965).

Broad production of one crop seriously lessened one's ability to spread risk. The farming system common in Mediterranean Antiquity, as in many regions of the world today, was 'mixed farming'. There are numerous types of this. But two regimes within mixed farming emerge clearly from the texts and material evidence as the most important for the late antique East: first, the production of various complementary crop types on the same land area, either simultaneously or in rotation; and second, the integration of animal husbandry with crop production. In order to render them fruitful, marginal landscapes required the devotion of heavy resources of labour, varied cropping strategies, and sometimes input of capital. Perhaps the most spectacular examples of the successful implementation of highly developed mixed farming regimes, combined with a strong specialized dependence on cash crops, are witnessed in the late antique Golan, the Limestone Massif, and the Negev.

Farming the Heights: Intensification and Mixed Farming in the Golan and Limestone Massif

The Golan comprises rugged, volcanic upland terrain, interspersed with a few fertile valleys around the northern reaches of the Jordan watershed. The Lower Golan, centred on Paneas (Banias–the early Roman Caesarea Philippi) receives abundant rainfall (700–900 mm) and had access to perennial water sources; whereas rainfall and water resources decrease as one proceeds east through the Upper Golan, where hills rise to 700–1000 m in elevation. From the Roman period onwards, Jewish, pagan, and Christian communities developed the agricultural potential of the region. Settlement steadily intensified. From the Roman period, 143 sites are known, and in Late Antiquity population apparently peaked: 184 Byzantine sites have been identified in survey.[2] The Golan was predominantly a land of villages and farmsteads: Urman's data include five very large settlements whose remains are over 12 ha in size; 14 settlements of 1.4–12 ha, which he terms 'towns'; 28 of 0.2–0.4 ha, called 'villages'; and 87 sites under 0.2 ha. In addition, he recorded 51 other sites of unspecified size but belonging to the Roman–Byzantine era.

The large settlements of the Golan were all in the lowlands. Paneas lay near the sources of the Jordan, not far from the foot of the Massyas valley. Hippos lay near Lake Tiberias, as did Hammat Gader. Gamala (al-Jamla), a large Jewish village famous for its role in the First Jewish Revolt against Rome during the first century AD, was situated on a tributary of the river Yarmuk,

[2] Ma'oz, 'Golan'. As noted above, this figure probably includes many sites that also lasted into the Early Islamic period. Urman, *The Golan*, 185–209.

around 20 km east of Lake Tiberias.[3] The final large site was Gaulon, which Eusebius (*Onom.* 64.8) noted as a large village (*komē megistē*), and Urman identifies with the modern Tell Jokhadhar.[4] The upland region, on the other hand, was a landscape of villages and dispersed settlement.

Jewish and Syrian settlers transformed the Golan from its previous state as a robbers'-nest to a thickly settled, agricultural landscape.[5] As so often, water management was the key to the enterprise. It was in the central heights of the Upper Golan that many dramatic landscape changes took place; the clearing of the forest cover there was already under way in the second and third centuries, but the fourth through seventh centuries witnessed a further marked decline in forest cover and the concurrent expansion of arable farming and stock-raising. In the Upper Golan, cereal production and stock-raising were the major agrarian activities. The volcanic rock in the uplands was generally impermeable, and the depressions in which water gathered formed *jibab* (sg. *jub*), natural reservoirs that had been utilized and expanded from the earliest times. At Sukeik, where the terrain and the finds of cereal processing installations suggest a focus on cereal culture, the inhabitants supplemented the natural lava reservoirs with numerous cisterns to supply water to livestock and crops.[6] Elsewhere in the Golan, farmers also exploited perennial water resources as never before; of the known Byzantine sites, 141 were located near springs, and many others were sited on other perennial water sources.[7] The concern for access to water and agricultural land was therefore paramount, and of far greater importance than other factors like security.

As in neighbouring Galilee and the Limestone Massif of North Syria, olive oil production formed a core part of the economy. This is especially true of the Lower Golan, where the majority of the 58 oil installations recorded by Urman were located.[8] As demonstrated by the numerous sites with multiple presses, oil was almost certainly produced, not just for local consumption, but also as a cash crop for regional and long-distance exchange. Late antique sites with multiple oil-presses include the villages of al-Manshiyye, 'Ain al-Taruk, Kafr al-Ma, Skufiyya, al-'Al, and Tlel.[9] The Lower Golan remained fertile into the medieval period: in the tenth century, al-Muqaddasi referred to the place

[3] On Gamala, see Josephus, *BJ* 3.3.
[4] Urman, *The Golan*, 88.
[5] Strabo (16.2.20) details the raiding activities of roving Arabs and Ituraeans.
[6] C. Dauphin, *La Palestine byzantine: peuplement et populations* (Oxford, 1998), 89–90.
[7] Urman, *The Golan*, 80.
[8] Ibid., 80ff.
[9] Al-Manshiyye: Urman, ibid., 202, site 146; 'Ain al-Taruk: ibid., 205, site 173; Kafr el-Ma: ibid., 202, site 174; Skufiyya: ibid., 206, site 177; Tlel: Dauphin, *La Palestine byzantine*, 90, 650, 53, 56, 58; Urman, *The Golan*, 208.

as the 'granary of Damascus'. However, the agrarian development of the Upper Golan proved ephemeral.[10]

The Limestone Massif (Massif of Belus), like the Golan, was a rugged land that received scant water and possessed marginal soil conditions. The only significant areas of unbroken land occur in the small valleys between the hills or on the floor of the nearby plains. Amidst the heavily degraded limestone, there are, however, pockets of good *terra rossa* soil. This agriculturally lack-lustre environment witnessed brisk growth in the fourth–seventh centuries when its inhabitants left an indelible mark in the archaeological record. More than 700 village settlements remain from Late Antiquity, attesting to an agricultural boom that in turn reflects the determination and abilities of Roman- and Byzantine-period Syrian cultivators. Given the rocky landscape and density of the massif villages, many of which are barely 1 km distant from one another, extensive cereal production was out of the question, and the region must have depended for the greater part of its cash and evident wealth on arboriculture and vineyards.

There is evidence of a flourishing and very diverse agricultural dimension to village life throughout the region, alongside more specialized production. Villagers worked gardens near to their homes, evidenced by archaeological remains of their walled enclosures. The dense scatter of winepresses, oil-presses, and animal troughs show the dominance of a mixed regime, where animal husbandry was combined with wine and oil production.

These regions exhibit 'limited specialization', in which agrarian production hinges on the production of cash crops with other products designed to mitigate risk and maximize output. In such a situation, plants and animals are raised in whatever way they can best complement cash crops within the farmed environment, or exploit ecological niches outside the cultivated zones.

Wine, Oil, and the Intensively Farmed Landscape

That wine and oil were produced within the limited lands of the small farm is well attested. Far to the south of the Massif of Belus, at Nahal Ela in the Negev, archaeologists recorded a farmstead with a winepress that exploited an area of a meagre 6 iugera.[11] The North Syrian village of Behyo preserves 37 presses; nearly all of them of the roller-type in which stone cylinders were used to process wine quickly. Given the large number of installations, the surplus was probably intended for market. These crushers served a village hinterland of approximately 250 ha (1,000 iugera), shared among perhaps as many as 77

[10] Al-Muqaddasi, *Best Divisions*, trans. Collins, p. 147.

[11] Haiman, *Map of Mizpé Ramon—Southwest (200)*, Site 59.

families. The average area per family, 3.2 ha, is far short of that required to sustain them, if grain alone was grown.[12] Likewise, we have already noted the finds of oil-presses in the village of Kafr Nabo and the thousands of others spread throughout other villages like it, particularly in the Limestone Massif, Negev, and Golan.

Perhaps more than any other crops, the relics of wine and oil exploitation provide the greatest insight into mixed farming and the methods which farmers used to enhance and extend the cultivated landscape. As we have seen in earlier chapters, both vine and olive were quintessentially Mediterranean crops and part of the basic diet. These plants were staples, but they also had deep social importance, and universal demand made them potentially highly profitable. As an intoxicant, wine had an innate appeal that deepened its value. Its status in trade was further enhanced by a developed cultural framework, in which the appreciation of wine was socially significant. Similarly, olive oil was an emblem of Mediterranean, especially Roman, cultural tastes, and it filled numerous needs in medicine, cosmetics, and industry. Both wine and oil exhibited considerable variation in taste and quality, and could be appreciably altered during processing, which added a further dimension to their potential marketability.

Since remote Antiquity, both the vine and olive had been bred or adapted to myriad conditions around the Mediterranean world, and they were available to farmers in the guise of hundreds, perhaps thousands, of cultivars. Each locale had dozens of varieties from which to choose. The modifications engendered by breeding, and the vigour of a developed marketplace in the late antique world, meant that grape and olives frequently offered better prospects for subsistence and profit than did cereals.[13] In short, the vine and olive were plants ideally suited for intensification and integrated farming practices.

The demand for wine and oil was also far more elastic than that for cereals and carried a greater risk. But with the risks came prospective reward in the market. A great deal of the activity in the late antique landscape was directed toward capturing this reward. By way of a quantified study of wine and oil production, I will examine methods of intensive production. The aim is twofold: first, to compare the yield potentials of wine and oil in the light of

[12] Callot, *Huileries*, 120, argues that there are only about 10 presses at Behyo, but as mentioned above, the roller installations in the north are most often best interpreted as wineries, and therefore I have followed Tchalenko's initial assessment of 37 presses at Behyo. Tchalenko, *Villages*, II.360. Behyo possesses the remains of 26 houses with 77 rooms. Tate, 'Les campagnes de la Syrie du nord', 207.

[13] Market conditions are discussed in Chapter 8.

subsistence requirements, and second, to determine the market value of any surpluses.

Palladius (*Op. Ag.* 2.12) shows that farmers often conceived of vineyard units as sections of 180 × 180 feet, an area equal to a iugerum of 32,400 square (Roman) feet, somewhat larger than the 28,800 square feet of the standard iugerum. The calculations that follow are based on the productive area of one of these plots of 28,800 square feet. Earlier data provided by Columella allow us to posit that an absolute minimum crop of 1 *culleus* (around 540 litres) of wine was expected from such an area. But yield was paramount: a vineyard that returned less than 3 cullei should, in Columella's opinion, be uprooted. The same concern with the sheer output in volume is expressed in the *Geoponica* (4.1; 5.2). We will choose the middle ground between Columella's absolute minimum and the minimum yield he personally considered necessary: 2 cullei per iugerum. Using this estimate, Columella's model vineyard plot of 7 iugera (1.75 ha) would have yielded at least 14 cullei of wine (circa 7,560 litres).[14] Given that a family of six annually required 1,095 litres, subsistence needs perhaps accounted for less than 20 per cent of the minimum proposed production of a 1-iugerum vineyard.[15]

We can thus estimate the production of both the Kurnub Farm and Michael's Farm, both in the Negev (discussed in Chapter 3 above), had they been exploited exclusively as vineyards. Labour requirements on vineyards were greater than those demanded for grain, and, according to early Roman authorities, one person was required for each 8 iugera of vineyard.[16] The small size of Michael's Farm (2.8 iugera) would likely have required no outside labour. The Kurnub Farm, which was 40–48 iugera in extent, was, as I mentioned above, about the size of farm that the Roman authorities proposed to give to discharged veterans; and thus it may indicate a holding considered suitable for an upper-tier smallholder. A vineyard of this size would have needed seven additional workers. As vinedressers possessed more skill than common labourers, they would likely have earned more than the labourer's wage of 3 solidi per year. Assuming therefore that the vineyard workers received 5 solidi per year, and owners paid rent of 10 per cent and taxes (noted in Chapter 3), we can approach a net value of the wine crop (see Table 7.1).

Whilst the prices of wine in Late Antiquity varied considerably, the data allow for plausible reconstructions of value. Sperber noted that the monetary relationship between bread, meat, and wine was approximately the following:

[14] Columella's 7-iugera vineyard's production of 14 cullei is equivalent to about 7,560 litres or 13,835 sextarii.

[15] See above, Chapter 4, for estimated wine consumption.

[16] R. Duncan-Jones, *Structure and Scale in the Roman Economy* (Cambridge, 1990), 206 tbl. 49. The necessary deduction for rent and taxes is assumed, as noted above in Chapter 3.

Table 7.1. Yields and major expenses of vineyards[2]

	Harvest sextarii/litres	Rent[1] sextarii	Subsistence sextarii	Labour costs	Taxes	Net yield sextarii/litres
Michael's Farm 2.8 iugera	5387/3035	539	1095	0	7 keratia or 168 sextarii	3585/1992
Kurnub Farm 40–48 iugera	76960/43360	7696	1095	25 solidi 7200 sextarii	8 solidi or 2304 sextarii	58665/32592

[1] Rent and subsistence are converted to cash in these calculations.
[2] Values calculated based on production of common wine.

8 denarii in the third–fourth century would buy one pound of meat, or a loaf of bread, or a sextarius of wine.[17] Diocletian's Price Edict offers three maximum prices for wines of type: first-quality one-year-old wine, second-quality one-year-old wine, and ordinary wine (*vini rustici*). The prices given are 24, 16, and 8 denarii respectively.[18] This set of figures does not reflect actual prices, but the cost relationship among the products is important. Since prices in the fourth through seventh centuries apparently remained fairly stable, a sixth-century price of 1 follis per sextarius is reasonable for common wine, while Aminean wine, fixed in the Price Edict at 30 denarii per sextarius, would have cost nearly four times as much as ordinary wine. If we convert these fourth-century ratios to sixth-century currency, 288 sextarii of common wine would have cost 1 solidus. The difference in cash potential between the two types, expressed in sixth-century gold currency, is reflected in Table 7.2.

Regardless of the type of grapes grown, vineyards offered a much greater potential return than did cereals. The yield from Michael's Farm could have produced wine that, once sold, would have generated enough cash to purchase the grain needed for subsistence (at a cost of approximately 4–8⅓

Table 7.2. Net value of wine crops

	Net Yield	Value (common wine)	Value (speciality wine)
Michael's Farm	3570 sextarii	12 solidi	45 solidi
Kurnub Farm	58665 sextarii	204 solidi	765 solidi

[17] Sperber, *Money and Prices*, 114–15. As I have noted earlier, the archive of the fourth-century traveller to Antioch, Theophanes (P.Ryl. 630–7) conforms with this view. It records very similar prices for food purchased at Antioch, with meat purchased for an average of 368.5 drachmas per pound, and wine at 350 drachmas per sextarius.

[18] Lauffer, *Diokletians Preisedikt*, 102.

solidi). In light of this, there is a strong likelihood that many small-scale producers depended heavily on cash crops. Wine production allowed for sustainable agriculture on smaller land areas than grain farming alone.

As with wine, we lack detailed accounts from Oriens that treat the production and yield of olive plantations. Early Roman handbooks give only perfunctory advice on planting densities. Varro recommended a spacing between trees of 25–30 feet (7.5–9 m), whilst Columella (2.2.24) counselled that on rich land suitable for intercultivation, olives should be planted 60 feet (18 m) in one direction and 40 feet (12 m) in the other; in poor soil, where cropping between the rows was not feasible, 25 feet (7.5 m).[19] In groves where grain was also sown (intercultivated), the *Geoponica* recommended that olive rows be placed 50 cubits (30 m) from one another, but also tells us that some were spaced more densely and still had cereals sown in between (*Geop.* 9.6.5). The text does not specify the distance between trees within each row. In Tunisia, aerial photography has revealed centuriation, Roman field boundaries, and orchards; in some cases individual tree pits are evident.[20] On the basis of this evidence, Mattingly proposes a density of 2,000 trees per *centuria*, approximately 40 trees per hectare. In modern Galilee, also a major oil producer, 10 m intervals are left between trees today, a density of 100 trees per hectare.[21] Forbes's data on modern Greece indicate, in areas where rainfall is around 400 mm, a density of 90 trees per hectare.[22] The Mishnah (redacted *c.*AD 200) attests to a very high density in plantations: 10–11 trees per *dunam* or 128–140 trees per hectare, which supports Forbes's assertion that over-planting in Antiquity was common.[23] *Geoponica* (9.9.10) states that farmers generally pruned olive branches to a maximum of 10 cubits (6 m) in length, which also hints that trees in eastern orchards were sometimes spaced about 10 m apart from one another.

One must then allow for three different types of groves. The first is that in which intercultivation was practised, where the spacing of the trees was greater and the olive density therefore lower: 30–40 trees per hectare. The second is the thicker plantation, where intercultivation was less likely, with plantings of approximately 100 trees per hectare. Finally, there may also have been dense plantations with up to 144 smaller trees planted at 8 m intervals.

[19] White, *Roman Farming*, 124.

[20] Mattingly, 'Oil for Export?', 45.

[21] Frankel, *Wine and Oil Production*, 37.

[22] H. Forbes, 'The Ethnoarchaeological Approach to Ancient Greek Agriculture: Olive Cultivation as a Case Study', in B. Wells (ed.), *Agriculture in Ancient Greece: Proceedings of the Seventh International Symposium at the Swedish Institute at Athens, 16–17 May 1990* (Stockholm, 1992), 97.

[23] Ibid., 97–8; Safrai, *Economy*, 122.

Yield has been variously estimated. Mattingly proposes 4 kg of oil per tree in Antiquity for Roman North Africa. (Unirrigated trees in Israel today yield 7–30 kg per tree.) Amouretti proposes 3 kg per tree, while Safrai calculated that oil yields in the Roman period were 2.64 kg per tree, and Brun a low 1.4–2 kg per tree.[24] The range is thus a minimum of 1.4 kg per tree to an absolute maximum of 30 kg per tree. Due to numerous factors which influenced the olive's fruiting, average yields were inherently variable. As noted earlier, the olive tended toward biennial cropping. This may or may not have been mitigated somewhat by planting varied cultivars, pruning, manuring, or other unknowable factors. Nor can the short- or long-term effects of climate be factored easily into the equation. Taking into account the lowest and highest figures for fruiting and averaging them, we obtain a low yield figure of 3.85 kg/tree and a high yield figure of 9.75 kg/tree. Given the habit of dense eastern plantings and the resultant small tree sizes, the lower figure of 3.85 kg/tree is perhaps a reasonable estimate, and is therefore used in the calculations below.

The example of St Simeon, discussed in pp. 149–50, does provide information on the upper tier of exploitation: 12,000 trees represented a large investment and would have occupied between 100 and 400 ha. Using the lower end of our calculated yield as a starting-point, Simeon's plantation would have produced 46,200 kg (49,434 litres) of oil, whereas the small farm used in this study, Michael's Farm in the Negev, if planted with olives at a density of 100 trees/hectare, would have supported only 70 trees, with a total yield of 270 kg (about 289 litres) of oil. Similarly, under an olive monoculture regime, the 40 iugera (10 ha) Kurnub Farm would have contained 1,000 trees and yielded 3,850 kg (4,120 litres) of oil.

Before calculating the market value of potential oil surplus for Michael's Farm, we must determine subsistence requirements for the farmer and his family (here assumed to be six members in total). Based on comparison with modern Mediterranean figures, Mattingly argues that an average per capita annual consumption of 20 litres of olive oil is reasonable.[25] Amouretti reaches similar conclusions, calculating that an urban ancient Greek household required 28.5 litres (52.1 sextarii) per person annually, whilst a rural household might use only 15 litres per person.[26] Using the latter figure for the rural household occupying our hypothetical allotment, the six household members would require in total 90 litres (162 sextarii) of oil per year. Table 7.3 summarizes the yield and principal demands placed upon the gross product of each plot.

[24] For cited estimated yields see: Amouretti, *Le pain et l'huile*, 204; Brun, *L'oléiculture antique*, 279 n. 25; Frankel, *Wine and Oil Production*, 37; Heichelheim, 'Roman Syria', 136–7; Mattingly, 'Oil for Export?', 45; Safrai, *Economy*, 122; White, *Roman Farming*, 392.

[25] Mattingly, 'Oil for Export?', 34.

[26] Amouretti, *Le pain et 'l'huile*, 181–3, 95–6.

Table 7.3. Yield and expenses of olive farms[1]

	Yield (sextarii)	Rent (10% of crop in sextarii)	Taxes (1 solidus/ 225 trees)	Subsistence (sextarii)	Labour	Net yield (sextarii)
Michael's Farm	520	52	4 keratia (64 sextarii)	162	None	242
Kurnub Farm	7415	742	4 solidi (768 sextarii)	162	9 solidi[2] (1728 sextarii)	4015
Simeon's Farm	90539	—	53 solidi (= 10176 sextarii)	unknown	none[3]	80363

[1] Oil-*solidus* values are based upon value in common oil of approximately 1 *follis* per sextarius.

[2] Assuming two full-time labourers working at a rate of 100 kg per day, harvest would take 39 days, well within the time available for the harvesting and processing season.

[3] Here monks provided the labour. If the olives were collected over a three-month period, at least 6 monks, working 80 days of the 90-day season, would be needed to harvest this olive crop.

Based on the available evidence, it appears that oil was the most expensive of the three major commodities which comprise this study. Certainly olive oil was significantly more costly than wine, ranging in price from approximately 1½; to 5 folles per sextarius, depending on local supply and the oil's quality.[27] If we consider that a sixth-century monk ate for 16 *nummi* (1/200 solidus) per day, this figure appears reasonable. In a normal diet, oil was likely used primarily as a dip; thus, a sextarius jar of common oil would likely last for several days. It is clear that oil was cheap enough to be used daily by many, but expensive enough to be denied to the poor during difficult times. In Leontius' *vita* of John the Almsgiver, the saint agonized over how much good he might do if he sold a valuable garment and used the money for the poor. 'How many are there who go one month and even two without tasting oil?' he

[27] Sperber, *Money and Prices*, 116, notes that, according to M Pe'ah 8.5, oil was approximately twice as costly as wine. According to Diocletian's Edict on Maximum Prices, oil ranged from 12 denarii per sextarius for ordinary quality to 40 denarii per sextarius for green oil. As common wine cost 8 denarii per sextarius, common oil was one-and-a-half times more expensive than common wine. Recorded prices from Egypt suggest an even higher differential in price: while 40–48 sextarii of oil could be bought for 1 solidus, the same price purchased 200 sextarii of wine. Jones, *Later Roman Empire*, 446–7. *CTh* 8.4.17 (AD 389 to the Praetorian Prefect of Oriens) states that *annona militaris* oil was purchased at 80 lb (approximately 52 sextarii) per solidus. The fact that a military solidus could purchase rather more oil (52 sextarii) than a civilian one (40–48 sextarii) is unsurprising, since military rations were always procured at a reduced price. B. H. Isaac, *The Limits of Empire: The Roman Army in the East* (rev. edn., Oxford, 1992), 286–7; A. Segrè, 'Essays on Byzantine Economic History, I: The *Annona Civica* and the *Annona Militaris*', *Byzantion* 16 (1942–3), 409.

Table 7.4. Approximated net value of oil crops

Farm	Net yield (sextarii)	Value as common oil (@ 1.5 folles /sextarius)	Value as first-quality oil (@ 5 1/2 folles /sextarius)
Michael's Farm	242	1 solidus, 75 folles	4 solidi, 7 keratia
Kurnub	4015	21 solidi	77 solidi
Simeon's Farm	80363	418 1/2 solidi	1535 solidi

lamented.[28] In Egypt, particularly, olive oil was generally an import and must have been the most expensive component of the common diet.

The value of the crops as calculated above may be approximated. As with wine, the value of the surplus would have varied considerably, depending on the quality of oil produced. Table 7.4 provides the minimum and maximum range for the crop value, the low figure, if common oil were produced, and the higher, if the orchard produced exclusively first-quality green (*omphakinos*) oil.

Our calculations are in themselves fairly conservative estimates of total possible income; they say nothing of other possibilities for profitable exploitation alongside the olive trees. But they do show that our smallest land unit under oleiculture could have met subsistence requirements. Given that the monetary value of grain required to feed a family ranged from 3 to 6 solidi, even a smallholding the size of Michael's Farm could have covered most of its grain needs through the profits from the sale of green (*omphakinos*) olive oil. Admittedly, the same land area planted in vines would have been much more profitable. The potential oil surplus of Michael's Farm's had a maximum value of approximately 4 3/4 solidi, whilst the same plot under viticulture produced three to ten times this amount (see Table 7.2).

Naturally, the greatest opportunities for gain from olives and other crops lay in the larger holdings; Simeon most of all would have reaped a windfall profit, as his direct labour costs were negligible. He employed monks in the harvest and therefore incurred no direct expenses for the most labour-intensive aspects of his olive enterprise.[29] The 418–1,535 solidi he received from his 12,000 trees alone would have made him a relatively well-to-do landowner, and the saint's monastic holdings were far more extensive than the olive groves. It is worth further noting that Simeon could have accommodated his entire olive grove on only 120 ha (480 iugera). Between the extremes of great and small, were farms such as that at the Wadi Kurnub. Despite the

[28] Leontius, ed. Festugière and Rydén, 19.35: '*πόcοι ἀρα ἔχουcιν ὅλον τὸν μῆνα ἤ και δύο μη γευcάμενοι ελαίον*'.

[29] Palmer, *Monk and Mason on the Tigris Frontier*, 164.

relatively small size of these properties, the returns from olives and wine were potentially substantial. If laid out for viticulture, many of these modest places could have provided livelihoods a long way above subsistence.

The Arbustum and Intensive Practices

The basic methods of vineyard production have been detailed in Chapter 4. What remains is to highlight the specifics of grape production within an intensive farming regime. To do so, we turn once again to the arbustum, the practice of growing grape vines on trees. The use of the arbustum is far older than Late Antiquity. The introduction of vines into the orchard illustrates three central aspects of mixed farming and its aims. First, it represented an attempt to diversify the farmed space of the orchard. Viewed as a closed system, the orchard is a microcosmic monoculture; the addition of vines into the system transformed the space to a mixed environment with the benefit of mitigating risk. Second, the vine was seen as a companion to the tree crops, and third, a substantial benefit of the arbustum was its elimination of the need for vine props. Tree-grown vines offered an alternative to the growth of vines unsupported by props, which were highly productive, but since they lay on the ground, were especially prey to vermin and rot.[30]

The creation of the arbustum required a good deal of planning, since trees on which vines were grown were started years in advance of the vineyard. Species were chosen that lacked excessive roots and canopy. The white and black poplar, ash, maple, and cherry (*Geop.* 4.1) were natural choices, and the *Nabatean Agriculture* adds to this list sycamore, stone pine, pomegranate, quince, apple, and olive as suitable varieties.[31] In good soil the trees were left unchecked. In Bithynia (*Geop.* 4.1.3–4), trees in the arbustum commonly grew to 12.5 m in height and sometimes reached 20 m. Poor soil, however, demanded that trees be pruned down to approximately 2.5 m, as smaller trees placed less burden on its fragile resources.[32] The distance between trees is

30 The nutrients and light taken by the vine in competition was considered to have no ill-effect on the yield of trees with which it was grown. Wood-bearing trees, such as elm, ash, and poplar, produced a whole series of products for the farm, including building materials, twigs for fuel, and withies for basketry and other implements. These products did have some market value, but in all likelihood the addition of vines gave a greater likelihood of a cash return.

31 The elm was particularly popular. M. A. O. Nemesianus and H. J. Williams, *The Eclogues and Cynegetica of Nemesianus* (Leiden, 1986), 473; Salinger, 'Kitab al-Filaha an-Nabatiya', 96, 99, 100. With the exception of the olive, all the aforementioned trees grow relatively quickly, especially the poplar. Trees in the arbustum had the added benefit of providing building material and fuel throughout the life of the vineyard. Poplars, in particular, grow extensive superfluous branches that would have yielded valuable wood for fuel after only a few years.

32 This also would make the harvest much easier, as many of the grape clusters were accessible from the ground.

expressed by Palladius (*Op. Ag.* 3.10.5), who specified: 'If it is a grain field where you place your vineyard, leave forty feet [12.3 m] between the trees, that it may be planted; in poor ground [leave] twenty feet [6.1 m]'.[33] Spacing and height were important: the larger the trees, the greater the size of the vines and their yield.

The arbustum represents a strategy of adaptation that was equally suited to the small-scale farmer or the estate. *The Geoponica* (4.1.1) makes it clear, however, that the arbustum was an important part of estate agriculture, and it is commended for the quality and durability of the wine that it produced. The same passage states that space between the trees could be sown within two years of creating the arbustum. The integration of trees, vines, and grain in one plot was thus an important concern for certain late antique estate owners whose first aim was commercial.

Sustainability, Intensification, and Mixed Farming: Animal Husbandry, Crop Rotation, and Agricultural Development

As illustrated by the integration of intensive practices in the creation of the arbustum, where space might incorporate orchard, vineyard, and grain field, mixed farming (or integrated farming) represents a broad category of practices rather than a specific method of managing the landscapes. It may, for example, include the growing of different plants within the same field, the growing of different plants in rotation in order to support animals, or the use of farmland and grazing in an integrated manner. Like many of the methods discussed here, mixed farming was certainly not an innovation of Late Antiquity, but there are strong indications that there were improvements or modifications in practice during this period that were economically significant. What must be stressed is that mixed farming approaches do not preclude specialization. On the contrary, careful integration of crops and livestock may be a vital component of intensification on the one hand, and of specialization on the other. Today integrated farming is often a symptom of bad economic conditions and the overuse of land, and is a poverty-induced option.[34] Poverty, however, need not have been the driving force encouraging mixed farming within the landscape of late antique Oriens; indeed the bulk of material evidence strongly suggests that it was not.

[33] *Sed si ager frumentarius fuerit ubi arbusta disponsis, quadragenos pedes inter arbores relinque, ut serri possit, in exili autem vicenos.*

[34] H. Schiere and L. Kater, *Mixed Crop–Livestock Farming: A Review of Traditional Technologies Based on Literature and Field Experiences* (Rome, 2001), 1–2.

Animals were, and continue to be, a major component of many mixed farming systems. In those marginal regions that I have examined, the Negev, Golan, and Limestone Massif, the archaeological record preserves abundant late antique material relating to animal husbandry, mainly in the form of sheepfolds, campsites, stables, troughs, and cisterns.[35] In order to see how these remains originally functioned within the framework of mixed farming, one must understand the zones of exploitation in the farmed environment. Simply stated, these may be divided into infield and outfield regions. The former represent spaces within 1 km of the village with the most intensively worked landscapes, whereas those in the outfields include more distant, less intensively worked landscapes. Where settlement is particularly dense, the entire territory of a given village might not extend beyond 1 km: the lands of the whole village are, in essence, 'infield' lands.

In the Limestone Massif, for example, sheep and goats were integrated into the agricultural regime primarily in two ways: first, by controlled grazing in the 'infield' land, and second, by less thorough grazing of the 'outfield' lands beyond. In return for wages in cash or kind, shepherds and cowherds from the local community watched over the animals for a number of different farmers; this was common practice in Antiquity and is later attested in the seventh–eighth-century 'Farmer's Law' (*Nomos Georgikos* §§23–9). Animals grazed the herbs and grasses among the olives and were watered from the innumerable cisterns found in the villages. They were folded, or kept in stables at night, where their dung accumulated and could be conveniently collected and used on the fields and the kitchen gardens of the infield area.[36] Additionally, while the animals grazed amongst the trees, they deposited dung that fertilized the orchards. Given that livestock pens and troughs are ubiquitous throughout the villages of the Golan, Negev, and Limestone Massif, infield grazing was certainly present. This arrangement was complemented by more extensive, outfield exploitation. In the case of the tightly packed villages of North Syria, the only 'outfield' was wasteland and steppe located at a considerable distance from the village territory proper. In the Negev, however, or the Golan, where less-dense settlement allowed wider movement of herds, waste areas were much greater in extent and much closer to hand. Whatever the region, exploiting the outfield lands meant herding flocks from the immediate environs of the village to outlying fields, to hilly wasteland, or to the steppe lands at the feet of the jebels. A transhumant model was also possible, with the

[35] Haiman, 'Agriculture and Nomad–State Relations'; Tate, *Campagnes*, 243–56; Urman, *The Golan*, 147.

[36] B. Rosen and I. Finklestein, 'Subsistence Patterns, Carrying Capacity and Settlement Oscillations in the Negev Highlands', *PEQ*, 124 (1992); Tate, *Campagnes*, 42.

flocks grazing away from the villages for extended periods of time, perhaps under the care of pastoralists who moved through the region on a seasonal basis.

Animal dung deposited in stables and on fields provided a source of fertilizer and fuel, a major advantage and one of the primary reasons stock often formed part of intensive agricultural regimes. Goats especially were able to convert shrubs and trees into energy; their browsing abilities made them ideally adapted to marginal landscapes where such vegetation, like the lower shoots of olive trees, predominated. With their ability to eat and eliminate shrubby and spikey plants, sheep and goats were also adept at clearing overgrown places in preparation for arable farming. Once the grain had been harvested, there was abundant stubble on which to feed them, and their grazing was welcomed by farmers who wanted their fields to be fertilized. Thus, sheep and goats fit well into the system, eating shrubbery, herbs, stubble, and weeds and, in return, providing wool, hides, milk, and meat. Since they were mobile and could be driven to market where their meat and hides were sold, flocks also offered an easy means of transporting valuable commodities to the cities.

The model just described is one form of animal and crop husbandry that fits well with the archaeological data collected from the most intensively settled regions of Oriens. From the agricultural handbooks and comparative study, other methods of using animals and crops together in an intensified agriculture manifest themselves. One is the use of what is presently termed 'ley farming'. Ley farming, or convertible husbandry, was a vital component of the enhancement of agriculture in northern Europe at the end of the Middle Ages, and its diffusion was a major factor of agricultural growth throughout much of Early Modern Europe. Ley farming involves the rotation of cropping and pastureland, in which several successive seasons of arable farming are followed by a number of years of rest (fallow). Ley lands are referred to in the handbooks as fallow, *novalis*, and their management, though most fully elaborated in the handbooks of the Early Roman period, continue to be detectable in the literature of the later Empire.[37] The *Geoponica*, for example, is probably referring to such long-fallowing when mentioning the *neos* land (*Geop.* 3.11.8) that was ploughed in August. Fallowing replenishes the nutrients needed by crops, such as phosphate and nitrogen, and, in one form or another, has long been recognized as a major feature of Roman farming. Only recently, however, have scholars begun to understand that the Romans utilized a sophisticated method of fallowing akin to the advanced systems of late medieval and early modern land management. Amongst the

[37] G. Kron, 'Roman Ley-Farming', *JRA* 13 (2000), discussing Varro, *De re rustica* 1.44.3; 2.7.11; 3.16.33; Columella, *De re rustica* 2.10.6; Virgil, *Ecl.* 1.70–1; Columella, *De re rustica* 7.3.9/19.

major features of Roman farming was the use of convertible husbandry. In the Roman system, *restibilis* lands were cultivated in successive years, followed by a period as ley lands.

In the pre-industrial era, when fertilizer was not abundant, the restoration of land through fallowing was vital to maintaining agriculture. At the same time, given the general shortage of land and the population pressures of Antiquity, farmers could ill afford to have land out of production for extended periods of time while it recovered for arable farming. Ley farming offered a solution to both difficulties. The introduction of animals into the farming system meant that fallow land could be converted into ley and restored to fertility. In its basic scheme, the system involved cropping of lands over several years using cereals or legumes, or some combination of the two. Following this, the land was allowed to revert to pasture, the ley component of the system. This pasture could consist of spontaneously generated indigenous grasses, herbs, and shrubs, but more often was a managed landscape in which fodder was sown and tended.

One of the keys to detecting and understanding the sophistication of the practice of ley farming in Late Antiquity is through textual references to the use of leguminous plants. Legumes, such as lupines, beans, and peas, were important components of the ancient diet. The use of legumes in conjunction with cereal crops was also critical to agricultural sustainability. Nitrogen, an essential element for production of cereal grains, is fixed into the soil via bacterial interaction with the root systems of leguminous plants. Of course the precise chemistry of this process was not understood, but the benefits to fertility of nitrogen fixation were explicitly recognized by ancient farmers, who frequently used legumes as 'green manure' to prepare for the cereal crop. As had their early Roman predecessors, both Palladius (*Op. Ag.* 1.6) and the *Geoponica* (2.39.6) demonstrate that legumes were sown, allowed to grow, and then ploughed under while still green in order to enrich the soil before grain was planted. Beyond this simple legume–cereal rotation, legumes were also a key feature within ley farmed systems of Late Antiquity.

The use of the *medic* (medicago) plant provides a glimpse into the ley farming practice that safeguarded soil fertility in the drylands of the eastern Mediterranean. As recently demonstrated by attempts to reintroduce these plants into regular cultivation there, medicago species are legumes ideally suited for the semi-arid dry-farming conditions of the Levant and North Africa.[38] One

[38] B. A. Chatterton and L. Chatterton, 'Medicago—Its Possible Rôle in Romano-Libyan Dryfarming and its Positive Rôle in Modern 'Dryfarming', *LibSt* 15 (1984); B. A. Chatterton and L. ChatterTon, 'A Hypothetical Answer to the Decline of the Granary of Rome?', *LibSt* 16 (1985); L. Chatterton, *Sustainable Dryland Farming: Combining Farmer Innovation and Medic Pasture in a Mediterranean Climate* (Cambridge, 1996).

common present-day dry-farming practice in Australia involves a medicago–cereal rotation. Around the time of the first autumn or winter rains, medic is sown. Through spring and summer, medic reaches full maturity and produces seed. Grazing is limited so that the plants may fully develop their seeds, and the seeds drop to the ground after maturation. In the following autumn, the grain crop is put into the field. Light ploughing ensures that the residual, hard-podded seeds of the medicago remain dormant, but near to the surface, as the grain crop matures and is harvested. After the grain is cropped, the residual medic seeds break open and germinate, allowing for the re-establishment of a rich medic pasture.

In antiquity, the process was much the same; the fact that ancient farmers utilized the light scratch-plough meant that the medicago seed was not buried too deep, so that it might germinate following the cereal crop. Medic rotation addressed several ancient farming concerns: first was the vital cereal crop needed for human subsistence; the second was the preservation of soil fertility through the use of nitrogen-fixing legumes; while the third element was the introduction of animal husbandry through convertible farming. In this last area, medic pastures provided rich grazing for livestock. Animals in turn added fertilizer and offered a way successfully to exploit the fallow, without infringing on the rather delicate restoration process needed to return the land to arable farming.

Medic varieties were an important and normal component of late antique agriculture, as indicated in both the material and written data. Archaeobotanical investigations have revealed ancient medicago seeds in cereal stores in Romano-Libyan contexts at Ghirza in the Tripolitanian pre-desert, on the late antique estate of Marinus at Raqit in Samaria, and in excavations of late antique deposits at Nessana in the Negev.[39] Likewise, references in Virgil's *Georgics* (215) and Palladius (*Op. Ag.* 3.6; 5.1) demonstrate not only that the plant was commonplace, but also that farmers understood that allowing the plant to shed its seed while it was grazed or utilized for hay was critical to replenishing the stand.[40] Based on the time of planting recommended in the *Geoponica* (3.2.4), which recommended a February sowing, it seems that medic was a normal part of the cropping cycle. More importantly, the handbook demonstrates the common use of medic (17.8.1 for cattle feed; 17.14.6 in veterinary medicine) and its seeds (16.9.4). In doing so it suggests both a full knowledge of the plant's life cycle, and a determination to cultivate

[39] Van der Veen and Hamilton-Dyer, 'A Life of Luxury in the Desert?', 45–8; Marinus: Dar, *Raqit*, 301; Negev: Mayerson, *Ancient Agricultural Regime*, 48.

[40] Palladius attests (5.1) that intensive pasturing and hay-making were performed under irrigation, and these irrigated pastures were then grazed by horses or cattle. Well managed, the pasture would last ten years.

it carefully. In the *Geoponica* there are indications that medic was part of a ley farming system. Sown in February (3.2.4), the plant would have matured and gone to seed in the summer, ready to be ploughed into the soil for an autumn planting of grain.

In the *Geoponica* (17.8.1; 18.2.6) medicago is sometimes paired in its usage with cytisus, a name that covers a large group of legumes. Cytisus varieties are generally fast-growing, hardy, shrubby legumes tolerant of poor soils; they are ideal for grazing in semi-arid, harsh landscapes. For these reasons, they are also attracting present-day agricultural interest.[41] That cytisus was often a companion crop of medicago in late antique Oriens once more suggests that late antique farmers sought plants that thrived on and improved poor earth. Their use also indicates long-term fallowing, during which time both medics and cytisus plants established, matured, and re-seeded themselves. In addition to their vouchsafed fertilizing properties mentioned in the handbooks, their links with animal husbandry are certain: the seeds were considered exceptional animal fodder. This exploration of particular plant types and their management reveals that there was clearly a conscious integration of animal husbandry with cereal culture, the aim of which was to sustain agriculture in the short term and over time.

It is worth considering the impact of medics and other leguminous plants on the landscape of the Levant in a period far removed from the present day, when similar plants are now regarded as potentially revolutionary. If the precise ways medic was used in Antiquity remain uncertain, the general place of legumes, and their widespread role as fertilizing plants and grazing material, are not in doubt. A seventh-century papyrus from Nessana (P.Ness. III.82), for example, records that 30 modii of vetch seed were sown and 97 returned from the fields of the Negev, perhaps in a ley-farming situation similar to that described above for medicago.

Intensive farming in Late Antiquity also included encroachment on the fallow period during the summer months. Summer cultivation is known from an earlier period: Columella (11.2.72) described millet as harvested in September, which implies a June or July sowing.[42] But in earlier Antiquity, conventional spring-sown cereal crops were not a prominent component of Mediterranean farming, since the summer season was too hot for them to thrive without irrigation. In the Classical period, it seems that many spring-sown crops were generally confined to cooler regions where autumn sown

[41] F. González-Andrés and J. María-Ortiz, 'Potential of Cytisus and Allied Genera (Genisteae: Fabaceae) as Forage Shrubs: 1. Seed Germination and Agronomy', *New Zealand Journal of Agricultural Research* 39 (1996). In some cases cytisus varieties, generally considered a weed, are being investigated as potential dry land crops.

[42] Spurr, 'Cultivation of Millet in Roman Italy', 7.

grains could not flourish. Spring-sown cereals typically were trimestrian, growing to harvest in just three months. Sometimes they were used as an emergency crop when the winter grains failed. Thus *Geoponica* (3.2.4) states that in February, one should sow three-month (*trimenaion*) wheat. In March, several types of wheat were sown, including, it seems, durum and Alexandrian (3.3.11).

It is tempting to see this 'Alexandrian wheat' as an altogether different grain from wheat, given that it was sown late (at the end of March) in light soils receiving full sunlight and in higher, and thus less moist, ground. A candidate for 'Alexandrian wheat' is sorghum, which thrives under these conditions, and is ideally suited to spring planting. Sorghum (*Sorghum bicolor*) is a member of the same family of grasses as millet and shares many of the latter's qualities, such as disease resistance and drought tolerance.

The wide range of sorghum finds from China to sub-Saharan Africa has defied attempts to determine the precise origin of the grain, but a starting-point in north-eastern Africa seems likely.[43] Though the early history of sorghum is disputed, Watson's claims that the diffusion of sorghum into the Mediterranean world on a significant scale was an Islamic endeavour appear to be false. A number of finds from the first centuries AD indicate that domesticated sorghum was a commonly cultivated crop in the Nile Valley and on the Red Sea coast of Egypt during the Roman–Byzantine period. Other finds, dating to the first centuries AD at the site of Qasr Ibrim in Nubia, indicate that domestic sorghum was an important crop.[44] Pliny mentioned a new type of *milium* about this time, and his reference bolsters the assumption that sorghum was a newcomer in the first century. In excavations at Berenike, an Early Roman period port on the Egyptian Red Sea coast occupied until the sixth century, sorghum has been found in some quantity during the first through fifth centuries. These finds perhaps represent local production at Berenike and nearby Wadi Shenshef or possibly import from the Nile Valley or abroad.[45] Sorghum grain has also been recovered at the Byzantine-period monastery of Kom el-Nana in northern Egypt.[46]

[43] P. Rowley-Conwy, O. J. Deakin, and C. H. Shaw, 'Ancient DNA from Archaeological Sorghum (*Sorghum bicolor*) from Qasr Ibrim Nubia: Implications for Domestication and Evolution', *Sahara* 9 (2000).

[44] P. Rowley-Conwy, 'Sorghum from Qasr Ibrim, Egyptian Nubia, *c.*800 BC–AD 1811: A Preliminary Study', in J. Renfrew (ed.), *New Light on Early Farming* (Edinburgh, 1991).

[45] R. T. J. Cappers, 'Archaeobotanical Remains', in S. E. Sidebotham and W. Wendrich (eds.), *Berenike 1995: Preliminary Report of the Excavations at Berenike (Egyptian Red Sea Coast) and the Survey of the Eastern Desert* (Leiden, 1996), 329; R. T. J. Cappers, 'Archaeobotanical Remains', in S. E. Sidebotham, W. Wendrich, and F. Aldsworth (eds.), *Berenike 1996: Report of the 1996 Excavations at Berenike (Egyptian Red Sea Coast) and the Survey of the Eastern Desert* (Leiden, 1998), 323–4.

[46] Rowley-Conwy, Deakin, and Shaw, 'Ancient DNA from Archaeological Sorghum', 28.

Due to the morphological similarities of millet and sorghum, it is possible that the late antique texts, including the sources of the *Geoponica*, mentioned sorghum under a name commonly taken to refer to millet or other grains. One such possibility is *kengchros*, generally translated as 'millet'. Similarly, *olyra* is generally equated with emmer wheat (discussed in Chapter 3 above). The high yield, mentioned in the *Geoponica* as characteristic of the grain, perhaps better describes sorghum, highly regarded for its prodigious production. The translators of a recent Spanish edition of the *Geoponica* (3.6) have indeed taken the mention of 'generous' olyra made into *tragum* (porridge), to refer to sorghum. André believed that both the Latin *milium* and *olyra* could have designated sorghum.[47] According to Cadell, who based her study on papyrological data, Greek olyra denoted sorghum.[48] Cadell concluded that olyra in many contexts was sorghum, and that the plant was widespread in Egyptian agriculture during the Roman period.[49] It is, of course, possible that in Late Antiquity, olyra signified different grains depending on the location, or that sorghum had simply assumed the name of a crop (emmer) that was now fading from the landscape, since it occupied the same niche as the crop that it replaced—namely, food for the poor.

The case of the Latin *milium* is more straightforward. While it is clear that milium signifies millet generally, it is an umbrella term, similar to the way 'wheat' today encompasses many distinct varieties. As mentioned, millet and sorghum share many characteristics and are often confused with one another.[50] The ancients regarded sorghum as a kind of millet, and therefore referred to both under the same imprecise word. In the Latin West, whence came the sixth-century *Piacenza Pilgrim*, sorghum was commonly viewed as a kind of exotic millet.[51] The *Piacenza Pilgrim* noted that 'millet' growing at Nazareth was 'abnormally tall, the stalks being bigger than the height of a man'.[52] Most millet varieties do not reach the height the *Pilgrim* described. Sorghum, its close relative, often does, and is depicted as such in a medieval Italian illustrated manuscript of the medical works of Ibn Butlan.[53] The *Pilgrim's* 'untrained eye' could easily have mistaken the unfamiliar sorghum plant for the more common millet. As we have already seen, there is ample archaeological evidence of sorghum being grown in Roman–Byzantine Egypt,

[47] J. André, *Les noms de plantes dans la Rome antique* (Paris, 1985).

[48] Ibid., 177.

[49] H. Cadell, 'Le vocabulaire de l'agriculture d'après les papyrus grecs d'Égypte: problèmes et voies de recherche', in D. H. Samuel (ed.), *Proceedings of the Twelfth International Congress of Papyrology* (Ann Arbor, 1970).

[50] Rowley-Conwy, 'Sorghum from Qasr Ibrim', 206.

[51] André, *L'alimentation*, 54.

[52] Wilkinson, *Jerusalem Pilgrims*, 81.

[53] Watson, *Agricultural Innovation*, 13.

and the recovery of a sorghum grain in the Byzantine fortress at En Boqeq in the Negev provides a trace, albeit a slender one, of its cultivation within our area in Late Antiquity.[54] The semi-arid climate, too harsh for dry-farming of wheat, was an environment to which sorghum was admirably suited and possibly led to its introduction there and its diffusion throughout Palestine.[55]

Much more palaeobotanical work is needed before we can adequately gauge the role of sorghum in the agrarian life of the late antique Levant. On the present evidence, it seems likely that the crop was fairly widespread. The plant apparently was brought from Egypt into Palestine from AD 100; it is probably attested in Palestine and Mesopotamia in the body of Rabbinic literature from the second century onward. As a new crop that spread over late antique Oriens, sorghum is emblematic of the efforts of that age to overcome the considerable challenges of climate and soil that confronted agriculture in the region. It shows conscious adaptation, willingness to experiment, and a perceived place for a grain that was both high-yielding and drought-resistant. Sorghum was one of many crops whose use in highly developed management, intercultivation, and rotational schemes left a prominent and durable mark in the agriculture of the day and was a component of the expansion into the fringes of the settled landscape.

The spring sowing season to which sorghum belonged, a planting time attested in the handbooks, such as the *Geoponica* (2.3), has important implications for late antique agriculture. It signifies the suppression of fallow land, and thus efforts to crop land continuously. In many cases, summer cereal cultivation also entailed the irrigation of plants throughout the hotter months and thus betokens widespread artificial watering, with all its attendant labour and equipment. The growing of well-adapted plants potentially extended not only the physical space that one could farm, but the temporal space as well. While sorghum represents a plant well-suited for management of lands during the scorching summer heat when other cereals would fail, it was relatively new to the lands of the eastern Roman empire. There were, however, other older summer cereals, whose importance scholars have generally failed to recognize.[56] Among these were certain fast-growing emmer varieties, which

[54] Lipschitz and Waisel, 'Dendroarchaeological Remains', 394.

[55] A final, intriguing scrap of evidence supports the view that sorghum was entrenched within late antique agriculture in the Levant. The sixth-century P.Ness. III.82.9 lists yields from fields, and mentions wheat, barley, and *arakos* (*ἄρακος*), an unknown plant that has been identified variously with emmer wheat and the weed wild chickling, but may instead represent sorghum. N. Jasny, 'Competition among Grains in Classical Antiquity', *AHR* 47/4 (1942), 126, 32–3, proposes emmer, whilst Liddell and Scott define *ἄρακος* as *Lathyrus annuusi*, wild chickling. Pliny (*H.N.* 18.92), however, equates the Latin word *arnica* (*ἄρακος*) with olyra.

[56] Such as Watson, whose insistence that the Arabs were the first to crop year round flies in the face of a vast quantity of ancient evidence; Watson, 'Arab Agricultural Revolution'.

apparently formed an important crop component in some cooler regions of the Mediterranean (Oribasius 1.2.7), such as the mountainous regions of the Taurus and Anti-Taurus or the Hermon, where the longer, colder winters shortened the growing season.

One exception to the rule that spring-sown wheats were restricted to cooler areas is the 'black wheat' (*sitos melanather*) of the *Geoponica* (3.3.11). This black wheat was a hardy, heat-tolerant crop that flourished throughout the eastern provinces as a spring cereal. It was sown in elevated, light land amongst vine trenches, a planting location which implies that it was viewed as an alternative to barley, and expected to yield comparably. Evidently *sitos melanather* was a late-sown variety cultivated opportunistically, when and where conditions favoured its production. Jasny suggested that black wheat was a variety of durum wheat, and a possible progenitor of one or several of the modern cultivars well-known throughout Greece and Anatolia, since these are also early ripening wheats sown in the spring.[57] Whatever the case, it appears that throughout the eastern Mediterranean of Late Antiquity, spring-sown wheats were widely available to farmers.[58] There were also spring-sown barleys, such as the trimestrian varieties mentioned by Isidore of Seville (*Etym.* 17.3.10), and a type of barley, called 'small barley' (*leptididas*), mentioned in the *Geoponica* (3.3). Planted in April, *leptididas* matured in the middle of the summer, probably in July.

Further support for the notion that spring planting was more common than is generally assumed is found in Palladius (*Op. Ag.* 4.3), who recommended March sowing for both *panicum* and *milium*, in warm, dry places; in cooler regions, it was sown in May (6.1). Further, *Geoponica* (3.3.12) placed the sowing of millet in March. In view of the dominance of wheat and barley, both principally winter-sown crops, and the need to find crops which could tolerate Mediterranean summers, it is unsurprising that late antique farmers exploited millet during the hot months. That millet was a summer crop in the Byzantine Near East is attested by Pseudo-Joshua's (§38) description of the famine at Edessa during AD 499/500 in which locusts devoured the grain crop of the city's hinterland.[59] Out of desperation, farmers sowed millet in June and July, but it performed poorly, and the famine continued. The agricultural historian may draw several important inferences from this passage. It is clear that millet was a familiar crop on the scene in Mesopotamian agriculture: the

[57] He associates the black colour of this ancient grain with the black awns that are today common on durum wheat throughout the Mediterranean: Jasny, *Wheats*, 92–3.

[58] They were a natural extension of farming practices that harked as far back as the Classical period, when Theophrastus described Pontic spring-sown wheat (that some scholars believe to have been durum). J. Cotte and C. Cotte, *Etude sur les blés de l'antiquité classique* (Paris, 1912), 28.

[59] Ps.-Joshua, *Chron.* 38–40.

farmers naturally turned to it in the summer months, since it was the only cereal which they could sow so late and still expect a crop. The months of June and July are beyond Palladius and the *Geoponica*'s recommended sowing dates of March–April. But by July, the success or failure of the wheat and barley crops was known. Thus, millet seems to have been a natural emergency crop and one to which farmers turned when the major cereals failed. Further proof of this comes from other textual sources. During a famine in the early sixth century, Cassiodorus ordered the release of millet stocks from granaries at Tortona and Pavia.[60] Similarly, the tenth-century historian Movses Dasxuranci (3.12), described millet as a famine food in Caucasian Albania. The crop was probably widespread throughout Upper Mesopotamia and the mountainous regions of eastern Anatolia.[61]

It is noteworthy that the Edessenes had millet seed on hand in sufficient quantity to hope that its planting would alleviate famine, a fact that indicates it was habitually cultivated. In all likelihood it represented a stock of seed from the previous year, or an emergency cache. However, it may have been that the farmers had already harvested one crop: millet can ripen in as little as ninety days, and thus, if planted in March, would have been ready for the harvest by June. It is also conceivable that the cultivators around Edessa generally planted millet in the summer, rather than the spring months. But, if so, this is likely to have required field irrigation.

This discussion introduces the hypothesis that the age-old regime of mixed farming, long dominant in the Mediterranean, witnessed several developments in Late Antiquity. The integration of crops, hydraulic features, and animals within a host of different mixed farming regimes had at its core a cadre of cash crops, particularly olive and vine. Catalysts for development of intensive regimes were not only the dense settlement found in many rural areas, but a host of other internal and external factors which can only be touched upon. Some of these stimuli are obvious and have already been mentioned: the monetization that accompanied state exploitation of taxes and a thriving market; and the possibilities that peace and technological developments opened up, including an overseas demand for eastern foodstuffs as both staples and luxuries. In the face of these pressures and inducements, farmers at all levels adapted their strategies. Sometimes they will have

[60] Cassiodorus, *Ep.*: *PL* 69.876–7. The situation of both Tortona and Padua, in the Po Valley in northern Italy, is telling; the abundance of water, and a shorter growing season, would have made the plant especially attractive to farmers there.

[61] M. Dasxuranci, *The History of the Caucasian Albanians by Movses Dasxuranci* (London, 1961), 203. I am grateful to Dr Timothy Greenwood for bringing this reference to my attention and for giving me access to his own translation of the text.

reacted in pursuit of profit, but in other instances they must have altered their practices simply to survive.

The late antique peak of integrated farming and specialization left an enduring legacy, founded on a fabric woven from the complex strands of population, crops, technology, and the market. In the end, the result was a late antique landscape that was overall more intensively farmed and more market-centred than had previously been the case. Most of the crops and techniques used in these transformations were not really new. But it appears that in Late Antiquity some of the methods, and especially the degree and frequency in which certain technologies and crops were employed, went beyond what had occurred before. Many of these changes persisted into the Early Islamic period. But, although Islamicists have sometimes attributed the origins of these changes to the Early Islamic period, many of them in fact originated earlier. Numerous regions of the late antique Levant had already undergone several centuries of agricultural enhancement. The Early Islamic period saw continued efforts by the caliphate and its landowners to develop farmland in the margins, to rehabilitate and expand canal networks, and pursue the production of crops, like cotton and sorghum, that were generally under-represented throughout the Mediterranean. The role of landowners of early Muslim times as disseminators of agrarian practice and technology is widely recognized as of tremendous import. But, rather than viewing the upward trajectory of intensive agriculture in the Islamic world as a decisive break with the past, it is best viewed as continuing long-standing ancient development and adaptation.

8

Trade, Agriculture, and the Economy of the Late Antique East

An abundance of archaeological work now offers considerable insight into the range, extent, and durability of food exports from the eastern Mediterranean to surrounding areas. While scholars now generally recognize that the East possessed a flourishing urban life and general prosperity, uncertainties about the nature of this prosperity and its duration remain. Did it, for example, entail a monetized economy, significant agricultural surpluses, and widespread travel and trade? Or was it, rather, an 'artificial' economy, dependent on an inflow of state funds or money from pilgrimage, and thus unsustainable without these external inputs? How long did the vitality of the urban life and economy of the East persist? Did it end in the later sixth century or with the Muslim conquests, or do we sense broad continuity through the seventh century in material culture and patterns of consumption and exchange?

THE MARKET IN THE LATE ANTIQUE EAST

Some, such as Horsley, have challenged the very existence of a significant market in Roman–Byzantine Palestine.[1] He believes, for instance, that evidence of grain traders in Palestine refers only to shipments contracted by the wealthy to supply their urban dwellings. In many ways, Horsley's view of Galilee is typical of wider 'minimalist' conceptions of the ancient economy. Past 'primitivist' scholars, notably Moses Finley and A. H. M. Jones, have exerted substantial influence in the discussion of the ancient economy.[2] For these material minimalists, trade was predominantly confined to luxuries, and large cities were not real centres of complex economic activity, but were primarily comprised of consumers whose elites were supported by land rents. All transport was ponderous and expensive, while land transport was

[1] Horsley, *Galilee: History, Politics, People* (Valley Forge, Pa., 1995), 201–7.

[2] M. I. Finley, *The Ancient Economy* (2nd edn, London, 1985), 159; K. Hopkins, 'Taxes and Trade in the Roman Empire (200 B.C.–A.D. 400)', *JRS* 70 (1980); Jones, *Later Roman Empire.*

prohibitively slow, underdeveloped, and marginal. The greatest movement of low-value goods occurred through government-controlled provisioning. Government-controlled provisioning, however, was not an economic activity *per se*; state compulsion and the collection of taxation in kind for feeding the army and the populations of Rome and Constantinople was not commerce.

Richard Duncan-Jones nuanced this view by arguing that the large-scale imperial movement of commodities facilitated trade in cheap goods, since these could fill out empty space in the holds of the state's ships.[3] For Late Antiquity, Jean Durliat envisioned a similar seaborne movement of *le grand commerce des subsistances*,[4] which generally fits with the models proposed by Duncan-Jones, Jones, and others. The movement, especially overland, of low-value commodities was largely eliminated from the ancient economic picture as it was seen as prohibitively expensive.[5] In support of their theories of underdeveloped markets and poor overland communications, scholars have pointed to a number of food crises that occurred in the East during Late Antiquity. During such crises, rural dwellers came to the city seeking grain, as in 502 and 582 at Edessa.[6] Episodes like these and the major famine in Cappadocia in 368 have been viewed as proof positive for the fragility of the economies of the inland cities and the anaemic market they sustained.[7]

Though many of the broad themes of this discussion have been superseded by new views on production and trade, there is still considerable need to clarify certain points, among them the view that the market was feeble and that inland cities did not rely on trade to any degree to feed themselves.[8] Evidence suggests, indeed, that real self-sufficiency, however much it may have been in the mindset of the ancient farmer, was rarely attained, and people such as priests, tradesmen, merchants, and government officials relied on the market to provide or supplement their food supply.[9] Even monks often bought grain

[3] Duncan-Jones, *Structure and Scale*, 49.

[4] J. Durliat, *De la ville antique à la ville byzantine. Le problème des subsistances* (Rome, 1990), 513.

[5] Jones, *Later Roman Empire*, 841. See also P. Garnsey and C. R. Whittaker, 'Trade, Industry and the Urban Economy', in A. Cameron and P. Garnsey (eds.), *The Cambridge Ancient History*, vol. XIII: *The Late Empire, A.D. 337–425* (Cambridge, 1998), 332–5.

[6] Patlagean, *Pauvreté économique*, 80.

[7] Durliat, *De la ville antique*, 445.

[8] New views about a more developed urban economy of production and exchange include: D. J. Mattingly and J. Salmon, *Economies Beyond Agriculture in the Classical World* (London, 2001); J. Paterson, 'Trade and Traders in the Roman World: Scale, Structure, and Organisation', in H. Parkins and C. J. Smith (eds.), *Trade, Traders and the Ancient City* (London, 1998); Ward-Perkins, 'The Cities', 403–7; B. Ward-Perkins, *The Fall of Rome and the End of Civilization* (Oxford, 2005), 88.

[9] T. W. Gallant, *Risk and Survival in Ancient Greece: Reconstructing the Rural Domestic Economy* (Cambridge, 1991), 143–69; P. Garnsey, *Famine and Food Supply in the Graeco-Roman World: Responses to Risk and Crisis* (Cambridge, 1988), 55–63.

after they sold their own products.[10] In fact, there were apparently many who, whilst they may have owned small parcels of land, could not depend on these plots to meet their subsistence requirements.[11] Their needs were not met primarily through the state dole or patronage, but rather through their access to the marketplace. Grain was widely available for sale in cities throughout the empire. In 362/3 at Antioch, Julian regulated the prices demanded for grain during the food crisis there. The market had reacted to shortage with skyrocketing prices, a situation repeated during the famine at Edessa in 502, as reported by pseudo-Joshua (§39). When Symeon the Holy Fool wished to procure the finest wheat flour, he had no trouble purchasing it in the markets of Emesa.[12] In a sixth-century story that John of Ephesus tells of the saints Mary and Euphemia, he describes the markets of Amida as filled with wheat and other agricultural produce.[13]

A great deal of local traffic involved no middlemen, since many farmers sold their own goods directly. Galen, writing in second-century Rome, noted how rustics brought grain in wagons from the countryside into the cities of the empire. He also related the fraud propagated by some of these farmers, who placed water jars in the midst of the grain, from which the latter drew moisture, thereby increasing in weight and bulk.[14] According to the *Digest* (50.11), farmers who transported their own produce into town often disrupted production by leaving their fields for extended periods of time; they were urged to dispose of their grain rapidly and return to work—better still, they should leave such activities to traders.[15] Farmers brought their wheat and barley to fourth-century Antioch according to Libanius (*Or.* 50.28), and John Chrysostom likewise described goods brought into the city, including animals, wheat, and barley.[16] Zachariah of Mitylene (7.5), writing in the sixth century, recalled that the farmers around Amida carried their produce of wheat, wine, and other foodstuffs to a market held outside the walls of the city, where they were sold to both Persians and Romans.

Markets primarily served urban dwellers who could not supply their own grain and other essentials. But they also almost certainly supplied farmers from the city territory with goods not necessarily available in their villages.

[10] George of Choziba, *The Life of Saint George of Choziba and the Miracles of the Most Holy Mother of God at Choziba*, trans. Vivian, p. 58.25.

[11] A. Bowman, *Egypt After the Pharoahs: 332 BC–AD 642 from Alexander to the Arab Conquest* (London, 1986), 112; Safrai, *Economy*, 209–21.

[12] *Symeon the Holy Fool: Leontius' Life and the Late Antique City*, trans. Krueger, 156.

[13] John of Ephesus, *Lives*, trans. Brooks, p. 173.

[14] C. Galen, *Galen on the Natural Faculties*, with an English translation by A. J. Brock (London, 1979), 87 1.14.

[15] A. Watson, *The Digest of Justinian* (rev. English language edn, Philadelphia, 1998), 50.11.

[16] W. Mayer and P. Allen, *John Chrysostom* (London, 2000), 186–7.

Rabbinic sources make frequent mention of markets; according to these, there was a well-known meat market at Emmaus, while in the markets at Bet She'an, one could purchase cattle, fruit, clothing, money, and metal vessels.[17]

The archaeological remains of well-developed market centres, and districts containing substantial store fronts, are known throughout the East: the main thoroughfares of every city contained stalls and shops lining the *cardo* and *decumanus*, where tradesmen and merchants sold their wares. One of the most prominent is that excavated at Scythopolis (Bet She'an), where the late antique market centre continued to be active well into the Umayyad period.[18] Textual references and archaeological remains provide other examples in the East, such as those at Antioch, Ephesus, Apamea, and Berytus.[19] At Sardis, the craftsmen of the Byzantine period produced an array of wares sold from the shop fronts in the centre of town.[20] The *agorai* and settings for other city markets are also known at Corinth, Constantinople, and Gerasa (Jerash), to name but a few.[21] We hear of the marketplaces of northern Mesopotamia at Dara and Melitene.[22] Even small cities, like Mampsis in the Negev, had market spaces at their centre.[23]

Besides the *agorai* and *stoai* that generally spring to mind when thinking of the classical city, there were other types of market space in cities and villages throughout the East. Several purpose-built marketplaces called *macella* are known from eastern cities, including Constantinople, Corinth, Alexandria, and Dura Europus (in eastern Syria).[24] The *macellum* served the same economic functions as the *agora* (though without any of its political or juridical roles), but was generally smaller. In Oriens, Coracesium in Cilicia and Antioch both had macella known from texts.[25]

17 Emmaus: TB Ḥul. 48a; Bet She'an: TB 'A.Z. 13a.

18 Y. Tsafrir and G. Foerster, 'Urbanism at Scythopolis-Bet Shean in the Fourth to Seventh Centuries', *DOP* 51 (1997), 138–40.

19 Antioch: Libanius, *Antioch as a Centre of Hellenic Culture as Observed by Libanius*, trans. A. F. Norman (Liverpool, 2000), *Or.* 11; J. S. Crawford, *The Byzantine Shops at Sardis* (Cambridge, Mass., 1990); Ephesus: C. Foss, *Ephesus after Antiquity: A Late Antique, Byzantine and Turkish City* (Cambridge, 1979), 63; Apamea, *IGLS* 4.1318, mosaic pavement denoting an *emporion*; Berytus (Beirut): L. J. Hall, *Roman Berytus: Beirut in Late Antiquity* (London, 2004), 66.

20 Crawford, *Byzantine Shops at Sardis.*

21 Corinth: R. L. Scranton, *Mediaeval Architecture in the Central Area of Corinth* (Princeton, NJ, 1957), 6–26; Constantinople: M. Mango, 'The Commercial Map of Constantinople', *DOP* 54 (2000). Constantine originally equipped his new city with a circular *agora* and expanded and improved the porticoes. He also added a large marketplace with four porticoes: Zosimus, *Hist. Nov.*, ed. Paschoud, 2.31.2; Gerasa: A. Uscatescu, *La cerámica del Macellum de Gerasa (Ŷaraš, Jordania)* (Madrid, 1996).

22 Procopius, *De Aed.* 2.2.15; 2.10.22; 3.4.18.

23 A. Negev, *The Architecture of Mampsis Final Report*, vol. I: *The Middle and Late Nabatean Periods* (Jerusalem, 1988), 163–6.

24 C. de Ruyt, *Macellum: marché alimentaire des Romains* (Louvain-la-Neuve, 1983), 263–6.

25 Ibid., 265.

Villages often served as market centres and were particularly important in the less urbanized areas of the empire, notably along the frontiers at places like Androna, and in the Hauran, where a sizeable village like Umm el-Jimal has reasonably been supposed to have functioned as a local and regional exchange centre.[26] These are the sorts of places Libanius described in fourth-century northern Syria in his *Antiochikos* (230–2).[27] Some of these markets probably consisted of ephemeral booths, huts, and tents, like the wine sellers' tents mentioned in pre-Islamic poetry, but others may have been more permanent structures.[28] There is, however, some debate over whether permanent shop-buildings existed within the rural communities of Late Antiquity. Although Butler and Tchalenko identified several buildings in the Limestone Massif as shops associated with marketplaces, Tate remains largely unconvinced, and considers that the only certain shops of the region are those at Deir Sim'an, adjacent to the processional path to the St Simeon monastery.[29] However, John Chrysostom noted that many wealthy patrons provided *agorai* in villages, by which he probably meant 'built' markets.[30]

In addition to the more or less durable urban features of shops, *macella*, or *agorai* and the rural networks of roads and rest houses, many features of trade were ephemeral and have left no obvious trace in the physical record. There may have been a rotational cycle of rural fairs, such as De Ligt hypothesizes for Syria, while evidence from Chrysostom, noted above, suggests that villages were served by some combination of retail outlets, regular markets, and seasonal fairs, with these various types of outlets complementing and competing with each other. Temporary market fairs, the *panēgyreis* and *nundinae* of the East, were common occurrences, and they appear frequently, especially in the Rabbinic literature. Libanius (*Or.* 11) offers a glimpse of the working of the panegyreis within Syria; these were held within the larger villages and apparently shifted from one important village to another throughout the calendar year. In a well-known passage, Libanius noted that the rural centres served by the *panēgyreis* had little need for the city. To my mind, this implies that these market meetings were regular, well-attended, and well-supplied with a range of goods.[31]

[26] De Vries and Betlyon, *Umm el-Jimal*, 232.

[27] M. F. Hendy, *The Economy, Fiscal Administration and Coinage of Byzantium* (Northampton, 1989), 10; Libanius, *Antioch as a Centre of Hellenic Culture*, trans. Norman, 54.

[28] A. Wensinck, 'Khamr', in E. Van Donzel, B. Lewis, and C. Pellat (eds.), *Encyclopaedia of Islam* (2nd edn., Leiden, 1960), IV.994.

[29] Tate, 'Les campagnes de la Syrie du nord', 81.

[30] *In Act. Apost.*, *PG* 60.147.8.

[31] L. De Ligt, *Fairs and Markets in the Roman Empire: Economic and Social Aspects of Periodic Trade in a Pre-industrial Society* (Amsterdam, 1993), 127–8.

Periodic markets could be frequent, occurring weekly or monthly, or seasonally. Some were celebrated annually on a saint's feast day, or during the festival of a local deity, and these undoubtedly played an important role within the agrarian economy. They supplemented the activities of the more regular markets and other retail outlets available to villagers. It is thus unsurprising to find fairs even in areas where permanent, built markets are attested. Ammianus Marcellinus described fairs at Batnae and Amida in Mesopotamia, both of which certainly possessed permanent market centres.[32] Local fairs of the sort mentioned by Ammianus often provided goods over extended periods of time: the famous one at Aegae in Cilicia lasted forty days and was visited by western merchants whose goods included slaves.[33] Several important fairs are known in Palestine, at Hebron, Scythopolis, Acre, Tyre, Gaza, and Jerusalem. The one at Jerusalem commemorated the consecration of the Church of the Holy Cross and was particularly large. It continued to be held from at least the fourth or fifth centuries into the late seventh century.[34] In the fifth century, a panēgyris in honour of St Zachariah was held at al-Bassa (Bizet) in Phoenice. At Imma, a large and prosperous North Syrian village and staging post on the road from Antioch to Beroia, a prominent fair was held over a number of days, and a fair at Hierapolis in Syria attracted merchants from all quarters: Phoenicians, Indians, Greeks, and people from Asia Minor congregated there.[35]

Merchants

Merchants were a universal feature of late antique Mediterranean society and many traded far beyond the confines of the Roman world. We gain some insight into the merchant world of Late Antiquity from the fourth-century merchant's guide, the *Expositio totius mundi*, offering glimpses into the merchants' world, where regional specialities were known and sought after and where merchant communities operated around the whole of the Mediterranean and sought information on ports of call and local products. One of the best known merchants of Late Antiquity, Cosmas Indicopleustes, made several voyages in the Red Sea region, something that was commonplace among Christian

[32] *SEG* VIII.18.9; Ammianus Marcellinus, *Rerum gestarum*, ed. W. Seyfarth, L. Jacob-Karau, I. Ulmann, 14.3.3;18.8.13.

[33] Theodoret, *Ep.*, ed. Azéma, no. 70; *Itinera hierosolymitana saeculi IIII–VIII*, ed. Geyer, 150.

[34] De Ligt, *Fairs and Markets in the Roman Empire*, 72–3.

[35] Theodoret, *Historia Religiosa*, ed. P. Canivet and A. Leroy-Molinghen, 7.1–3; J.-P. Sodini, 'L'artisanat urbain à l'Epoque Paléochrétienne (IV–VIIe S.), *Ktèma* 4 (1979), 113; Procopius of Gaza, *Procope de Gaza, Priscien de Césarée, Panégyriques de l'empereur Anastase Ier*, ed. A. Chauvto, §18.

merchants.[36] During the Prophet Muhammad's lifetime, timber from the wreck of a Greek merchant ship at Judda (Jidda) was used to refurbish the Kaʿaba.[37] At least some merchants were apparently specialized retailers; from Rabbinic sources, we hear of wax sellers, clothes vendors, and those who sold pots.[38]

I have noted above that some grain was sold from the field directly to the purchaser, and that farmers also sold their produce themselves in the market. However, middlemen also played a prominent role in the grain trade in Late Antiquity, as they had from at least the Hellenistic period.[39] Many traders dealt in subsistence goods. Grain dealers who bought in bulk and sold in both city and village were commonplace throughout Syria and Palestine. A particularly unsavoury grain merchant makes an appearance in the seventh-century *vita* of St Spyridon of the city of Tremithus in Cyprus.[40] We know of grain dealers in Egypt and in Palestine who were accused of adding water to the grain they sold in order to increase the bulk and weight.[41] In the fourth century, Themistius told of grain merchants who sailed to Egypt, Thrace, and the Chersonese, whilst Chrysostom mentioned the corn merchants of Antioch, and *Digest* 50.11.12 attests farmers selling their produce to merchants. The move to smaller, cheaper ships, like the well-known seventh-century example at Yassi Ada, hints at the ubiquity of smaller-scale merchant operations where cabotage was the rule. These smaller ships, however, were also capable of sailing in deep water and over longer distances than previously believed.[42]

Wine merchant and wine vendor were common occupations, and they were controlled by the legislation of the emperor Theodosius.[43] Cilicians seem to have been particularly engaged in the wine trade; wine sellers are frequently found among the late antique epigraphic record of Corycus. The lower tax rate accorded their wine cargoes in the fifth–sixth-century tariff inscription from Abydos led Dagron to identify Cilicians as specialized wine merchants.[44]

[36] Cosmas Indicopleustes, ed. Wolska-Conus; John of Nikiu, *Chronicle*, trans. Charles, XC.72.

[37] Ibn Ishaq, *Life of Muhammad*, trans. Guillaume, 84.

[38] Wax merchants: TB B.M. 63b; clothes merchants: TB Pes. 26b; pot sellers: TB Pes. 30a.

[39] Brooks, *Peah*, 1.3; Rathbone, *Economic Rationalism*, 52.

[40] TJ Dem. 2.4; Theodorus, *Vita Spyridon*, 2; Ps.-Joshua, *Chron.* 39.

[41] Duncan-Jones, *Structure and Scale*, 149; Sperber, *Money and Prices*, 126.

[42] Themistius, *Orations*, trans. Penella, 170; G. F. Bass and F. H. Van Doorninck, *Yassi Ada* (1st edn., College Station, 1982); S. A. Kingsley, A. Johnson, and D. Williams, *A Sixth-Century AD Shipwreck off the Carmel Coast, Israel: Dor D and Holy Land Wine Trade* (Oxford, 2002), 91–4; Mayer and Allen, *John Chrysostom*, 186–7.

[43] John of Nikiu, *Chron.* XC.72.

[44] Dagron and Feissel, 'Inscriptions inédites', 454. More likely, like other eastern or 'Syrian' merchants, Cilicians traded in a whole range of goods, though with wine and oil perhaps the most prominent amongst them. See Decker, 'Wine Trade of Cilicia in Late Antiquity'.

Egyptian and Palestinian wine traders were also prominent. According to the *vita* of Porphyrius of Gaza (58.7), Egyptian wine traders were found in the fourth century both in the Black Sea coastal city of Tomis and at Gaza. In the seventh century, Egyptians still plied wine to Constantinople; an unnamed Egyptian *oinemporos* appears in the *Life of St Artemios*.[45] A fine Syrian mosaic depicts a donkey laden with wineskins, and donkey caravans that passed from village to village to buy and sell wine and oil are common in the Rabbinic sources.[46] These sources mention wine merchants who bought in bulk from numerous sellers, as well as smaller-scale pedlars, such as innkeepers and wine bar owners, whose establishments were ubiquitous in the cities of the late antique East.[47] Jewish wine merchants were especially prominent and operated extensively in the late antique world, especially in the Arabian Peninsula, where they are mentioned in pre-Islamic poetry.[48] At Gaza, two Jewish merchants, who may have traded in the famous wine of the city, used their wealth to construct a new synagogue.[49]

Similarly, epigraphic and other evidence supports the notion of a vigorous oil trade. At Palmyra, a second-century inscription reveals that merchants were taxed according to whether their camels carried loads of two or four goatskins of oil, and similar charges applied to asses carrying oil.[50] These shipments were not all for provisioning the city; the tariff suggests that many of the items it taxed were exported.[51] In Late Antiquity, a document from the reign of Anastasius (491–518), the 'Abydos Tariff', attests a similar transport of oil to the imperial capital by eastern merchants, perhaps as part of the annona. Taxes are stated for wine and oil shippers, from which Cilicians are exempt.[52] In necropolis inscriptions at Cilician Coracesium and Corycus, oil merchants appear in eleven instances, and their prominent place may indicate the involvement of those settlements in long-distance oil trade.[53] Cilician

[45] *Corpus Inscriptionum Judaicarum* 681b3; Artemios, ed. and trans. Crisafulli and Nesbitt, 164–5.

[46] Pentz, *Invisible Conquest*, 23, fig. 7; Safrai, *Economy*, 234–5.

[47] TB B.M. 51a; 60a: wine merchants who buy wine from five presses.

[48] P. Crone, *Meccan Trade and the Rise of Islam* (Oxford, 1987), 105.

[49] A. Ovadiah, 'Excavations in the Area of the Ancient Synagogue of Gaza (preliminary report)', *IEJ* 19 (1969).

[50] J. Matthews, 'The Tax Law of Palmyra: Evidence for Economic History in a City of the Roman East', *JRS* 74 (1984), 176.

[51] Bowersock, 'Social and Economic History of Syria', 74.

[52] Dagron and Feissel, 'Inscriptions inédites', 425–55. The tax inscription from Anazarbus in Cilicia does not list olive oil, but does list cuttings used in grafting. The transport of such stock has been described above. In view of Cilicia's reputation as an oil producer, we may assume safely that at least some of these cuttings were olive shoots.

[53] Keil and Wilhelm, *Denkmäler aus dem Rauhen Kilikien*, nos. 114, 39, 40, 62, 64a, 72, 245, 350, 468, 768; Trombley, 'Corycos in Cilicia Trachis'.

ports were admirably placed to receive oil from the hinterlands of Tarsus, Anazarbus, and Seleucia-in-Isauria, as well as from North Syria and Cyprus. In some cities, at least, speciality oil markets existed: the ninth-century Muslim historian Balādhuri reported that there was such a market in Damascus, where the Muslim commanders met after taking the city in AD 635.[54] From the above evidence, it is clear there were merchants who tended to specialize in one product, and that grain, wine, and oil were each commodities on which merchants could focus their livelihood.

There are many indicators that late antique overseas trade often comprised small-scale individual ventures. Many merchants apparently developed contacts through ecclesiastical circles or through émigré communities. In the sixth-century West, dozens of cities possessed eastern communities of some social and economic importance. Here, the term 'Syrian' became almost synonymous with the word 'merchant'.[55] Colonies of 'Syrians' or other easterners are known to have operated throughout the late- and post-Roman West, mainly in Gaul and Spain.[56] In both places, easterners were influential, especially so within the church, where many rose to prominent bishoprics including that of Rome. The *Lives of the Fathers of Mérida* in Spain provide insight into how such trade networks worked in the post-Roman world. The sixth-century bishop Paul, himself '*natione Grecum*' and a physician, was visited by 'Greek' merchants (*negotiatores Grecos*). These merchants had with them a boy named Fidelis whom Paul discovered was his nephew. In exchange for a sum of money, Paul persuaded the merchants to leave Fidelis with him. Fidelis eventually followed his uncle as bishop of Mérida. The story of Paul and Fidelis shows the church as a natural meeting point for foreigners, especially easterners, who apparently found it worth their while to call on the bishop in places where they hoped to do business. But there is no hint that bishop Paul was engaged in any kind of commerce himself, rather he, and those like him, were probably consumers and contacts for those who might wish to purchase eastern goods. As can be seen by these examples from the West, the late antique East had an active and far-ranging merchant community whose success is best exemplified by their being synonymous with commerce.

[54] Hitti, 'The Origins of the Islamic State', 189.

[55] L. Bréhier, 'Les colonies d'Orientaux en Occident au commencement du moyen-âge', *BZ* 12 (1903) remains a fundamental study; for Spain see L. A. García Moreno, 'Colonias de commerciantes orientales en la península ibérica S. V–VII', *Habis* 3 (1972) and for Italy L. Ruggini, *Ebrei e orientali nell'Italia settentrionale fra il IV e il VI secolo* (Rome, 1959). For the West generally see K. R. Dark, 'Early Byzantine Mercantile Communities in the West', in C. Entwistle (ed.), *Through a Glass Brightly: Studies in Byzantine and Medieval Art and Archaeology Presented to David Buckton* (Oxford, 2003), and A. Harris, *Byzantium, Britain and the West: The Archaeology of Cultural Identity AD 400–800* (Stroud, 2003).

[56] *Vitas sanctorum patrum Emeretensium*, ed. Maya Sánchez, 25, 31–3.

As noted, the merchants left the boy Fidelis with Paul in exchange for money. While there is evidence for direct ecclesiastical involvement in trade, for example the stories preserved in the *vita* of John the Almsgiver, most evidence does not tie the traders to the immediate service of the church or state; rather some explicitly describe them as driven by a desire for profit. Thus, Jerome, in the fifth century, portrayed the activities of Syrian merchants as compelled by love of money; they are 'those greediest of men'.[57] In roughly the same period, Salvian noted the unruly 'merchants and Syrians' who filled the cities of Gaul, and described them as profligate liars who only cared to gain by deceit.[58] The fourth-century Asterius, in a precious passage about the accumulation of wealth in Late Antiquity, was clear as to the motives of contemporaries:

> No-one sails the sea for its own sake, nor spends his life toiling in farming for its own sake, but it is clear that they both persist in their pains, the one to profit from the land, the other so that he might receive the wealth of maritime trade.[59]

Late Roman Amphora and Late Antique Trade

Did these contacts, for which we have scattered textual testimony, involve eastern agricultural goods on a significant scale? Were they indicative of a pattern of trade that was sustained and economically important to both the producers and consumers? The evidence suggests that we should answer the question in the affirmative. Eastern dominance of the Mediterranean involved substantial movements of agricultural cargo, on a scale and with a regularity that rendered them economically and culturally significant. The cargoes flowed from at least the fifth until at least the mid-seventh century. The peak of trade occurred in the sixth century, but there is growing evidence to support the probability that Levantine goods, especially some of those commodities described here, continued to be important trade items, especially in the eastern half of the Mediterranean, into the eighth century.[60]

[57] Jerome, *Epistulae*, ed. Hilberg and Kamptner, 184.

[58] Salvian, ed. Lagarrigue, 288: '*Negotiatorum et Syricorum omnium turbas quae maiorem ferme ciuitatum uniuersarum partem occupauerunt, si aliud est uita istorum omnium quam meditatio doli et tritura mendaci*...'

[59] Asterius, *Hom.*, ed. Datema, 3.7.2; the same sentiment is echoed in Theophylact Simocatta, *Ep.* 10.17–21.

[60] This interpretation is obviously based on the belief that the Umayyad period was one of general continuity rather than disruption, and the evidence increasingly points to this being the case. See R. M. Foote, *Umayyad Markets and Manufacturing: Evidence for a Commercialized and Industrializing Economy in Early Islamic Bilad al-Sham* (Cambridge, Mass., 1999); Pella: Walmsley, *Early Islamic Syria*; P. Watson, 'Change in Foreign and Regional Economic Links with Pella in

Amphorae provide critical evidence for agricultural trade within the Levant and to the wider ancient world. Although we are far from having a complete picture—and many methodologies of survey, excavation, and publishing remain underdeveloped or ignored—we nevertheless possess a usable corpus of ceramic finds that produces a clear picture of late antique eastern trade in wine, olive oil, and other commodities. The geographical spread of the discoveries is impressive. They extend from India in the south-east, to Britain and Ireland in the north-west. The limit of the distribution directly to the north seems to be the Crimea, while Ethiopia and the Sudan mark the southern extremes.

In many ways, the science of ceramic studies is in its infancy. Quantified studies of amphora finds within archaeological excavation still remain relatively rare, particularly in the eastern Mediterranean. Whatever the shortcomings of our present state of knowledge, the data are sufficient to sustain some burden of economic hypothesis. While a detailed survey of amphora studies is not possible here, a brief outline of the present state of knowledge of the major eastern amphora types is necessary, in order to relate their production and distribution to the agrarian economy of the East.[61]

The main types of transport jars that concern us are the Levantine types Late Roman 1 (LR1), 4 (LR4), and 5 (LR5) (Figures 8.1, 8.2, 8.3). It is now well-established that these containers were manufactured at multiple locations along the eastern coast of the Mediterranean. Unlike the jars associated with individual classical and Hellenistic *poleis*, but akin to many of the well-known types of the Roman period, the LR transport amphorae of Late Antiquity belonged to regions rather than individual cities.[62]

Perhaps the best known transport jars of Late Antiquity are the Palestinian types LR4 and LR5. Kingsley noted type LR4 ('Gaza'), jars that Stephen of

the Seventh Century A.D: The Ceramic Evidence', in P. Canivet and J.-P. Rey-Coquais (eds.), *La Syrie de Byzance à l'Islam: VIIe–VIIIe siècles: actes du colloque international Lyon—Maison de l'Orient méditerranéen, Paris—Institut du monde arabe, 11–15 Septembre 1990* (Damascus, 1992); Magness, *Early Islamic Settlement.* Much of the survey data from rural areas in the south of former Oriens contradict this, for one example see Parker and Betlyon, *The Roman Frontier in Central Jordan*, 49–50. The view of largely continuous site habitation and economic continuity also seems incorrect in the case of Mesopotamia and Cilicia, the borderlands between Byzantium and the Caliphate. See Decker, 'Frontier Settlement and Economy in the Byzantine East'.

[61] In-depth studies of the distribution and quantification aspects may be found in Kingsley, 'Sumaqa Pottery Assemblage'. The distribution of LR4, LR1 and LR3 has been addressed by Pacetti, 'Appunti su alcuni tipi di anfore orientali', though another such study is needed. There are numerous other amphora varieties about which data continue to accumulate: Samos cistern type, Beirut amphorae, Robinson M334, Sinop jars, Cyprus type (LR1 similis), North Syrian bag-shaped jars, hollow-foot amphorae.

[62] This phenomenon finds precedent in the Roman period, when regional amphora types become common.

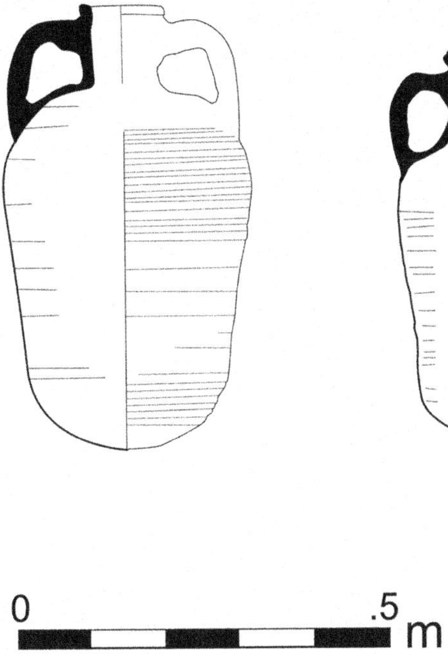

Figure 8.1. LR1 amphora types A and B (after Peacock and Williams, *Amphorae and the Roman Economy*, fig. 104).

Byzantium (*Ethn.* 1.10) in the sixth century referred to as *keramoi Gazitai*, on eighty-four sites throughout modern Israel.[63] These were filled primarily with the famous wine of Gaza, though papyri inform us that the *gazition* carried a range of products, including grain and nuts.[64] LR5, the so-called Palestinian 'bag-shaped' amphora, is a ridged, ovoid container. Large-scale winepresses and dozens of LR5 production centres are known from the territory of Ascalon. But, as with the LR4, the LR5 also served as a multi-purpose jar whose typical contents could include fish brine, dried figs, and olive oil, as well as wine. The distribution of these jars is impressive: Kingsley noted LR5 at 180 sites in Byzantine Palestine and found LR4 and LR5 at a further 123 sites around the Mediterranean. Together they represent a significant portion of amphora exports from the East.[65]

63 Kingsley, 'Specialized Production', 153.

64 P. Mayerson, 'The Gaza "Wine" Jar (*Gazition*) and the "Lost" Ashkelon Jar (*Askalonion*)', *IEJ* 42 (1992), 79.

65 Kingsley, 'Specialized Production', 153, fig. 31.

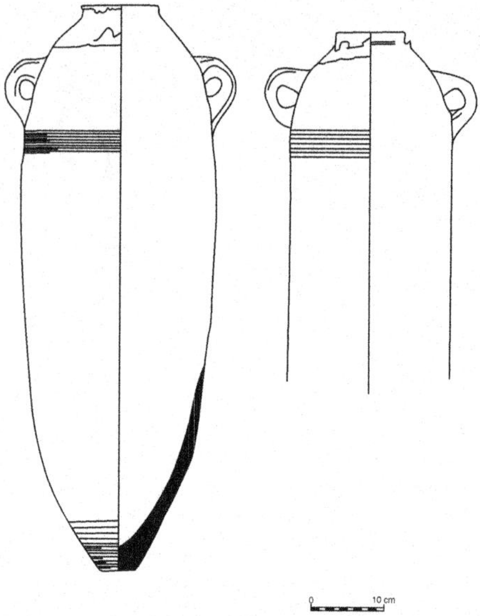

Figure 8.2. LR4 amphora (after Hirschfeld, *Ramat Hanadiv Excavations*, pl. XIV. Reproduced by permission of The Israel Exploration Society, Jerusalem).

At present, production sites of LR1 amphorae are known on Rhodes, the south-east Anatolian coast, Cilicia, Cyprus, and on the north shore of Syria.[66] Its purpose was probably first as a wine carrier, and this is especially true of the smaller sub-varieties of LR1.[67] Larger types carried wine and oil, and likely other regional products as well. LR1 was the most widely exchanged jar in the late antique Mediterranean; it has been found on nearly three hundred sites, and several quantified assemblages illustrate the importance of the type in a wide range of times and places, from Arikamedu in India to Britain. Even at western sites, LR1 represent a substantial, and sometimes a high, percentage of the late antique amphorae discovered.[68]

[66] Empereur and Picon, 'Les régions de production'.

[67] P. G. Van Alfen, 'New Light on the 7th-c. Yassi Ada Shipwreck: Capacities and Standard Sizes of LRA1 Amphoras', *JRA* 9 (1996).

[68] From India (Arikamedu) to Britain: Decker, *Agricultural Production and Trade in Oriens*, Appendix IV. LR1 represented 15–20 per cent of the total amphora deposits of the sixth–seventh centuries in excavations at Saraçhane, Istanbul. Hayes, *Excavations at Saraçhane in Istanbul*, II.64. In the Balkans, at Svetinja near Viminacium, LR1 comprised 54 per cent of the total

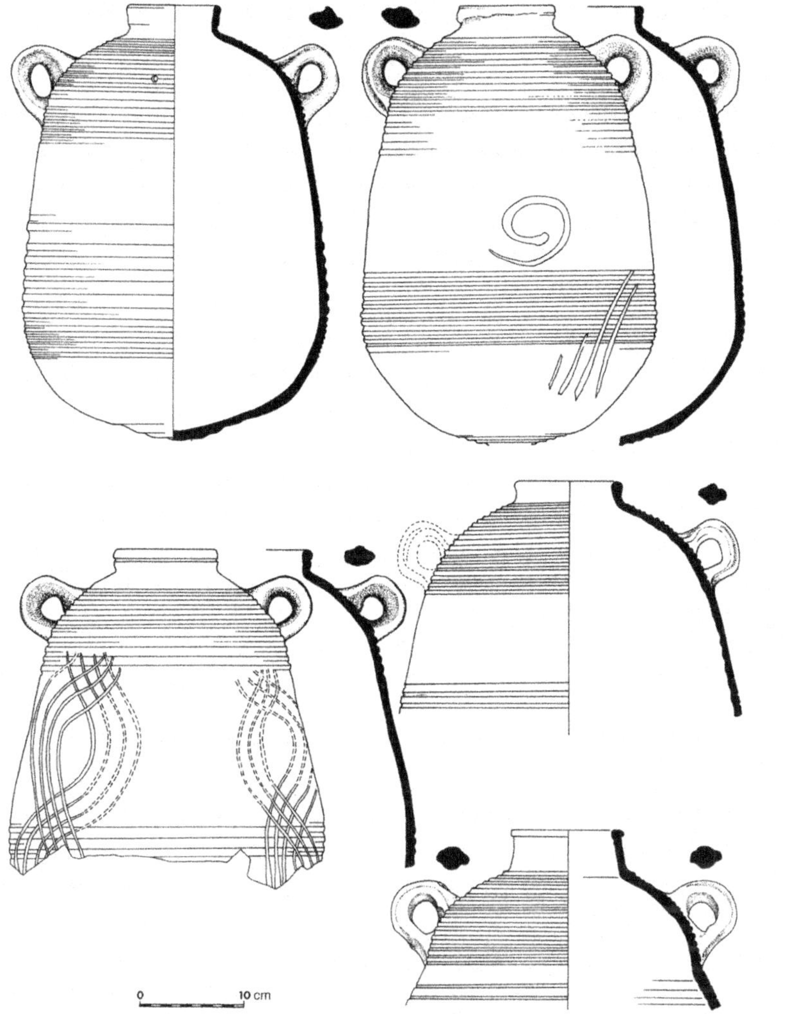

Figure 8.3. LR5 amphora (after Hirschfeld, *Ramat Hanadiv Excavations*, pl. XVII. Reproduced by permission of The Israel Exploration Society, Jerusalem).

The Scale of the Transport and Marketing of Agricultural Commodities

Up to this point, I have made a case for a market for agricultural goods and for a presence of the agents involved in this process, and have suggested that there existed a steady trade in agricultural commodities both within the empire and abroad. But what of the scale of the trade? The recovery of substantial quantities of amphorae in urban excavated contexts in places like Carthage, Rome, and Marseilles indicates regular contact, with merchants selling eastern products to citizens. But how can we translate a high proportional presence of eastern amphorae into overall quantities of foodstuffs imported? One can get some idea of this by examining the case of the metropolis of Syria I, Antioch. Although by no means typical, the city illustrates the considerable task that ancient peoples had in feeding large concentrations of urban dwellers.

The circa 150,000 inhabitants of the urban area of Antioch may be divided into 25,000 of our hypothetical families of six discussed in Chapter 3. If each family of six required about 94 modii of wheat annually, the land required to fulfil this demand would be about 13 iugera (3.25 ha) per family. This is the equivalent of the total output of 81,250 ha (about 13.5 per cent of the whole of the Antiochene hinterland of 600,000–700,000 ha). Obviously, in addition to the land and labour required to feed these citizens, the logistics of moving a

amphorae sherds of the deposits dated AD 567–96: O. Karagiorgou, 'LR2: A Container for the Military *annona* on the Danubian Border?', in S. A. Kingsley and M. Decker (eds.), *Economy and Exchange in the East Mediterranean during Late Antiquity* (Oxford, 2001), 134. Reynolds provides the best summary of western data, noting that in the Schola Praeconum I excavations at Rome, LR1 comprised 19.2 per cent of all amphorae by count in the contexts dated AD 430–40. Reynolds, *Trade*, 180–1. At Carthage, LR1 accounted for up to 47.5 per cent of all amphorae finds across a range of fifth–sixth-century contexts, becoming more frequent from the early fifth century and peaking in the mid-sixth century. Excavations in Marseilles yielded a presence of LR1 accounting for 27 per cent of total amphorae (by count of rims, bases, handles, and sherds) in the *c.*AD 400 contexts, increasing to 75 per cent by AD 450. Reynolds, *Trade*, 180–1. Inland sites also show a strong presence: at El Monastil in Alicante, Spain, LR1 accounts for 16.6 per cent of all possible amphorae vessels (based on rim, handle, and base count) in the period AD 400–75. Reynolds, *Trade*, 180–1. At Berenike in Libya, Riley's study showed that LR1 accounted for 20–45 per cent of all amphora diagnostics. J. A. Riley, 'The Coarse Pottery from Benghazi', in J. A. Lloyd (ed.), *Excavations at Sidi Khrebish Benghazi (Berenice)* (Tripoli, 1979), II.212–13, fig. 41. In eastern contexts, less quantified material is available, but Kingsley's synthesis shows that LR1 accounts for 44 per cent of all imported amphorae distributed through the three Palestines. Kingsley, 'Specialized Production', 167. In excavations of the Egyptian Red Sea port of Berenike, LR1 is present throughout fifth–sixth-century contexts, and represents up to 50 per cent of all amphorae by weight, whilst at nearby Wadi Shenshef, Tomber reports that LR1 represents 32–57 per cent of all amphorae. R. S. Tomber, 'The Pottery', in S. E. Sidebotham and W. Wendrich (eds.), *Berenike 1996: Report of the 1996 Excavations at Berenike (Egyptian Red Sea Coast) and the Survey of the Eastern Desert* (Leiden, 1998), 146–7. The use of amphora remains as proxies for pre-industrial economic activity has received mixed reactions from scholars; for a dim view, see Horden and Purcell, *Corrupting Sea*, 372–5.

minimum 30,550 metric tons of grain into the city was a considerable undertaking. Much of the grain undoubtedly reached the city by river craft, and the whole process was spread throughout the year. But grain that arrived overland demanded the efforts of thousands of beasts of burden. Cities, for the most part far smaller than Antioch, generated considerable traffic, as farmers and merchants sought to fill their need for grain and other foodstuffs. The supposed inability of ancient peoples to transport large quantities of material overland seems highly implausible when we consider the number of sizeable inland cities in Oriens. Urban centres like Hierapolis, Edessa, Chalcis, and Bostra must have depended on local and regional overland transport of food.

Like grain, wine circulated widely around the Mediterranean periphery as well as inland. Mentioned above, but worth repeating, is the compelling argument for substantial long-distance trade in wine from Oriens in the LR1, LR4, and LR5 jars. Their frequent use as wine containers helps to explain both their wide circulation and why they are found in quantity in oil-producing regions like Tripolitania. Present evidence attests the movement of wine from Oriens to every coastal region of the Mediterranean, from Spain to Turkey. The penetration of LR1 along the western British coast, around the Black Sea, and deep into Nubia hundreds of miles away is explained by the universal demand for wine and the activities of eastern merchants who sought products like tin and ivory. The *Codex Justinianus* (4.41) prohibited the traffic of wine and other goods to northern barbarian regions, thus making it clear that there was regular trade in wine throughout the Black Sea region. The appearance of LR1 at least thirty-two sites around the Crimea and the Danube shows the extent of a northern trade that is scarcely considered in secondary literature. Inland areas within the empire took part in this trade as well. The Cappadocians, who came to Mesopotamia to buy wine from the holy man Addai, had to travel at least 150 km, as Melitene was the nearest city to Cappadocia, and the wine they carried back had to be taken overland.[69]

Evidence from Vandal Carthage and from Rome affords the opportunity to quantify eastern wine exports and assess the role of LR1 amphorae in the trade. Both cities have produced quantified ceramic data, and, for both, some notion of population numbers is possible. Starting with these figures, and knowing roughly how much wine ancient town dwellers consumed, Sean Kingsley has proposed possible total consumption figures for Carthage of up to 782,000 litres per annum of Palestinian wine carried in LR4 and LR5.[70]

69 Though administratively in Armenia, Melitene was often considered part of Cappadocia. In all likelihood, the Cappadocians came from even further afield.

70 Kingsley, 'Economic Impact'; Kingsley, Johnson, and Williams, *A Sixth-Century AD Shipwreck*, 84; S. Kingsley, 'The Dor D Shipwreck and Holy Land Wine Trade', *IJNA* 32/1 (2003). I must stress that, however precise the figures appear, they are illustrative only, used to support an argument that eastern agricultural goods were regularly traded and that in

Using the same methodology, we can attempt a rough estimate for the total quantities imported into Carthage of another eastern product, the wine of Syria, Cilicia, and Cyprus, carried in LR1 amphorae. Carthage and its environs flourished in the fourth and fifth centuries and probably had about 100,000 inhabitants. Pre-industrial Mediterranean wine consumption rates were, compared to the present, quite high. At a consumption rate of 73 litres of wine a year on average for each citizen, the city would have required 7.3 million litres annually. The British excavations at Carthage record that LR1 comprised, on average, 11 per cent of the total amphora presence from selected contexts for the years 475–500. Taking 11 per cent to represent the maximum of annual LR1 wine shipments to Carthage (wine rather than oil is assumed, due to the wealth of the African provinces in the latter), the total volume of eastern wine carried to Carthage in LR1 may have amounted to some 800,000 litres a year. This trade would have filled approximately 100,000 LR1 amphorae, and required around 100 vessels the size of the seventh-century Yassi Ada freighter to transport.[71]

The population of mid-fifth-century Rome was perhaps 350,000.[72] Assuming once again that LR1 jars in Rome carried wine—and estimating from quantified data that they represented 6 per cent of the city's wine imports in the later fifth century—1.78 million litres a year (3.2 million sextarii) may have flowed into Roman ports. This is the equivalent of 222,500 LR1 amphorae, or 202 shiploads, with a total value of 11,125 solidi.[73] These calculations provide only notional maximums, but they illustrate the potentially considerable demand generated by the largest Mediterranean cities. Urban centres, like Carthage and Rome, remained dynamic and important in Late Antiquity, to say nothing of the hundreds of minor cities with populations of a few thousand. Servicing even a tiny portion of their demand for foodstuffs generated sustained and economically significant trade; several wine and oil ships per week arrived in the harbours of metropoleis like Rome and Carthage, and probably monthly during the sailing season in lesser settlements.

aggregate their quantity and value were economically significant. An obvious problem, in seeking to estimate total volume of imports from percentages of amphora sherds, is the varying size of different types of amphora.

[71] The Yassi Ada ship carried 850 mixed LR1 and LR2 amphorae of varying capacities. LR1 B jars have been assumed for this calculation, as they were commonly used to carry wine; this was the main type found in the Yassi-Ada wreck, and usually holds about 8 litres, only about a third of the capacity of large varieties.

[72] Durliat, *De la ville antique*, 117.

[73] LR1 deposits from the Temple of Magna Mater on the Palatine, deposit (Saggio) P and earlier, ranged from 2.4 per cent for Saggio A (*c*.AD 390–420), through 4.3 per cent in Saggio I-L (*c*.AD 420–40), to 6.0 per cent in Saggio P (*c*.AD 440–80). LR1 at the Crypta Balbi made up 10.5 per cent of all amphorae in deposits dated AD 410–80. Reynolds, *Trade*, 327–36.

Neither the movements of annona grain fleets, which had long ceased to serve Rome, and anyway had never come from Oriens, nor the supply of eastern imperial soldiers (who were wholly absent before the Justinianic conquests), can explain this exchange activity. Rather, the ceramic evidence is a fossil of an active commerce in the fifth and early–mid-sixth centuries.

As I have noted, a taste for olive oil generated a demand similar to that for wine. In the East, Constantinople was obviously a prime destination for both wine and oil exported from Oriens. Greece and Bithynia provided significant quantities of oil, and after the fall of Carthage to the Vandals, some North African oil still reached the East.[74] But most of the produce that flowed to the capital was from the south-eastern Mediterranean. The supply of oil and other foodstuffs, however abundant in a given year, was always a cause of imperial angst. In its prohibiting oil exports to northern countries outside the empire, the legislation, promulgated by Anastasius (*CJ* 4.41), probably betrays worries over the capital's oil supply. It also indicates that merchants found a market for oil as far away as the Crimea. In 524, a few years after the dearth in Alexandria had caused a major civil disturbance, an oil shortage afflicted Constantinople.[75] These notices are all the evidence we have of disruptions in supply to the major cities, and suggest that the supply was generally well maintained.

In seventh-century contexts at Saraçhane in Istanbul, LR1 (for purposes of illustration here assumed to be an oil carrier only) represented 15–20 per cent of all amphorae. Imported across the city at such a rate, LR would have supplied perhaps 1 million litres of Levantine oil from an annual demand in the capital for around 5 million litres of olive oil to a city whose population can reasonably be placed at a conservative total of 250,000.[76] Transportation of these 1 million litres of eastern oil may have required as many as 45,000 LR1 amphorae and perhaps seventy-five ships the size of the Yassi-Ada wreck. The total cash value of such shipments would have been something like 40,000 solidi (around 550 lb of gold).

In addition to Black Sea and Mediterranean trade, the first-century *Periplus Maris Erythraei* testifies to the circulation of olive oil along the East African coast in the Early Roman period.[77] This trade persisted in Late Antiquity. At the Egyptian port of Berenike on the western side of the Red Sea, a port of

[74] Rossiter and Freed, 'Excavations at Domuztepe', 172–3.

[75] Marcellinus, *Chronicle*, ed. and trans. Croke, 42 (AD 542); A. A. Vasiliev, *History of the Byzantine Empire, 324–1453* (2nd English edn., Madison, 1958), 374.

[76] Hayes, *Excavations at Saraçhane in Istanbul*, II.64. Assuming consumption of 20 litres of oil per person per year, based on Mattingly, 'Oil for Export?', 34.

[77] L. Casson, *The Periplus Maris Erythraei: Text with Introduction, Translation, and Commentary* (Princeton, NJ, 1989), 52–3.

call in the Roman–Byzantine Red Sea and Indian trade, finds of olive stones show importation of the fruit and probably also of olive oil. The presence of LR1 in significant quantity from the first- through fifth-century contexts at Berenike and late antique levels at Axum, strengthens the view that there were regular contacts between the eastern Mediterranean and the Red Sea coast.[78]

Other evidence for the export of Syrian oil beyond the empire comes from the Arabian Peninsula. As has been noted, the Hauran was especially well positioned to profit from this trade, since its communities had frequent contacts with the inhabitants of Arabia. Islamic period textual evidence suggests that in the time of the Prophet, Arabian caravans regularly operated to and from Syria, whence they carried grain and olive oil.[79]

Although port cities were easily nourished by seaborne supplies, scholars have taken a dim view of the capabilities of the ancients to transport bulk goods overland, where the potential costs seem to far outweigh the value of the goods if any distance was covered.[80] Such a view precludes the inclusion of inland areas into a wider Mediterranean economy of long-distance trade in basic goods. At its most drastic, such a theory limits the Mediterranean economy and its material culture to a narrow, coastal, urban freckle of cities. Clearly the inability of the ancients to ship olive oil and wine to the interior of regions like Mesopotamia would have considerably curtailed the diet in the interior of the Semitic east, and hence one of the basic expressions of the Roman lifestyle. The major consequence of this model is that the majority of rural areas suffered from a lack of market integration and communication, and thus were, economically at least, undeveloped. The social and cultural repercussions of this theoretical position are clear, since only the coastal cities could be fully integrated into the Roman–Byzantine economy.

The ability of grain merchants to supply large quantities of cereals to cities, especially those far from the sea, is a problem that needs closer examination, and the presence or absence of a land-locked grain trade says a great deal about the capacity of ancient people to move heavy, low-value goods long distances. The textual evidence for the long-distance movement of grain is

[78] J. W. Hayes, 'The Pottery', in S. E. Sidebotham and W. Wendrich (eds.), *Berenike 1994: Preliminary Report of the 1994 Excavations at Berenike (Egyptian Red Sea Coast) and the Survey of the Eastern Desert* (Leiden, 1995), 33; S. E. Sidebotham, W. Wendrich, and F. Aldsworth, *Berenike 1995: Preliminary Report of the 1995 Excavations at Berenike (Egyptian Red Sea Coast) and the Survey of the Eastern Desert* (Leiden, 1996), 443.

[79] Crone, *Meccan Trade*, 150.

[80] Finley, *Ancient Economy*, 32, 126; Jones, *Later Roman Empire*, 841–2; M. I. Rostovtzeff and P. M. Fraser, *The Social and Economic History of the Roman Empire* (2nd edn, Oxford, 1957), 146; R. P. Saller, 'Framing the Debate Over Growth in the Ancient Economy', in W. Scheidel (ed.), *The Ancient Economy* (Edinburgh, 2002), 254.

somewhat ambiguous. At a glance, textual evidence suggests that large shipments moved long distances only with imperial help, and that the grain trade was a local affair. One such example of the latter is the reviled grain merchant in the *c.*seventh-century *Life of Spyridon*, who bought corn cheaply and stored it for years while waiting to cash-in during a food shortage. When he did so, he became a hated figure in the city of Tremithus in Cyprus.[81] This example suggests that the grain supply was in local hands, and that, if crop failure affected an extensive area, long-distance supply was not readily available. Such was the case when the city of Edessa was pressed by hunger in 502. This was not a local famine, but part of a regional disaster alleviated only by the importation of grain under state directive. In such an extreme case, imperial invention was necessary.

A similar case of a suffering metropolis occurred in Caesarea of Cappadocia in eastern Anatolia in 368/9. One of the principal texts used to support the lack of overland transport, and the inability of the late antique city to be supplied overland across any distance, is a famous passage from Gregory of Nazianzus (*c.*330–90) in his funeral oration on Basil of Caesarea (*Or.* 43.34), describing the background to a famine which Basil helped alleviate:

> There was a famine, the most severe within the memory of man. The city was in distress, but there was no help forthcoming from any quarter, nor any remedy for the calamity. The maritime cities support without difficulty times of want like these, since they can dispose of their own products and receive in exchange those which come to them by sea. But we inland can make no profit on our superfluous products, nor procure what we need, having no means of disposing of what we have or importing what we lack.[82]

Both Hendy and Durliat use this passage to support their view that the ancients could not move large quantities of low-valued products, including food, overland.[83]

Other evidence, however, exists that indicates a regular overland flow of goods, even bulky staples. It is clear, for example, from a letter attributed to Basil of Caesarea (*Ep.* 345), that Cappadocia depended on grain imports from the neighbouring provinces of Galatia, Paphlagonia, and Pontus.

> For since it [the River Halys] is never crossed on foot at any time or in any manner, it does not permit our indispensable and profitable homelands to transport across it the provisions we have for sale. I mean, namely, the lands of the Galatians and the

[81] Theodorus, *Vita Spyridon* 2.

[82] Trans. L. P. Macauley: *Funeral Orations by Saint Gregory Nazianzen and Saint Ambrose* (Washington, 1968), 57.

[83] Durliat, *De la ville antique*, 516–17; M. F. Hendy, *Studies in the Byzantine Monetary Economy, c.300–1450* (Cambridge, 1985), 296.

Paphlagonians, and Hellenopontians, through which and from which we have our necessities, especially abundance of bread, since the land all around is subject to frosts, and fettered by the surrounding climate...[84]

The situation described is therefore the opposite of what Gregory implies in *Oration* 43: during the fourth century, staples arrived in Cappadocia by overland passage, but at the time of writing, this trade was impeded by the lack of a crossing on the Halys. Since the previous structure had been swept away in a flood, the writer petitioned the imperial authorities for a new bridge. The distances involved in this trade were considerable: the one-way journey from Caesarea to the border of Galatia was about 170 km, whilst Caesarea to Gangra in Paphlagonia was 270 km, and to Amaseia in Pontus 217 km. All these distances are as the crow flies, and would have been considerably longer on the ground.

One way to test the cost of the movement of goods is by examining a case in which we know a large quantity of food was moved overland, which we will do below. Libanius complained that the *translatio annonae* was ruinous to the Antiochene curials involved in supplying the distant garrison of Callinicum. In order to meet the expenses of the annona, claimed Libanius, councillors were forced to sell their estates.[85] The sophist's picture of overland victual transport as burdensome fits very well with the general scholarly consensus that such transport of bulk goods overland was prohibitively expensive. It thus offers the opportunity to test the still-current dogma of 'A wagon load of wheat... would be doubled in price by a journey of 300 miles; a camel load by a journey of 375 miles.'[86] Did land transport really represent 'the most decisive failure of ancient technology'?[87]

Overland Costs and the Market

In order to reconsider these views we must first assess the evidence for the existence, and frequency, of the traditional components of overland trade: beasts of burden; their use in caravans; and, finally, the infrastructure normally associated with land movement.[88] We have already seen that merchants throughout the Levant dealt in bulk goods, often as wholesalers, such

[84] *Ep.* 345, ed. Courtonne, 3.226–7.

[85] Liebeschuetz, *Antioch*, 163.

[86] Jones, *Later Roman Empire*, 841.

[87] Quote from: P. A. Brunt, 'Review of *Roman Farming*', *JRS* 62 (1972), 156; K. D. White, *Greek and Roman Technology* (London, 1984), 131.

[88] The overland passage of caravans in the Levant and North Africa during the medieval and early modern periods was a major pillar of commerce and is perhaps suggestive of the scale of the pre-Islamic Levantine caravan trade. E. W. Bovill, *Caravans of the Old Sahara: An Introduc-*

as those mentioned in the sixth-century collection of the Babylonian Talmud, where we find men who bought from local producers and sold their farm products elsewhere.[89] These grain and wine traders operated inland, where there was no recourse to water transport, and thus relied upon caravans to move their goods. Certainly the Rabbinic sources take for granted the prevalence of caravans of various sorts—some of them Jewish, some of them Arab. They also attest the variability of rural communities' access to such trade goods: there were inland areas where visits from travelling merchant parties were common, and others where they were not.[90] It seems probable, however, that all but the remotest villages, far removed from the larger cities, were places of call for local or long-haul caravans.

The caravans were of two sorts: the donkey train and the camel train. Donkeys were ubiquitous in the Roman East, as they had been for centuries, and there are plentiful references to donkey drivers in Rabbinic sources and papyri.[91] A Syrian mosaic depicts an ass transporting wineskins, and while individual farmers frequently depended on this animal as their only means of transport, teams of donkeys with professional drivers were also common. Typically these donkey caravans were used by itinerant merchants for short-haul trading as they moved from village to village.[92] In the days of the Old Assyrian Empire, donkey trains were the major caravan type, and apparently some late antique merchants who travelled far afield continued to rely on donkeys. The late-sixth- or early-seventh-century Nessana accounts of a merchant group (P.Ness. III.89) documents the activity of a cadre of merchants who used both donkeys and camels to haul dates and other commodities from the Negev to Egypt, a one-way journey of perhaps 400 kilometres. Across the East, a variety of contractual situations existed; sometimes merchants owned their own animals, but often individual donkey drivers were hired, typically along with their own animals.[93]

We can gain a rough cost estimate of operating a donkey caravan by using comparative data. In a calculation of this kind, there are numerous assumptions that have to be made. It is therefore important to stress once more that,

tion to the History of the Western Sudan (London, 1933); E. W. Bovill, *The Golden Trade of the Moors* (London and New York, 1958); D. Carruthers, 'The Great Desert Caravan Route: Aleppo to Basra', *Geographical Journal* 52/3 (1918).

89 TB B.M. 48a.

90 TB B.M. 18a.

91 *Geoponica* 10.90.3; 16.20.4; 18.15.3; TB Ḥag. 20a; TB Kid. 82a; TB 'Er. 35a; TB Sanh. 111b.

92 Safrai, *Economy*, 234–5.

93 Like their modern counterparts, ancient donkey drivers probably owned between two and five animals. F. Gebreab et al., 'Donkey Utilisation and Management in Ethiopia', in P. Starkey and D. Fielding (eds.), *Donkeys, People and Development: A Resource Book of the Animal Traction Network for Eastern and Southern Africa (ATNESA)* (Wageningen, 2004), 46–52.

like the other quantified test cases in this work, the present discussion is useful only in establishing some possible relative costs, not in providing absolute transportation costs.

We can now return to the case cited by Libanius, in which curials of Antioch were obliged to supply the *annona militaris* to the garrison at Callinicum, just over 250 km away. Both the Theodosian (7.5.2) and Justinianic Codes (12.38.2) reveal that the duties of the *translatio* were undertaken unassisted, though in some instances the distances involved were very limited. The Theodosian Code (7.4.9) tells us that imperial troops carried fodder from the twentieth milestone of a city; presumably provincials were obliged to provide transport over the first 20 miles. However, Libanius' complaints must have been inspired by an onus far more severe than moving food 20 miles. For the purposes of this study, I assume that the curials paid for movement of the annona for the entire journey from Antioch to Callinicum. This would not have been unprecedented, and was probably in fact the norm: in 504–5, for instance, the citizens of Edessa engaged in the long-haul supply of soldiers under state compulsion and in difficult circumstances.[94]

The fifth-century *Notitia Dignitatum* records an *ala* of *equites promoti Illyriciani* stationed at Callinicum, a unit which at full strength perhaps numbered 500.[95] If the food shipment from Antioch was intended to supply wheat to this force for a year, 196,500 kg would have been required. This quantity of wheat was worth between 440 and 1,000 solidi (6–14 lb of gold). The transport of such a shipment would have required a train of 3,023 donkeys, each able to carry 65 kg. The movement of the annona grain would have thus required about 750 drivers, who over a month would have been able to complete two trips between Antioch and Callinicum. A driver's wages in Late Antiquity, which included the cost of his animals, was about 4.5 keratia per month. The Antiochenes would have thus paid something like 225 solidi, a little more than three pounds of gold.[96] This represents half the value

[94] ps.-Joshua, *Chron.* 99–100.

[95] *Notitia Dignitatum*, ed. Seeck, 35.16; W. T. Treadgold, *Byzantium and its Army, 284–1081* (Stanford, 1995), 89.

[96] According to Zachariah of Mytilene (trans. Hamilton, p. 166) when Anastasius built Dara, he paid exorbitant sums to labourers to induce large numbers of workers to come to the site. Workers at Dara who came with their donkey received 8 keratia per day, about double what other evidence suggests that donkey drivers earned for an entire month. Egyptian evidence suggests a wage of 4–4.5 keratia per month, whilst Palestinian donkey-drivers would cover the 250 km from Antioch to Callinicum in approximately four stages and earn wages of 4 keratia: Sperber, *Money and Prices*, 124. The keration is an accounting value and is used here for the purposes of illustration. The relationships obtained are only approximate, since substantial regional variation through time is to be expected; 15 keratia equals one solidus. We should further reflect that this shipment thus hypothesized, represents a worst-case scenario, in which there is no possibility of profit by trafficking goods on the return.

of the grain that they transported, not a doubling of the price of goods carried this distance overland, as suggested by Jones. The fact remains, however, that the enterprise was not cheap compared to the value of the goods transported. Given the paucity of data and their imperfect nature, these figures could also easily be doubled, in which case they would uphold the tenets of those who consider overland trade to have been prohibitively expensive.

It is also very important to bear in mind the price of grain was not a fixed one, and that in reality it varied greatly according to place and to circumstance. A merchant who bought grain in the quantity discussed above for 6 lb of gold, and paid a donkey caravan another 6 lb of gold, would, to make a reasonable profit, have had to sell the grain for some 14 lb. This was at the high end of the estimated value of the grain noted above. But, as we have seen, regional and seasonal price differences normally created opportunity for profit. In the Early Roman period, wheat was typically twice as expensive in Palestine as in neighbouring Egypt, and Egyptian traders frequently brought grain to Palestine in order to profit from the more competitive markets there.[97] The movement of a large quantity of a bulky low-value food staple by donkey caravan was expensive, but not outlandishly so.

Camels, Caravans, and Late Antique Trade

Over long distances, donkeys were perhaps a relatively costly option and thus more likely to be used for local or regional transport. On longer journeys of more than a few days in duration, asses were clearly outclassed by camels. Camels, or (more accurately) dromedaries, ranged widely over the Levant and North Africa and were certainly important in Late Antiquity.[98] The ability of these beasts to survive in marginal environments, common in the Levant and the Arabian Peninsula, recommended them and facilitated their spread: dromedaries often went for days without requiring water (*Geop.* 16.22) and fed on rough grazing and shrubs that cattle and equids found unpalatable.[99] Diocletian's Price Edict (§14.9) suggests that camels typically carried 600 Roman pounds (196 kg), three times the 65 kg load of the donkey. These advantages help explain the extensive use of camels in Palmyra, where a second-century edict demonstrates the animal was a common vehicle. Palmyrenes

[97] Sperber, *Money and Prices*, 126–7; D. Sperber, 'Objects of Trade Between Palestine and Egypt in Roman Times', *JESHO* 19 (1976).

[98] R. W. Bulliet, *The Camel and the Wheel* (2nd edn, New York, 1990), 22. In this discussion 'camel', refers to one-humped dromedaries, not to Bactrian two-humped camels, though Bactrian camels were present in much of the empire and surrounding lands in Antiquity.

[99] In Provincia Arabia, camels were fed on thistles and thorns left in the grain fields, a practice forbidden by Jewish religious custom: TB B.B. 156b; TB Shab. 144b.

conducted long-distance trade in the region of the Persian Gulf by dromedary caravan (συνοδία), witnessed by epigraphic data and stone reliefs depicting the loaded animals.[100]

In Late Antiquity, camels were a common beast of burden of the landscape in most of the empire. They were frequently found around Antioch and were even ubiquitous in Anatolia: in Cappadocia, they provided meat and milk, and in Galatia, the presence of camel herders is attested by the record of their visit to the seventh-century St Theodore of Sykeon (d. 613).[101] A terracotta camel with its wine amphorae, and the mosaic of Kissufim in Palestine, where the merchant or farmer Orbikon is shown moving his wine to market, are reminders that these animals were a standard feature of Late Antiquity.[102] Several Talmudic passages refer to the donkey and camel caravans that stayed within cities and camped in the countryside on their route of march. In the Babylonian Talmud, we find them in the company of Arab merchants, carrying flax, in breeding herds, or being treated for ailments.[103]

From the fourth through the sixth centuries, when the *limes* from the Hauran to Chalcis reached a peak of agricultural development, the interplay between semi-nomadic groups and settled farmers intensified. Trade generated along the meeting-places of the desert and the sown was one driven by mutual need.[104] As farming settlements expanded into traditionally stock-raising districts, herders played an even greater part in the life of sedentary communities. Frequent movement and mixing was conducive to trade, all of which was conducted overland. As had traditionally been the case, transient herding groups traded pastoral products for goods from the settlements, provided labour during the harvest, and grazed their flocks on the stubble of harvested fields. A certain symbiosis was thus maintained, as Haiman proposes for the Negev Highlands, and Banning envisions for the entire frontier zone, where semi-nomadic groups became dependent on settled farmers and sometimes joined them by forming semi-permanent or perman-

[100] M. Gawlikowski, 'Palmyra as a Trading Center', *Iraq* 56 (1994); M. Gawlikowski, 'Palmyra and its Caravan Trade', *AAAS* 42 (1997); F. Millar, 'Caravan Cities: The Roman Near East and Long Distance Trade by Land', in M. Austin, J. Harries and C. Smith (eds.), *Modus Operandi: Essays in Honour of Geoffrey Rickman* (London: Institute of Classical Studies, School of Advanced Study, University of London), #690; Matthews, 'Tax Law of Palmyra'; Waddington, nos. 2589, 2590, 2596.

[101] R. Teja, *Organización económica y social de Capadocia en el siglo IV, según los padres capadocios* (Salamanca, 1974), 31. Theodore of Sykeon, *Vie de Théodore de Sykeôn*, ed. Festugière, 3.

[102] A. Ovadiah and S. Mucznik, 'The Mosaic Pavement of Kissufim, Israel', in H. Stern (ed.), *Mosaïque: recueil d'hommages à Henri Stern* (Paris, 1983).

[103] TB Sanh. 32b; TB Ḥul. 76a; TB B.B. 74a, 85b, 93a; TB Shab. 21b, 154b; TB B.B. 74a.

[104] Crone, *Meccan Trade*, 151.

ent settlements.[105] Farmers, in turn, relied on the state to provide security along the frontier zone from periodic upsurges of violence by these volatile nomads. In this, the Roman authorities from Diocletian to Heraclius seem to have been, for the most part, successful. Despite occasional violence, most of the desert margin, from the Euphrates to Palestine, remained reasonably stable, first under direct Roman control and then under that of their clients, the Ghassānids.[106]

Land communications all over Oriens were generally good, facilitated by major roads that continued to be maintained throughout Late Antiquity. As city demography altered, the road system changed to reflect the new human geography of the East. The dense network of settlements, whose closeness rendered them easily accessible to one another, is nowhere better reflected than in the mosaic masterpiece, the Madaba Map, from the province of Arabia.[107] Throughout the coastal regions of the diocese, north–south highways like the Via Maris were linked with other branch roads of only regional and local importance. On the eastern margins, the developed road system around Bostra linked with the Strata Diocletiana, which carried traffic the length of the frontier to the Euphrates. Bostra was also the terminus of the southern trunk road, the Via Nova Traiana that connected the Hauran to the Red Sea. The east–west networks of roads, criss-crossing the frontier region of Chalcis and Mesopotamia to as far west as the Phoenician coast, are well known, especially in northern Oriens, where the most famous of these was the military route that led from Antioch to the Euphrates via Hierapolis.[108]

The frequency of Late Roman forts and way-stations, as well as the high density of settlements, offered rest and provisions to passing caravans, outlets for their goods, and the opportunity to procure other goods. The Roman roads and other facilities that rendered the movement of merchants and other travellers easy are well attested.[109] Numerous inns and way-stations are recorded in the textual and archaeological record, including places of refuge

[105] E. B. Banning, 'Peasants, Pastoralists and Pax-Romana, Mutualism in the Southern Highlands of Jordan', *BASOR* 261 (1986); Haiman, 'Agriculture and Nomad–State Relations', 34–5.

[106] Parker, 'Peasants, Pastoralists, and Pax-Romana: A Different View', 35–51.

[107] D. Graf, 'Roman Roads East of the Jordan', in M. Piccirillo and E. Alliata (eds.), *The Madaba Map Centenary, 1897–1997: Travelling through the Byzantine Umayyad Period: Proceedings of the International Conference Held in Amman, 7–9 April 1997* (Jerusalem, 1999); I. Roll, 'The Roads in Roman–Byzantine Palaestina and Arabia', in M. Piccirillo and E. Alliata (eds.), ibid., 109–13.

[108] M. Fisher, B. H. Isaac, and I. Roll, *Roman Roads in Judaea*, II: *The Jaffa–Jerusalem Roads* (Oxford, 1996); B. Isaac and I. Roll, *Roman Roads in Judaea*, I: *The Legio-Scythopolis Road* (Oxford, 1982); Mouterde and Poidebard, *Limes de Chalcis*; Poidebard, *Trace de Rome*.

[109] Fisher, Isaac, and Roll, *Roman Roads in Judaea*, II; Y. Frankel, 'Roads and Stations in Southern Bilad ash-Sham in the 7th–8th Centuries', *Aram* 8 (1996); Graf, 'Roman Roads'; Isaac and Roll, *Roman Roads in Judaea*, I; Mouterde and Poidebard, *Limes de Chalcis*; R. Paret, 'Les

(*phrourion*). Some of these were built by monks whose monasteries lay close to the roads. Military stations (*mētata*), hostels (*xenodocheia*), and inns or lodgings (*pandocheia, katagōgia, katalumata*) are also attested. Inscriptions commemorating the construction of *phrouria* are known throughout Syria; in the Chalcidike, they have been found at Tarutia Emporon and Ma'an.[110] These refuges undoubtedly served as rest-stations along the roads which they fronted, as did the *mētata* or *mitata*, billeting or military staging-posts. The *mētaton* at Raphanea was dedicated to St Sergius. Others are known from inscriptions at al-Burj, Jawr, and probably Deir Solaib.[111]

Epigraphic data attest *xenodocheia* (hostels) at Umm al-Khallahil in northern Syria, Soada (Suweida) in the Hauran, Ma'in in Arabia, and Jerusalem.[112] Cyril of Scythopolis records the presence of several Palestinian xenodocheia, some clearly belonging to monasteries.[113] The remains of a *xeneon*, a way-station that perhaps served as a hospice for the sick as well as for travellers, is known from Umm al-Khallahil in northern Syria not far from Hama.[114] The courtyard building, about 60 m long and 44 m wide, is analogous in form to later *khans*. At Antioch, there were many similar structures that served the sick and travellers, and the emperor Julian noted them as important in the underpinning of Christian success.[115] These institutions were a universal feature along the highways and byways of the late antique East.[116] There are two probable Byzantine–Early Islamic khans at Palmyra and Rusafa, both serving caravans travelling the Euphrates–Bostra corridor.[117] In Orshoene, the third-century governor Aurelius Dassius built a *pandocheion*, an inn, near Edessa, and during the year 397/8, several men dedicated a similar pandocheion at Harran near Bostra.[118] At Rimat 'Azim another is known from an

villes de Syrie du Sud et les routes commerciales d'Arabie à la fin du VIe siècle', *Akten des XI. Internationaen Byzantinischen Kongresses 1958* (Munich, 1960).

110 *IGLS* 4.1631, 1809.

111 Raphanea: *IGLS* 4.1397.1. The dedication to St Sergius makes it possible that this was a Ghassānid installation. El Burj: *IGLS* 4.1610.1 included a *pyrgos* and was dedicated to the Archangel Michael and St Longinus; Jawr (uncertain): *SEG* VII.110; Deir Solaib: *IGLS* 4.1385.

112 Umm al-Khallahil: *IGLS* 4.1750; Soada: Le Bas–Waddington, no. 2327, undated but Christian (naming St Theodore); Ma'in: *SEG* XXXV.1581, dated AD 598/99; Jerusalem: *SEG* XXXVI.1350.

113 Cyril of Scythopolis, *Vit. Sab.* = *Kyrillos von Skythopolis*, ed. Schwartz, 109.14–15, 16.10, 21; 86.19, 28; 94.3.

114 Lassus, *Inventaire archéologique*, 67.

115 Sozomen, *Historia Ecclesia* = *Kirchengeschichte*, ed. Bidez, 5.6.19; Julian, *Ep.*, ed. Wright, 84.23; the ecclesiastic origin and charitable function of the installations is further underscored by Justinian: *Nov.* 57.38; 62.36; 319.27.

116 Eusebius, *HE*, ed. Bardy, 7.22.4 (*pandocheion*).

117 Palmyra: J.-M. Dentzer, 'Khans ou casernes à Palmyre? À propos de structures visibles sur des photographies aériennes anciennes', *Syria* 71 (1994); Rusafa: W. Karnapp, 'Der Khan in der syrischen Ruinstadt Resafa', *AA* 93/2 (1978), where pilgrims were probably the major clientele.

118 C. A. Mango, 'A Late Roman Inn in Eastern Turkey', *OJA* 5 (1986); Waddington, nos. 2462–3; Butler et al., *Syria*, III.A, no. 794.

undated dedicatory inscription. Better known are the traces of two pandocheia at Deir Simʿan built for pilgrims as a shrine to St Simeon, both dated to AD 479.[119]

Caravans also travelled through much rougher territory that often lacked way-stations and other facilities. Groups like those which plied the desert roads and trackways of Parapotamia through the Hauran and Palmyra are attested in early Islamic sources and are probably indicative of the fourth–seventh centuries. The eighth-century writer Ibn Isḥaq preserves a story in which a caravan of the Kalb tribe travelled from Nisibis to the Arabian Peninsula.[120] Natural stopping points along the desert route were the forts and inns of the Strata Diocletiana and cities like Resafa. These features in the human geography of Oriens did not make overland trade cheaper than waterborne transport, but they did facilitate the process of moving goods. Byzantine merchants did not sail to Medina, just as Arab tribesmen did not sail to Bostra. As Crone observes for Bedouin and itinerant merchants, travelling was simply another way of living.[121]

In Late Antiquity, Bostra and Petra remained important centres at the end of the long-distance trade routes which traversed the Arabian Peninsula. During the Early Islamic period, Syrian grain supplied the interior of Arabia: in some instances, cereals were carried part of the way by camel, then floated down the Euphrates to Lower Mesopotamia, and beyond.[122] This is not unlike the movement of goods during the Ottoman period, and immediately after, when caravans journeyed from northern Mesopotamia and Arabia to Iraq and, after selling their goods, made the return march up the Euphrates.[123] Modern travellers have observed that eastern caravans frequently carried low-value items such as soap, from Antioch to Caesarea and along other routes, like that from Aleppo to Ankara, as well as to more distant points like Amaseia, Diyarbakir, Mardin, and Mosul. Trade among these cities was, as Weurlersse noted, predicated upon the low cost of transporting freight.[124] Common, heavy goods were also a frequent item of trade; Ottoman-period quarries in the Leja of south-eastern Syria exported millstones as far as Aleppo.[125] In the same period, caravans carried apricots, raisins, wine, nuts, olives and olive oil, barley, wheat, tobacco, potash, and wool, not to mention sizeable quantities of raw silk, from

119 Butler et al., *Syria*, III.B, nos. 1154–5.

120 Ibn Isḥaq, *Life*, ed. Guillaume, 96.

121 Crone, *Meccan Trade*, 158.

122 M. Lombard, *The Golden Age of Islam* (Amsterdam, 1975), 163–4.

123 C. P. Harris, *The Syrian Desert: Caravans, Travel and Exploration* (London, 1937), 131–53.

124 J. Weulersse, 'Antioche, essai de géographie urbaine', *Bulletin d'études orientales* 4 (1934), 66–8.

125 Merrill, *East of the Jordan*, 24–5.

Berytus to Damascus, Mosul, and other eastern cities.[126] As demonstrated below, pre- and early-Islamic sources indicate that similar low-valued products, including food, were exchanged along these routes during Late Antiquity.

The outfitting and travelling expenses of a caravan were potentially relatively low.[127] A second-century Palmyrene inscription commemorates the euergetism of Thaimarsas, who provided 300 gold *aurei* (perhaps 360 later solidi) for the expenses of a caravan from Palmyra to Spasinou Charax on the Persian Gulf, a journey of about 1,000 km.[128] This cost of a Palmyrene caravan is quite close to the amount of money a captain needed to outfit and load a merchant vessel in seventh-century Alexandria.[129] Unfortunately, we have no information as to how the sum, noted from Palmyra, was spent, nor do we know the size of the caravan (though it must have been sizeable to have warranted the thanks of the entire guild of caravaneers).

Many long-distance caravans numbered over 1,000 camels, and it is on this scale that we should envisage the operations of a merchant metropolis like ancient Palmyra and probably the large cities of Late Antiquity as well.[130] Large-scale animal trains were certainly involved in the overland grain trade between pre-Islamic Arabia and Syria. ʿAbd al-Rahman b. ʿAwf exported skins, cottage cheese, and clarified butter to Syria. The sale of these products provided him with the means to import Syrian grain with a caravan of 700 camels. ʿAbdallah b. Judʿan purportedly sent 2,000 camels to Syria, seeking clarified butter and honey to distribute in Mecca. During the lifetime of Muhammad, Byzantine merchants and 'Nabataeans' from Syria carried on active exchange with the Arabian Peninsula. They sold grain and oil to the settlements in the Arabian Peninsula—among them, Medina, the north-western Arabian city of Dumat al-Jandal, and Yamama (about 70 km south-east of present-day Riyadh)—and bought dates for the return trip.[131] The regular merchant activity and food deliveries to Mecca, approximately 1,150 km from Bostra, were certainly exceeded between sedentaries and nomads living close to one another along the frontier. Although the role of the state in this trade is uncertain, the Byzantine government continued to maintain and police this frontier. It is striking that, at the time of the Muslim conquests, the Byzantines maintained a post at Tabūk, some 700 kilometres south of Damascus.[132]

[126] Ibid., 96.

[127] Crone, *Meccan Trade*, 158.

[128] Heichelheim, 'Roman Syria', 210.

[129] Leontius, ed. Festugière and Rydén, 8.1–77.

[130] Harris, *Syrian Desert*, 144.

[131] Crone, *Meccan Trade*, ʿAbd al-Rahman b. ʿAwf: 98; ʿAbdallah b. Jud'an: 104; grain and oil to Dumat al-Jandal and Yamama: 39–41.

[132] H. Seyrig, 'Antiquités syriennes: postes romains sur la route de Médine', *Syria* 24 (1941), 218–23.

Returning to the scenario of the Antiochene transport of annona grain to Callinicum, it is useful to discuss a minimum-cost scenario for the transport of such a huge quantity of grain. As camels in the Roman period could carry 600 Roman pounds—and positing one driver per camel and two round trips between the cities over a single month of hire—500 camels and drivers would be required. Assuming the same conditions of hire and distances travelled for camels as for donkeys, transport of the annona from Antioch to Callinicum would have cost 150 solidii, about 2 lb of gold. Thus, utilizing camels to transport the grain could have reduced the transport costs to approximately 14–33 per cent of the value of the cargo.

A substantial body of evidence suggests, and these economic estimates support, the notion that low-value goods commonly circulated around the interior of the Levant during Late Antiquity. Land transport inland was performed by caravans of donkeys over short distances and, more often, by camels over longer routes of travel. Caravan traffic was neither expensive nor rare. The donkey and camel caravans of the Rabbinic sources carried relatively low-value, bulk goods: grain, wine, olive oil, and clothing. The long-distance caravans of pre-Islamic Arabia carried similar goods. It is time to rethink the nature of overland trade: it was neither dominated by luxury goods, nor was it infrequent.

Conclusion

Rarely, if ever, in the history of the pre-industrial Mediterranean have levels of agrarian development, intensity of settlement, and a combination of security, easy communications, and monetization coalesced in the way that they did in the late antique East. The relative peace and stability of centuries of Roman rule continued over large areas of the Levant through the fourth to seventh centuries. Plague and warfare in the sixth century dampened commerce, but they destroyed neither the deep external connections, nor the impetus of movement and trade, that social conditions had instilled within the peoples of the Levant. The degree to which the material culture of the eastern Mediterranean flowed into the void left by the disintegration of the Roman West is reflected in major cultural markers: the penchant for eastern goods abroad, the style of church building, and the continued observance in western chroniclers of eastern regnal years. This was a brief moment, between the collapse of the western state and the coming of Islam, when there was no ideological or commercial alternative to the Roman East. The balance of Mediterranean trade shifted momentarily but decisively to the Levantine seaboard and its outlying colonies of Greeks, Cilicians, Syrians, and Egyptians throughout Italy, Gaul, and Spain.

Contemporaneously, in Oriens, there were important changes within the fabric of the rural landscape. The isolated farmsteads of the Negev or the Hauran were novel in a land that traditionally belonged to villages and towns, but they sprang up by the hundreds and persisted for centuries. Whatever the security concerns of those wealthy farmers who built the fortified farms at al-Tuba and Stabl Antar, there were clearly others who felt the press of neither Persians nor predatory nomad. In part, the dusting of farms over the wider countryside reflected a real need for more land, as populations increased at least until the mid-sixth-century Justinianic Plague. The rural dwellers of Late Antiquity built up villages, farmsteads, and monasteries into the heart of the drylands in the Negev, Judaea, the Biqāʿ, and the plains of Chalcis, and transformed a patchwork rocky desert or burning steppe into fertile land. The resultant country owed its existence to a whole range of different actors. Much of this landscape belonged to imperial estates (and we recall that large areas were worked by tenants of the state under emphyteusis), to the church, or to aristocrats who had come to possess tracts of former public lands. So inhabitants, like the benefactor of Androna, Thomas, were apparently weal secular or ecclesiastical estate owners. Others, however, like the inhabitants of the villages of the Limestone Massif, or those of the Hauran, were tenant- or independent, small-scale farmers. The settlements of Aramaic and Arabic-speakers, in places like Nessana or the Chalcidike, no doubt enlivened relations with the empire of Persian Mesopotamia and the Arabian Peninsula.[1] Places like Bostra and Chalcis fronted a sea of steppe to the East, with which they dealt equally, or more intently, than with the more distant Mediterranean.

Beyond the need for land, insufficient in itself to explain the colonization of the marginal regions of the empire, desire and know-how made such exploitation possible. Intensive farming practices were not new to Late Antiquity. Indeed, many of the components that formed the integrated farming systems that I have discussed throughout this work were known before or during Classical Antiquity.[2] Floodwater farming, qanat, and saqiya were technologies far from novel; but they were new in their range of use and spread among the people of the late antique Levant. These technologies joined centuries-old means of exploiting the land. In a carefully managed countryside, old patterns of dry-farming, crop selection, grafting, and intercultivation mingled with newer forms of working the countryside, like ley-farming, the use of the screw-press, and the sowing of new crops like sorghum. Something of this efflorescence is owed to the class of men who produced the last Roman agricultural handbooks, men like Vindanius Anatolius and Cassianus Bassus (whose works form the bulk of the *Geoponica*).[3] Roman bureaucrats and estate owners crafted manuals that, like those of their Hellenistic predecessors, expressed a desire for better ways of doing things, not in the cities or court, but on their own land.[4] They experimented in their gardens and orchards and competed with one another over the best ways to graft a tree or breed cattle. They ordered the land and profited from it.

Nor were their efforts the mere musings of armchair farmers. Vernacular directives on agriculture, like the Coptic almanacs, that provided farmers with the best planting times and ordered the agricultural year, and the translation of Roman farming manuals into Syriac, suggest a deeper penetration of these methods into the local elite and probably further down the social ladder.[5]

[1] For example, Sergius Jammal, whose death in 572/573 is commemorated in an inscription near Androna: R. Hoyland, *Arabia and the Arabs: From the Bronze Age to the Coming of Islam* (London, 2001), 237, fig. 8.

[2] Theophrastus, *Hist. Pl.*, ed. Wimmer, 8.7.7; *Caus. Pl.*, ed. Wimmer, 4.3.4.

[3] Appendix.

[4] Vindanius Anatolius was probably Prefect of Illyricum in the fourth century. See Appendix.

[5] T. G. Wilfong, 'Agriculture among the Christian Population of Early Islamic Egypt', in A. K. Bowman and E. L. Rogan (eds.), *Agriculture in Egypt from Pharaonic to Modern Times* (Oxford, 1999).

Cassianus Bassus boasted of his experiments, while his tenants and hired hands worked using their master's methods, then went home to their own small plots, orchards, and screw-presses. Farming in this period was frequently far from new or scientific, but sometimes, when compared with previous methods, it was progressive, rational, and persistent.

The face of the landscape, changed by rich and poor alike, in response to need or desire, meshed with the wants of an eastern empire framed by the great metropolitan cities of Constantinople, Antioch, and Alexandria. The hunger and thirst of New Rome was sated by a vast hinterland, once the garden of Old Rome, and spread across the coastlands and uplands from Anatolia to Egypt. Even as the farms and fields of the Roman West foundered under the disruption and dislocation of the fifth century, the integration and dominance of the East rose, confident in its ideas and ideology, and in its economic foundations.

Wine and oil from the presses did not flow only to the sea, but to the steppe as well. The landlocked cities of the East—Bostra, Chalcis, Edessa, and Hierapolis—were orbited by a constellation of small new cities and large villages linked by a network of roads and tracks which ran throughout Mesopotamia and into Arabia. In these regions, the worlds of the pastoralist and the sedentary farmer met, and fairs, *pandocheia*, highways, and markets served not only soldiers and the rich, but very ordinary Greek-, Aramaic-, Armenian-, and Arabic-speaking populations. Their goods arrived in donkey and camel caravans, whose frequency, if not their speed, rivalled the merchantmen that put to sea from Gaza, Caesarea, and Seleucia to Rome, Carthage, and Marseilles. Perhaps more than ever before, the Roman East was close to the Arabian Peninsula: when the Christians at Najran, far to the south, suffered martyrdom in the sixth century, the news spread with powerful effect over the whole of the East.[6]

Modern scholarship has yet to reach a satisfactory understanding of the impact of the Justinianic Plague on the society of the eastern Mediterranean and to agree whether or not the later sixth century saw a marked falling off of the prosperity of the fifth and early sixth centuries.[7] New research must be undertaken before we can advance beyond the limits that our texts to date have imposed on us. To my mind, the impact of the plague was severe, and in combination with a number of other disasters (such as earthquakes and Persian invasion), the cocktail of calamities reduced the prosperity and vivacity of eastern life. But the East was not the West, which was indubitably

[6] I. Shahîd, 'Byzantino-Arabica: The Conference of Ramla, A.D. 524', *JNES* 23/2 (1964).

[7] L. K. Little, *Plague and the End of Antiquity: The Pandemic of 541–750* (New York and Cambridge, 2007) is a start, but much more needs to be done. Specialist studies on mortality rates and the spread of the plague are needed, like that of M. McCormick, 'Rats, Communications, and Plague: Toward an Ecological History', *JInterdiscipHist* 34/1 (2003).

in severe decline: when the armies of Islam came, they found Damascus full of people, Jerusalem likewise, and places like Palmyra vigorous enough, and perhaps 'Roman' enough, to resist.[8]

With the coming of the armies of the Prophet and decades of warfare, the northern regions of Oriens became a border zone. The Limestone Massif villages, once the hinterland of Syria's greatest city, and the steppelands around Chalcis, formerly the scene of a flurry of late antique building and farming, now lay astride the battlegrounds between Byzantium and Islam. We know little of the conquests themselves, much less about the movements of people who were displaced or fled the Muslims, or those removed by either state. What is certain is that the northern territories in Mesopotamia and around Antioch were the threshold of the *Dar al-Harb*, the House of War. Unlike the southern territories of Oriens, where building inscriptions continued through the Persian period and the Muslim invasions, the swift decline in north Oriens of Greek building inscriptions hints at a precipitous deterioration of the economy. Unmistakable, however, is the cultural continuity in the landscape over much of the Umayyad caliphate: Greek continued to be spoken in many places, and Christians remained the overwhelming majority. This cultural continuity was probably built on broad economic continuity; although some cities like Caesarea suffered prolonged siege and sack, cities like Scythopolis and Gerasa got on with their affairs, just as settlement persisted in marginal areas like the Negev and the Balkh in Transjordan. If, as some claim, the Muslims did ravage the land and sever old connections, they also built new ones. Indeed, the Muslims felt most at home in the area that had once been the contested periphery between desert and sown, between Persia and Byzantium, and later between Muslim and Christian. The garland of Umayyad 'desert castles', with their water installations, gardens, and *khans*, mirror the world that preceded them. On balance, it seems that some vibrancy of the economy of the Levant continued well into the seventh century, and quite probably beyond it.

Indications of material continuity and general prosperity are especially evident in the south of Oriens. The epigraphic record there indicates broad continuity across the sixth and seventh centuries. Evidence for widespread destruction and immediate decline are lacking: Scythopolis-Bet She'an, for example, continued to flourish in the Umayyad period and had its public areas substantially renovated under the early caliphate.[9] The end for that city seems to have come not from the Muslim conquests, but from an earthquake

[8] Ward-Perkins, *The Fall of Rome and the End of Civilization.*

[9] E. Khamis, 'Two Wall Mosaic Inscriptions from the Market Place in Bet Shean/Baysān', *BSOA* 64/2 (2001); Tsafrir and Foerster, 'Urbanism at Scythopolis-Bet Shean'.

in 749.[10] Similarly, Pella, Bostra, and other cities in southern Oriens continued on, though not unchanged, and perhaps not unscathed.

There is little better expression of this persistence than the mosaic from Deir al-Adas in southern Syria. Created in AD 621, when the eastern Mediterranean was throttled by the war between Byzantium and Persia, the bottom of the panel depicts olive trees clad with vines, whose fruit is dutifully harvested by the estate's workers. Just above them, the pleasures of the hunt are on full display, with the beaters herding animals into nets and towards hides that conceal the wealthy huntsmen now lost from view. Here we witness both the toils of the estate, and the pleasures that it sustained. Finally, we view the scene that the patron could not have imagined omitting: a figure leads a string of camels laden with sacks and skins on the journey from press and granary to market and money.

[10] Y. Tsafrir and G. Foerster, 'The Dating of the "Earthquake of the Sabbatical Year" of 749 C.E. in Palestine', *BSOAS* 55 (1992).

Appendix: The Value of the *Geoponica* as a Source for Late Antique Agricultural Practices

The *Geoponica* was compiled during the reign of Constantine VII Porphyrogenitus (emperor AD 913–59), to whom it was dedicated.[1] Constantine's scholars created a medieval smash hit: 50 *Geoponica* manuscripts survive.[2] In the late nineteenth and early twentieth centuries, the text was closely examined and subject to a flurry of scholarly attention that has yet to be replicated. In addition to Heinrich Beckh's edition, Wilhelm Gemoll, Eugen Oder, and Eugen Fehrle undertook detailed source studies of the *Geoponica*.[3] Their work revealed that it contains virtually no medieval material, although it was itself widely applied to agrarian practice in the post-Roman world, especially in the eastern Mediterranean and Slavic world. Although the excerpting and recycling of material in such a fashion may seem strange to classicists and ancient historians, those familiar with Byzantine civilization are aware of the latter's predilection for encyclopaedism, which especially characterized the scholarship of the ninth to eleventh centuries. The Macedonian Dynasty desired to cultivate and render more meaningful its connection to the Roman past, and to this end encouraged research into the administrative and cultural practices of antiquity. One facet of this revival of interest in classical work was the perceived need to systematize, preserve, and apply classical knowledge, including technical works, and to shape it into useful handbooks like the *Geoponica*.[4]

Although the *Geoponica* and its source documents have a long and complicated history, with some questions unresolved, there is broad scholarly agreement about several major items of importance for our study. First, that the *Geoponica*, as it stands, is a reworking of earlier material, in particular the writings of Cassianus Bassus described below. Secondly, that this reworking was done in the tenth century in the reign of Constantine VII Porphyrogenitus (ruled AD 913–59), to whom a dedication is preserved in the manuscripts from the 'F' family.[5] During Constantine's reign, Byzantine scholars

[1] E. Oder, 'Beiträge zur Geschichte der Landwirtschaft bei den Griechen I', *Rheinisches Museum für Philologie* 45 (1890), 58–61.

[2] These range in date from the tenth through the sixteenth centuries. H. Beckh, 'De Geoponicorum codicibus manuscriptis', thesis (Erlangae, 1886); F. W. K. Gemoll, *Untersuchungen über die Quellen, den Verfasser und die Abfassungszeit der Geoponica* (Berlin, 1883). Additional manuscripts were discovered after Beckh's edition was completed: A. D. Wilson, 'A Greek Treatise on Agriculture', *British Museum Quarterly* 13 (1939), 10–11.

[3] E. Fehrle, 'Studien zu dem Griechischen Geoponikern', *Στοικεῖα* 3 (1920); Gemoll, *Quellen*; Oder, 'Beiträge I'; E. Oder, 'Beiträge zur Geschichte der Landwirtschaft bei den Griechen. II', *Rheinisches Museum für Philologie* 45 (1890); E. Oder, 'Beiträge zur Geschichte der Landwirtschaft bei den Griechen. III', *Rheinisches Museum für Philologie* 48 (1893).

[4] P. Lemerle, *Byzantine Humanism* (Sydney, 1987), 121–346.

[5] Included in the proem of Beckh's Teubner edition, the dedication is to Constantine *τὸ τερπνὸν τῆϲ πορφύραϲ ἀπάνθιϲμα*. The allusion is to one born in the purple: though there were

undertook a flurry of activity in epitomizing and preserving classical knowledge.[6] Thirdly, that there is some, but only a little, medieval material included in the *Geoponica.* A chapter that provides a monthly planting guide for Constantinople (*Geop.* 12.1) contains late vocabulary and has been cited as proof for a medieval date of compilation on the one hand, or for a late date for the source material on the other. Also early medieval must be the references to the destruction of the Arabs in *Geop.* 1.10. This passage probably makes most sense post-seventh century but certainly would have had little meaning by the eleventh century, when Turks, rather than Arabs, were the threat.

In the main, the conclusions of the nineteenth- and early-twentieth-century scholars mentioned above, which naturally vary on some issues, broadly concur that the medieval encyclopaedists relied most heavily on two works: the *Collection of Agricultural Practice* (*Synagōge geōrgikōn epitēdeumatōn*) (*Cυναγωγὴ γεωργικῶν ἐπιτηδευμάτων*) of Vindanius Anatolius, and the *Farming Extracts* (*Peri geōrgias eklogai*) (*Περὶ γεωργιάς ἐκλογαί*) of the *scholastikos* (*scholasticus*) Cassianus Bassus. In the case of the *Synagōge* of Anatolius, the belief that his work forms a broad portion of the material included in the *Geoponica* is based on comparisons of the latter with surviving portions of eastern sources that preserve Anatolius, namely a Syriac agricultural work (referred to somewhat confusingly as the 'Syriac Geoponica') and Arabic manuscripts mentioned below. The attribution of a considerable part of the *Geoponica* to Cassianus Bassus is a function of the mention of Cassianus in some of the *Geoponica* manuscripts, personal comments in chapters which the medieval Greek lemmata have labelled as belonging to Cassianus (though these are not always reliable), and finally, the considerable references within the Islamic agricultural and botanical tradition of 'Qustus', possibly an Arabized form of 'Cassianus'.

Anatolius' *Synagōge* survived as late as the ninth century, when the patriarch Photius noted it in his *Bibliotheca* (§163). It is also possible that the editors of the *Geoponica* knew other texts not yet lost in the tenth century, like the fifteen-book *Geōrgika* of Didymus of Alexandria, mentioned above. Based on what we know of the nature of farming manuals that do survive, it is extremely likely that there was a good deal of repetition and overlap in the sources used by Anatolius and Cassianus, so that these authors would naturally find themselves repeating multiple authorities on the same subject, with the origin of the specific knowledge sometimes as opaque to them as it is to us. It is worth pointing out that a similar compilation, the ninth-century Arabic *Nabatean Agriculture*, was also apparently constructed from earlier sources.[7] Unlike the *Geoponica*, which makes no claims to authority based on its age, the *Nabatean Agriculture* does explicitly stand on such authority, a claim that scholars have met with considerable scepticism. The editor of the critical edition, Toufic Fahd

eleven emperors named Constantine, the only one who this description fits is Constantine VII Porphyrogenitus, so called because he was born in the purple chamber of the imperial palace.

[6] See A. E. McCabe, *A Byzantine Encyclopaedia of Horse Medicine: The Sources, Compilation, and Transmission of the Hippiatrica* (Oxford, 2007) for the history of the closely related *Hippiatrica,* the collection of horse veterinary treatises that parallels the *Geoponica* in both its original late antique source materials and medieval compilation.

[7] T. Fahd (ed.), *al-Filahah al-Nabatiyya* (Damascus, 1993–8).

holds the *Nabatean Agriculture* to be what it purports to be: an Arabic translation of a Syriac compilation, which, Fahd believes, took shape in Late Antiquity.[8] It is interesting then to recognize that the encyclopaedic movement, which saw Greek-speakers collecting and editing their farming texts, had parallels in medieval Mesopotamia. The *Nabatean Agriculture* itself probably draws on this Hellenistic tradition that flourished in the fourth through seventh centuries in the Levant.[9]

The earliest Roman agricultural tradition does not survive, but beginning with the late Republic and early Empire a handful of collections do come down to us, in particular works by Cato (234–149 BC), Varro (116–27 BC), and Columella (fl. AD 50). These early Latin authors depended on the highly esteemed Punic treatise of Mago the Carthaginian (3rd–2nd c. BC?), whose twenty-eight books were translated into Latin as the *Rusticatio*, by order of the senate following the destruction of Carthage in 146 BC.[10] Roman authors also drew considerably upon a Hellenistic tradition that is only sketchily known to us, but was perpetually reworked and added to by later Roman writers like Florentinus (early/mid-third century), about whom we know almost nothing, save that he wrote a *Geōrgika*, in Greek, in eleven or more books. Also in the early third century, Sextus Julius Africanus (fl. AD 220) compiled agricultural precepts, along with many other items, in a work called *Kestoi*; the surviving fragments are not different from earlier material and add little insight into farming practice. Probably about the same time, an Alexandrian named Didymus wrote a *Geōrgika* in fifteen books, it too largely a compilation of earlier material.[11] In the Latin West, Quintus Gargilius Martialis (early/mid-third century) wrote *De Hortis*, part of which survives, as well as works on human and possibly on veterinary medicine, though the authorship of the latter is disputed.[12] Martialis seems to have been more original than his Greek-speaking contemporaries, and his work demonstrates attempts at new methods that continued uninterrupted into the later empire.[13]

The only complete late antique handbook is the mid-fifth-century Latin work of Rutilius Taurus Aemilianus Palladius, who probably wrote in Italy or Sardinia, where he held estates.[14] The *De Agricultura* of Palladius (*Opus Agriculturae* in its modern

[8] T. Fahd, *Matériaux pour l'histoire de l'agriculture en Irak: Al-filâha n-nabatiyya* (Leiden and Cologne, 1977); T. Fahd, 'Traductions en Arabes 'd'écrits geoponiques', *Ciencias de la naturaleza en al-Andalus: textos y estudios* (Granada, 1990); T. Fahd, 'Botany and Agriculture', in R. Rashed (ed.), *Encyclopedia of the History of Arab Science* (London, 1996), III.817. Others do not share this view, e.g. T. Nöldeke, 'Noch Einiges über die "nabataïsche Landwirtschaft"', *ZDMG* 29 (1875), 445–55.

[9] Rodgers, 'Hail, Frost, and Pests in the Vineyard'.

[10] J. A. Greene and D. P. Kehoe, 'Mago the Carthaginian', in M. Hassine and M. Ghaki (eds.), *Actes du IIIe congrès international des études phéniciennes et puniques: Tunis, 11–16 novembre 1991* (Tunis, 1995).

[11] M. Wellmann, 'Didymus 7', in *RE* IV (Stuttgart, 1895), 445.

[12] E. Christmann, 'Gargilius 4', in H. Cancik and H. Schneider (eds.), *Brill's New Pauly*, V (Leiden, 2004), 700.

[13] M. S. Spurr, 'Palladius (1)', in S. Hornblower and A. Spawforth (eds.), *OCD*[3] (Oxford, 1995).

[14] J. Svennung, 'De auctoribus Palladii', *Eranos* 25 (1927), 164–9; J. Svennung, *Untersuchungen zu Palladius und zur lateinischen Fach- und Volkssprache* (Uppsala, 1935).

Teubner edition) is an instructional manual and a practical guide, not a theoretical work. It shows extensive practical experience, although Palladius also depended on Columella and Martialis, and was especially dependent on the work of Vindanius Anatolius, whose work, discussed below, was the linchpin in the transmission of Greek agricultural writing into the *Geoponica*.[15]

Vindanius Anatolius

During Late Antiquity, the last Greek agricultural manuals were produced. Their completion marked the end of a thousand-year activity in the Graeco-Roman world of the production of didactic technical treatises aimed at improving agriculture. The *Synagōge Georgikōn Epitēdeumatōn* was a compilation of Vindanius Anatolius of Berytus.[16] Anatolius probably lived in the fourth century and may be the Anatolius who was a friend of Libanius and later prefect of Illyricum under Constantius.[17] Since Anatolius drew on the *Kestoi* of Julius Africanus, he must have worked after the early third century, and the use of the *Synagōge* by the mid-fifth-century Latin agronomer Palladius provides a *terminus ante quem*.[18]

Anatolius hailed from the Phoenician coastal city of Berytus, according to a notice in the *Bibliotheca* (§163) of the polymath patriarch of Constantinople, Photius (*c*.820–93). Photius informs us that the *Synagōge* comprised fifteen books and was among the most useful agricultural works he knew, though it contained pagan beliefs that were better shunned by Christian agriculturists.[19] The *Bibliotheca* further informs us that the collection of Anatolius derived its material from a range of writers in the Greek agricultural tradition, specifically the work of Democritus, Africanus, Tarentinus, Apuleius, Florentinus, Valens, and Leo, as well as the *Paradoxa* of Diophanes of Bithynia. Very little is known about any of these authors, although a text ascribed to Democritus survives in Arabic.[20] The *Synagōge* thus accessed Hellenistic authors whose work underlay Roman handbooks, including those of Varro and Columella; their similarities indicate not only shared sources, but also the durability of agricultural practice and the wide dissemination of agricultural literature.

Anatolius, called variously 'Anatolius', 'Berytius', or 'Vindanonius' (*sic*) in the *Geoponica*, is cited twice within the text of the work and additionally named no

[15] R. Browning, 'Minor Figures', in E. J. Kenney (ed.), *Cambridge History of Classical Literature*, II (Cambridge, 1982), 770–3; Svennung, *Untersuchungen zu Palladius.*

[16] Or, probably more correctly, 'Vindonius': Oder, 'Beiträge I', 67 n. 1.

[17] M. Decker, 'The Authorship and Context of Early Byzantine Farming Manuals', *Byzantion* 77; Beckh, 'De Geoponicorum codicibus manuscriptis', 261–70; Gemoll, *Quellen*, 221–5; A. H. M. Jones, J. R. Martindale, and J. Morris (eds.), *The Prosopography of the Later Roman Empire*, I: *A.D. 260–395* (Cambridge, 1971); R. H. Rodgers, '*Κεποποΐα*', in A. Littlewood, H. Maguire, and J. Wolschke-Bulmahn (eds.), *Byzantine Garden Culture* (Washington, DC, 2002), s.v. Anatolius 3.

[18] Oder, 'Beiträge I'.

[19] Some idea of such superstitions may be seen in Rodgers, 'Hail, Frost, and Pests in the Vineyard'.

[20] Paris Bibliothèque Nationale *Ms. Arabe 2802*; a summary (*mukhtasar*) that has much material in common with that preserved in the *Geoponica.*

fewer than twenty-eight times in the chapter headings of the modern editions, which are themselves culled from the manuscript lemmata. Each lemma tells the subject of the section, names its author, and, less frequently, indicates additional sources that the primary author used. Comparisons of the Greek *Geoponica*, the Syriac version ascribed to Anatolius mentioned below, and Arabic and Armenian versions, have established the *Synagōge* as the foundation for much of the agronomic tradition of the medieval East.[21] In addition to its eastern receptions, the text of Anatolius was reworked by the second principal source of the *Geoponica*, Cassianus Bassus (discussed below), who is traditionally assigned a sixth-century date. Thus, the *Synagōge* entered the *Geoponica* in two ways: via the original work that survived into the medieval period as attested by Photios, and via the work of Cassianus which, re-edited by Byzantine scribes in the tenth century, formed the main basis of the *Geoponica*.[22]

Though the Greek original of Anatolius' text is lost, fragments of it are known from the papyrus Parsinus gr. 2313 fol. 49v and also from a papyrus of the late sixth or early seventh century, P. Vindobonensis G 40302.[23] The *Synagōge* was clearly highly regarded; in addition to the heavy use made of it by Cassianus and the testimony of Photius, the work proved of immense influence in the agricultural tradition of the medieval Middle East. The *Synagōge* was translated into Syriac in the sixth century, a version of which survives in a unique manuscript of the eighth or ninth century (British Library, Additional 14662), which is in rather poor condition and preserves the name of neither author nor scribe.[24]

Arabic translations of an epitomized version of Anatolius' text survive, but these are of limited value; for instance, a rather poor Arabic summary of the Anatolius text survives in a Spanish manuscript in Madrid (Madrid Gayangos 30). While this text confirms the original form of Anatolius' work, in books arranged by topic, it offers little help in fully reconstructing the text nor in detailing issues of practical knowledge.[25] One Arabic version, which preserves the text in its entirety, does perhaps

[21] Fehrle, 'Studien'; J. Ruska, 'Cassianus Bassus Scholasticus und die arabischen Versionen der griechischen Landwirtschaft', *Der Islam* 5 (1914); G. Sprenger, 'Darlegung der Grundsätze nach denen die syrische Übertragung der griechischen Geoponika gearbeitet worden ist', doctoral thesis (Göttingen, 1889).

[22] Our understanding of these rather complicated reworkings is assisted by the existence of the *Hippiatrica*, a tenth-century compilation of horse medicine whose sources and transmission history are extremely similar to the *Geoponica*. See McCabe, *A Byzantine Encyclopaedia*.

[23] Beckh, 'De Geoponicorum codicibus manuscriptis', 268–70. This fragment appears both in the *Geoponica* (1.14.8) and in Palladius, *Op. Ag.* 1.35: see A. A. Carrara, 'Geoponica and Nabatean Agriculture: A New Approach into their Sources and 'Authorship', *ASciPhi* 16 (2006), 107; A. Papathomas, 'Das erste antike Zeugnis für die veterinärmedizinische Exzerptensammlung des Anatolios von Berytos', *Wiener Studien* 113 (2000).

[24] De Lagarde's modern edition of this work has been heavily criticized. A. Baumstark, *Lucubrationes syro-graecae* (Leipzig, 1894); P. A. De Lagarde, *Geoponicon in sermonem Syriacum versonum quæ supersunt. P. Lagardius ed* (Leipzig, 1860); P. R. Duval, *La littérature syriaque* (Paris, 1899), 279–81.

[25] See M. C. Vázquez De Benito (ed.), *El manuscrito no. XXX de la Colección Gayangos (fols. 1–98)* (Madrid, 1974).

survive in two manuscripts in Iran (Meshed, Riḍa 5672 of the fourteenth century and Teheran Mīlli 796, a modern copy of Riḍa 5672). Without access to this text I am unable to comment on its utility in reconstructing the *Synagōge*, though Carrara, who has read a copy of Teheran Mīlli 796, confirms that it contains both various versions of the name of Anatolius and fourteen books—two more than the original twelve mentioned by Photius. An Arabic version was in turn translated into Armenian, a version on which little work has been devoted. Though the Armenian treatise has many misreadings of the Arabic, it may prove useful in reconstructing the original text.[26] These translations are sure indicators that the work of Anatolius thus occupied a fundamental place within the Greek, Semitic, and Caucasian agricultural traditions.[27]

Cassianus Bassus

One manuscript, the thirteenth-century Venice Marcianus gr. 524, has led scholars to identify it as representing an earlier stage in the development of the *Geoponica* text. The Marcianus work is virtually identical in order and content to other *Geoponica* manuscripts but it differs in several key ways. In its incipit Marc. gr. 524 names the title and author of the whole: the *Eclogues* or *Selections on Agriculture* by a certain Cassianus Bassus (*Ἀρχὴ cὺν θ[ε]ῷ τῶν περὶ γεωργίας ἐκλογῶν: Καccιανοῦ Βάccου cχολαcτικοῦ*). This version has no dedication to Constantine VII, possibly indicating that it predated the reworking of the text into the mature *Geoponica*, presumed to have occurred during the tenth century. Furthermore, books 7, 8, and 9 in this same manuscript are dedicated to Bassus, son of Cassianus Bassus.[28]

Several scholars have, through comparison of the *Geoponica* and its eastern relatives, concluded that Cassianus relied heavily upon Anatolius' text.[29] In addition, these comparisons have shown that there is additional material in the *Geoponica*, which does not belong to the material coming from Anatolius. Thus, it has been assumed (though this remains uncertain and, as Rodgers has pointed out, can never

[26] Aspects of the textual history of the eastern language tradition of the Anatolius treatise are explored by Fehrle, 'Studien'; E. H. F. Meyer and F. Verdoorn, *Geschichte der Botanik* (Amsterdam, 1965); F. Sezgin, *Geschichte des arabischen Schrifttums* (Leiden, 1967), IV.314–15. Unsurprisingly, given the number of languages involved, the eastern tradition has yet to receive full scholarly treatment. Carrara, 'Geoponica and Nabatean Agriculture', 132, notes that the Armenian version depends on the fuller Arabic text of MS Teheran Mīlli 796, and thus it may help to unravel some of the complexities of the eastern and Greek traditions.

[27] The transmission into Arabic is widely known, but studied in little depth: Sezgin, *Geschichte des arabischen Schrifttums*, IV.303–18; M. Ullmann, *Die Natur- und Geheimwissenschaften im Islam* (Leiden, 1972), 427–39.

[28] Oder, 'Beiträge III', 24–36; Rodgers, '*Κεποποΐα*', 162.

[29] Beckh, 'De Geoponicorum codicibus manuscriptis'; E. Fehrle, 'Richtlinien zur textgestaltung der griechischen geoponica', *SBHeid* 11 (1920); E. Fehrle, *Studien zu dem griechischen geoponikern* (Leipzig, 1920); Gemoll, *Quellen*; Oder, 'Beiträge I'; Oder, 'Beiträge II'; Oder, 'Beiträge III'; E. Oder, *Ein angebliches Bruchstück Democritus über die Entdeckung unterirdischer Quellen* (Leipzig, 1898); Sprenger, 'Darlegung der Grundsätze'.

be proved until every single source of the *Geoponica* is somehow verified) that the bulk of this additional material belongs to Cassianus Bassus. That the material added to that of Anatolius belongs to Cassianus is supposed for three reasons. First, because of the Marcianus manuscript dedication, and the close relationship between this work and the fully evolved Geoponica. Secondly, a handful of passages, for instance *Geop.* 5.6 which speaks about the Maratonomo village or estate, cannot be found in the Syriac text belonging to Anatolius, nor in any of the other surviving fragments of the western agricultural tradition. It has been thus supposed (though not proved) that this voice is that of Cassianus Bassus scholasticus. Thirdly, the Arabic agricultural handbook tradition preserves a branch of tradition attributed in various forms to 'Qustus', which has been generally accepted to be a corruption of Cassianus Bassus scholasticus. Since no thorough survey of the Arabic or other eastern language agricultural handbook literature has been conducted, we can not isolate genuine references to Cassianus' work at this point.[30]

A mere sixteen sections (out of 621 sections in the Teubner edition) are said to belong to Cassianus, and these are hardly beyond dispute.[31] However, one of them, *Geoponica* 5.6 (which the lemma attributes to Cassianus), I have argued suggests that Cassianus was a Syrian. In this passage the author speaks of his estates at 'Maratonumo', which is none other than the Syrian town of Ma'arrat an-Nu'mān about 80 km south of Aleppo and right on the edge of the ancient Limestone Massif villages that witnessed such great development in Late Antiquity.[32]

The title *scholasticus*, meaning a lawyer or legal advocate, had largely gone out of use by the seventh century, providing one probable *terminus ante quem*. Another *terminus ante quem* is provided by a steady stream of citations to the work of the Arabicized 'Qustus' (probably a corruption of 'Cassianus'), that appear in various forms and texts, beginning with the philosopher al-Razi (*c.*AD 850–923).[33] From the above evidence it seems that anonymous Byzantine scholars re-edited the work of Cassianus and dedicated his encyclopaedia to Constantine Porphyrogenitus.[34] Others believe that Cassianus Bassus was himself responsible for the editing of the *Geoponica* in its current form, and that he thus belongs to the tenth, rather than the sixth, century.[35]

30 Rodgers, '*Κεποποιΐα*', 162.

31 Gemoll, *Quellen*, 258–61; cf. Carrara, 'Geoponica and Nabatean Agriculture', 112. These are: 4.1.3; 4.1.14; 5.2.12; 5.3.1; 5.6.6; 5.11.5; 5.12.1; 5.13.1; 5.17.4; 5.32.2; 5.36.3; 7.7.7; 7.18; 10.2.3; 10.34; 13.5.3; 16.22; 20.46. McCabe, *A Byzantine Encyclopaedia*, 73, notes that *Geoponica* 16 derives most of its material from Anatolius, and it is possible that it is Anatolius, not Cassianus, who is speaking in *Geop.* 5.6.

32 M. Decker, 'The Authorship and Context of Early Byzantine Farming Manuals', *Byzantion* 77, 113.

33 Carrara, 'Geoponica and Nabatean Agriculture', who, however, does not believe that Qustus is to be equated with Cassianus Bassus.

34 *Geop.* proem 11.

35 Gemoll, *Quellen*, 255–61, first proposed that Cassianus was the medieval editor and is followed by E. E. Lipshits, *Geoponiki: vizantiiskaia sel'skokhoziaistvennaia entsiklopediia X veka* (Moscow, 1960), 11, and more recently by Carrara, 'Geoponica and Nabatean Agriculture'.

Current scholarly consensus thus suggests that the *Geoponica* is almost entirely comprised of the works of Anatolius of Berytus and Cassianus Bassus, with the latter relying substantially on the former. Both authors, and the others included in the *Geoponica*, drew on a long tradition of Graeco-Roman farming manuals dating back to the Classical period and even earlier. The practices they describe were features of centuries of Mediterranean agricultural practice; but in the *Geoponica* (e.g. 5.6.6), as in Palladius (2.2.1; 4.10.24), the late antique authors add notices of their own efforts and of new ways of doing things. Different methods and experimentation continued into Late Antiquity.

The late antique agrarian handbooks were not simply armchair tomes for the intellectually curious, charmed by the thought of the country life while they lived as far away from it as possible. Nor were they simply the exercises of a disengaged, rural aristocracy. In agriculture, like few other human endeavours, theory and practice collide so frequently that it is impossible to conceive of a purely theoretical agriculture. In fact, the handbooks themselves give good evidence of their pragmatic application, and the authors, all of them estate owners, frequently allude to personal experience with the tried-and-true, or at least the tried, like the 'olive grape', a combination of the two fruits discussed in *Geoponica* 9.14. The fields and terraces of ancient estates were the place for the empirical—a realm where, if one could imagine it, one could try it. One might be sceptical of the results of grafting an olive onto a vine, but certainly not about the attempt. In a world in which every member of the elite was a landowner, an acquaintance with farms and farming was something all important men shared. Not all, of course, were as fond of the countryside as were the fourth-century Synesius of Cyrene and Themistius, but many apparently were. The texts that these interested estate owners produced continued to be read and talked about. We know, for example, that several prominent Byzantine intellectuals of the eleventh century, Michael Psellos and Michael Choniates amongst them, esteemed and made use of the *Geoponica*.[36]

It should be stressed that these handbooks do not shed light only on the methods of elite-owned properties. Before the industrialization of agriculture, the differences in tools and techniques available to rich and poor were quite narrow. While it seems that most of the important innovations in husbandry that spread throughout the Late Roman world were due to the interactions, capitalization, and competition among elites, new crops or farming techniques that did develop would have soon spread to interested rural communities outside of the direct control of these elites by way of hired labourers and tenants. If we take the hypothetical case of a particularly high-yielding variety of vine, when the harvesters descended on the vineyard, there would have been nothing impeding their acquiring and sharing knowledge of the superior plant, and, even if the owner or his or her agents were of a mind to stop them, of acquiring cuttings from the vineyard. In actual fact, there was no incentive for the owner to stop such activity, since the variety of vine was less a factor in his or her success than the economies of scale in land and equipment to which his or her

[36] J. Teall, 'The Byzantine Agricultural Tradition', *DOP* 25 (1971), 43.

operations had access. Nor do we have any indication that ancient and medieval people thought at all in such monopolistic ways. This is in no way to denigrate the importance placed on plant breeding and the production of specific narrow products from famous types of plants; quite the contrary. But any owner recognized that his or her Aminean wine was famous because he/she could produce it in such vast quantities, had amphorae kilns that fired thousands of jars in a week, and agents or connections in nearby cities who would both transport and sell that beverage. His or her tenants, over whom he/she held substantial control, in all likelihood tended the same Aminean grape, and lacked only the economies of scale available to the market-oriented estate. Support for the notion of a broader exchange in agricultural theory and a wide utilization of agronomic literature comes from Egypt, where vernacular texts in Coptic point to a wide dissemination of ideas aimed to regulate and manage farming practice. Almanacs and other practical guides, written in Coptic and often from monastic contexts, suggest that the wider, non-Greek-speaking population of Egypt managed their farming activities in ways analogous to those suggested by the agricultural handbooks.[37] It is likely that this situation was not restricted to Egypt, but rather indicative of the situation prevailing throughout the late antique and early Islamic Levant, where the local farming populations relied on not only a shared body of oral knowledge, but a written one as well.

[37] Wilfong, 'Agriculture Among the Christian Population of Early Islamic Egypt'; *Pachomian Koinonia*, trans. Veilleux, II.217–23.

Bibliography

PRIMARY SOURCES

Aetius of Amida, *Aetii Amideni libri medicinales*, ed. A. Olivieri (= Corpus Medicorum Graecorum 8) (Leipzig, 1935).

Alciphron, *Alciphronis rhetoris epistularum libri IV*, ed. M. A. Schepers (Leipzig, 1969); ed. and trans. A. R. Benner and F. Fobes, *The Letters of Alciphron, Aelian and Philostratus* (LCL, Cambridge, Mass., 1949).

Alexander of Tralles, *Alexander von Tralles*, ed. T. Puschmann (Vienna, 1878–9).

Ammianus Marcellinus, *Ammiani Marcellini Rerum gestarum libri qui supersunt*, ed. W. Seyfarth, L. Jacob-Karau, and I. Ulmann (Leipzig, 1978).

Anastasius the Persian: *Saint Anastase le Perse et l'histoire de la Palestine au début du VIIe siècle*, ed. B. Flusin (Paris, 1992).

Anthimus, *De observatione ciborum*, ed. M. Grant (Totnes, 1996).

Ps.-Apicius: *De re coquinaria*, ed. and trans. B. Flower and E. Rosenbaum (London, 1958; repr. 1974).

Aretaeus of Cappadocia, *Aretaiou Kappadokou ta sozomena. The extant works of Aretaeus, the Cappadocian*, ed. and trans. F. Adams (London, 1856).

Artemios, *The Miracles of St. Artemios: A Collection of Miracle Stories by an Anonymous Author of Seventh Century Byzantium*, ed. and trans. V. S. Crisafulli and J. Nesbitt (Leiden, 1997).

Asterius, *Asterius of Amasea. Homilies i–xiv*, ed. C. Datema (Leiden, 1970).

Athenaeus, *The Deipnosophists*, ed. and trans. C. Gullick (LCL, London and New York, 1927–41).

Al-Baladhuri, *The Origins of the Islamic State*, trans. P. Hitti (New York, 1916; repr. 1968).

Basil of Caesarea: *San Basilio. Commento al profeta Isaia*, ed. P. Trevisan (Turin, 1939).

——*Homélies sur l'Hexaéméron*, ed. S. Giet (Paris, 1968).

——*Lettres*, ed. Y. Courtonne (Paris, 1957–66).

Caelius Aurelianus, *On Acute Diseases and On Chronic Diseases*, ed. and trans. I. E. Drabkin (Chicago, 1950).

Cassius Felix, *Cassius Felix de la médicine*, ed. A. Fraisse (Paris, 2002).

Cato, *De agricultura*, ed. A. Mazzarino (Leipzig, 1962); *Marcus Porcius Cato, On agriculture; Marcus Terentius Varro, On agriculture*, ed. and trans. W. D. Hooper and H. B. Ash (LCL, Cambridge, Mass. and London, 1934).

Choricius of Gaza, *Choricii Gazaei opera*, ed. Richard Foerster and Eberhardus Richtsteig (Stuttgart, 1972).

Codex Theodosianus, ed. T. Mommsen and P. Krueger (Berlin, 1954); *The Theodosian Code and Novels and the Sirmondian Constitutions*, trans. C. Pharr (Greenwood, Westood, CT, 1969).

Columella, *L. Juni Moderati Columellae opera quae exstant*, ed. V. Lunström (Uppsala, 1897–1968).
Corippus, *In laudem Iustini Augusti minoris*, ed. A. Cameron (London, 1976).
Corpus Iuris Civilis[14], ed. T. Mommsen and P. Krüger (Berlin, 1954).
Cosmas Indicopleustes, *Topographie chrétienne*, ed. W. Wolska-Conus (SC 141, 159, 197, Paris, 1968–73).
Cyril of Scythopolis, *Vit. Sab.* = *Kyrillos von Skythopolis*, ed. and trans. E. Schwartz (Texte und Untersuchungen zur Geschichte der altchristlichen Literatur 49.22, Leipzig, 1939).
Digesta Iustiniani Augusti, ed. T. Mommsen, rev. P. Krüger, *Corpus Iuris Civilis* I (Berlin, 1928); *The Digest of Justinian*, trans. A. Watson (rev. edn, Philadelphia, 1998).
Diocletian, Price Edict: *Diokletians Preisedikt*, ed. S. Lauffer (Berlin, 1971).
Diodorus Siculus, *Diodori Bibliotheca historica*, ed. F. Vogel, C. T. Fischer, L. Dindorf, and I. Bekker (Leipzig, 1888–1906; repr. 1969–85).
Ps.-Dionysius of Tell-Maḥre, *The Chronicle of Zuqnīn, Parts III and IV: A.D.488–775*, trans. A. Harrak (Medieval Sources in Translation 36, Toronto, 1999).
Dioscorides, *Pedanii Dioscoridis Anazarbei De materia medica libri quinque*, ed. M. Wellmann (Bonn, 1906–7).
Eunapius, *Eunapii Vitae sophistarum*, ed. J. Giangrande (Rome, 1956).
Eusebius, *Histoire ecclésiastique*, vol. II, ed. G. Bardy (SC 41, Paris, 1952–71; repr. 1994).
——*Onomasticon*: *Eusebius Das Onomastikon der biblischen Ortsnamen*, ed. E. Klostermann (Hildesheim, 1966).
Evagrius, *Ecclesiastical History with the Scholia*, ed. J. Bidez and L. Parmentier (London, 1898); *The Ecclesiastical History of Evagrius Scholasticus*, trans. M. Whitby (Translated Texts for Historians 33, Liverpool, 2000).
Expositio totius mundi et gentium, ed. J. Rougé (SC 124, Paris, 1966).
Galen, *Galeni De sanitate tuenda, De alimentorum facultatibus. De bonis malisque sucis, De victu attenuante. De ptisana*, ed. K. Koch, G. Helmreich, K. Kalbfleisch, O. Hartlich, and W. John (Leipzig, 1923).
Geoponica sive Cassiani Bassi scholastici De re rustica eclogae, ed. H. Beckh (Leipzig, 1895).
George of Choziba, *The Life of Saint George of Choziba and, the Miracles of the Most Holy Mother of God at Choziba*, trans. T. Vivian (San Francisco, 1994).
George of Cyprus, *Georgii Cyprii Descriptio orbis romani* (Leipzig, 1890).
Grattius, *Poetae Latini minores*, ed. E. Baehrens, F. Vollmer (Leipzig, 1910–13).
Greek Anthology, ed. and trans. W. R. Paton (LCL, London and New York, 1916–18).
Gregory Nazianzus, *Discours 42–43*, ed. J. Bernardi (SC 384, Paris, 1992); *Funeral Orations by Saint Gregory Nazianzen and Saint Ambrose*, trans. L. P. McCauley (Washington, D.C, 1968).
Gregory of Nyssa, *Gregorii Nysseni opera*, ed. W. Jaeger (new edn, Leiden, 1960).
Gregory of Tours: *Gregorii episcopi Turonensis Libri historiarum X*, ed. B. Krusch and W. Levison (MGH 1, Hanover, 1937).

Hierocles, *Hieroclis Synecdemus accedunt: fragmenta apud Constantinum Porphyrogennetum servata et nomina urbium mutata*, ed. A. Burckhardt (Leipzig, 1893).
Ibn Isḥaq, *The Life of Muhammad*, trans. A. Guillaume (Oxford, 1955; repr. 1967).
Isidore of Charax, *Parthian Stations by Isidore of Charax: An Account of the Overland Trade Route Between the Levant and India in the first century, B.C.*, ed. and trans. W. H. Schoff (Philadelphia, 1914).
Isidore of Seville, *De l'agriculture: Livre XVII*, ed. and trans. J. André (Paris, 1981).
Itinera hierosolymitana saeculi IIII–VIII, ed. P. Geyer (CSEL 39, Vienna, 1898).
Jerome, *Epistulae*, ed. I. Hilberg, M. Kamptner (CSEL 54–6, Vienna, 1996).
——*Vita Hilarion: Trois vies de moines: Paul, Malchus, Hilarion*, ed. P. Leclerc, E. Martín Morales, and A. de Vogüé (SC 508, Paris, 2006).
John of Ephesus, *Lives of the Eastern Saints*, trans. R. Payne Smith (PO 17.1: 1–304; 18.4: 513–694; 19.2: 153–282; Paris, 1923–6).
John of Nikiu, *The Chronicle of John (c.690 A.D.) Coptic Bishop of Nikiu Being a History of Egypt Before and During the Arab Conquest*, trans. R. H. Charles (London, 1916; repr. Amsterdam, 1980).
Josephus, Flavius, *Jewish Antiquities*, ed. and trans. L. H. Feldman (LCL, Cambridge, Mass., 1965).
——*The Jewish War*, ed. and trans. H. S. J. Thackeray et al. (LCL, Cambridge, Mass., 1958).
Ps.-Joshua the Stylite, *The Chronicle of pseudo-Joshua the Stylite*, trans. F. R. Trombley and J. W. Watt (Translated Texts for Historians 7, Liverpool, 2000).
Julian, *Works*, ed. and trans. W. C. Wright (LCL, London and New York, 1913).
Julian of Ascalon, *Le traité d'urbanisme de Julien d'Ascalon: Droit et architecture en Palestine au VIe siècle*, ed. C. Saliou (Paris, 1996).
Julius Africanus, *Les 'Cestes' de Julius Africanus*, ed. J.-R. Vieillefond (Paris, 1970).
Leontius, *Vie de Syméon le Fou; (et) Vie de Jean de Chypre*, ed. A. J. Festugière and L. Rydén (Paris, 1974).
Libanius, *Opera*, ed. R. Foerster (Leipzig, 1903–27); *Selected Works*, partial trans. A. F. Norman (LCL, London and Cambridge, Mass., 1969–77); *Antioch as a Centre of Hellenic Culture as Observed by Libanius*, trans. A. F. Norman (Translated Texts for Historians 34, Liverpool, 2000).
Liber Pontificalis: The Book of the Pontiffs (Liber Pontificalis), trans. R. Davis (Translated Texts for Historians, Liverpool, 1989).
Longus, *Daphnis et Chloe*, ed. M. D. Reeve (Stuttgart, 1994).
Macrina, *Vie de sainte Macrine*, ed. P. Maraval (SC 178, Paris, 1971).
Macrobius, *Saturnalia, apparatu critico instruxit, in Somnium Scipionis commentarios selecta varietate lectionis*, ed. J. Willis (Leipzig, 1970); *The Saturnalia*, trans. P. Davies (New York, 1969).
Marcellinus Comes, *The Chronicle of Marcellinus*, ed. and trans. B. Croke (Sydney, 1995).
Michael the Syrian, *Chronicle: Chronique de Michel le Syrien*, ed. J.-B. Chabot (1899–1905).

Movses Dasxuranci, *The History of the Caucasian Albanians*, trans. C. J. F. Dowsett (London and New York, 1961).

Mu'allaqat, *The Seven Odes*, trans. A. J. Arberry (Oxford, 1957).

al-Muqaddasi, *The Best Divisions of the Knowledge of the Regions: A Translation of Ahsan al-taqasim fi ma'rifat al-aqalim*, trans. B. A. Collins (Reading, 1994).

Nabatean Agriculture: Kitāb al-filaha al-nabatiyya, ed. T. Fahd, *L'agriculture nabatéenne* (Damascus, 1993–7); partial trans. pt. 10: *Kitab al-Filaha an-Nabatiya* (sic): *Bab Dhikr al-Kurum*, G. Salinger (unpublished MS, Berkeley, 1963).

Nemesianus, *The Eclogues and Cynegetica of Nemesianus*, ed. and trans. H. J. Williams (Leiden, 1986).

Nomos Georgikos, *The Farmer's Law*, ed. W. Ashburner, *JHS* 30 (1910), 85–108; trans. W. Ashburner, *JHS* 32 (1912), 68–95.

Notitia dignitatum: accedunt Notitia Urbis Constantinopolitanae et latercula provinciarum, ed. O. Seeck (Berlin, 1876).

Ps.-Oppian, *Oppianus Apameensis Cynegetica*, ed. M. Papathomopoulos (Munich and Leipzig, 2003).

Oribasius, *Oribasii Collectionum medicarum reliquiae*, ed. J. Raeder (CMG 6.1, Leipzig and Berlin, 1926–30); *Dieting for an Emperor: A Translation of Books 1 and 4 of Oribasius' Medical Compilations with an Introduction and Commentary*, partial trans. M. Grant (Leiden, 1997).

Pacatus, *Panegyric of Theodosius*, trans. C. E. V. Nixon and B. S. Rodgers (Berkeley, 1994).

Pachomian Koinonia, trans. A. Veilleux (Kalamazoo, Mich., 1980–2).

Palladius, *Palladii Rutilii Tauri Aemiliani viri inlustris Opus agriculturae de veterinaria medicina de insitione*, ed. R. H. Rodgers (Leipzig, 1975).

The Periplus Maris Erythraei: Text with Introduction, Translation, and Commentary, ed. and trans. L. Casson (Princeton, 1989).

Peter the Iberian, *Petrus der Iberer: ein Charakterbild zur Kirchen und Sittengeschichte des fünften Jahrhunderts: syrische Ubersetzung einer um das Jahr 500 verfassten griechischen Biographie*, ed. and trans. R. Raabe (Leipzig, 1895).

Philaretos the Merciful, *The Life of St. Philaretos the Merciful Written by His Grandson Niketas*, ed. and trans. L. Rydén (Studia Byzantina Upsaliensia 8, Uppsala, 2002).

Philo of Alexandria, *Philonis Alexandrini Opera quae supersunt*, ed. L. Cohn and P. Wendland (Berlin, 1962–3).

Pliny the Elder, *C. Plini Secundi Naturalis historiae libri XXXVII*, ed. C. Mayhoff and L. Ianus (Leipzig, 1878–1906).

P.Nessana: Excavations at Nessana, vol. III: *Non-literary Papyri*, ed. C. J. Kraemer (Princeton, 1958).

P.Petra: The Petra Papyri, ed. J. Frösén, A. Arjava, M. Lehtinen, and Z. Fiema (Amman, 2002–).

P.Rylands: Catalogue of Greek Papyri in the John Rylands Library, Manchester, II: *Documents of the Ptolemaic and Roman Periods (Nos. 62–456)*.

Procopius of Caesarea, *Procopii Caesariensis Opera omnia*, ed. J. Haury (Leipzig, 1905–13); trans. E. B. Dewing, vol. I: *History of the Wars, books I and II. The Persian*

War; vol. II: *History of the Wars, books III and IV. The Vandalic War*; vols. III–V: *History of the Wars, books V–[VIII]. The Gothic War*; vol. VI: *The Anecdota, or Secret History*; vol. VII: *Buildings* (LCL, Cambridge, Mass., 1953–61).

Procopius of Gaza, 'Un'epistola inedita di Procopio di Gaza', ed. E. V. Maltese, *Parola del passato* 39 (1984), 53–4.

——*Panegyric*: *Procope de Gaza, Priscien de Césarée, Panégyriques de l'empereur Anastase Ier*, ed. A. Chauvot (Berlin, 1986).

Salvian of Marseilles, ed. G. Lagarrigue (Paris, 1971–5).

Sidonius Apollinaris, *Poems and Letters*, ed. and trans. W. B. Anderson (LCL, Cambridge, Mass., 1936–65).

Simeon of the Olives, 'The Life of Simeon of the Olives', trans. S. Brock, *Ostkirchliche Studiem* 29 (1979), 174–9.

Sopater Rhetor, 'Διαίρεϲιϲ ζητήματων', in *Rhetores Graeci*, ed. C. Walz, vol. VIII (Stuttgart, 1835; repr. 1968), 2–385.

Sozomen of Bethelia, *Hist. Eccl.*: *Kirchengeschichte*, ed. J. Bidez (Berlin, 1960).

Stephanus, *Stephani Byzantinii Ethnikoü quae supersunt*, ed. A. Meineke (Berlin, 1849; repr. Chicago, 1992); partial edn *Stephani Byzantinii Ethnica*, vol. I, *Α–Γ*, ed. M. Billerbeck, J. F. Gaertner, B. Wyss, and C. Zubler (Corpus fontium historiae Byzantinae 43, Berlin, 2006).

Strabo, *Strabonis Geographica*, ed. A. Meineke (Leipzig, 1852–3).

Symeon the Holy Fool, *Symeon the Holy Fool: Leontius' Life and the Late Antique City*, trans. D. Krueger (Berkeley, 1996).

Synesius, *Synésios de Cyrène correspondance*, ed. A. Garzya, trans. D. Roques (Collection des universités de France, Paris, 2000); *The Letters of Synesius of Cyrene*, trans. A. Fitzgerald (Oxford, 1926).

Syro-Roman Lawbook: *The Syriac Text of the Recently Discovered Manuscripts accompanied by a Facsimile Edition and Furnished with an Introduction and Translation*, ed. and trans. A. Vööbus (Stockholm, 1982); *Das syrisch-römische Rechtsbuch*, ed. and trans. W. Selb and H. Kaufhold (Vienna, 2002).

Themistius, *Orationes quae supersunt*, ed. H. Schenkl, G. Downey, and A. F. Norman (Leipzig 1965–74); *The Private Orations of Themistius*, Partial trans. R. J. Penella (Berkeley, 2000).

Theodore of Sykeon, *Vie de Théodore de Sykeôn*, ed. A. J. Festugière (Paris, 1970).

Theodoret of Cyrrhus, *Epistulae*: *Théodoret de Cyrrhus. Correspondance*, ed. Y. Azéma (SC 40, 98, 111, Paris, 1964–82).

——*Historia Religiosa*: *Théodoret de Cyrrhus. L'histoire des moines de Syrie*, ed. P. Canivet and A. Leroy-Molinghen (SC 234, 257, Paris, 1977–9).

Theodorus of Paphos, *Vita of Spyridon*: *La légende de S. Spyridon, évêque de Trimithonte*, ed. P. van den Ven (Louvain, 1953).

Theophanes, *Theophanes chronographia*, ed. C. de Boor (Leipzig, 1883–5); *The Chronicle of Theophanes Confessor*: *Byzantine and Near Eastern History, AD 284–813*, trans. C. Mango and S. Scott (Oxford, 1997).

Theophrastus, *Theophrasti Eresii Opera quae supersunt omnia*, vols. I–II: *Historia plantarum*; *De causis plantarum lib. VI*, ed. F. Wimmer (1854–62).

Theophylact Simocatta, *Theophylacti Simocatae Epistulae*, ed. J. Zanetto (Leipzig, 1985).

Varro, *Marcus Terentius Varro, Gespräche über die Landwirtschaft*, ed. and trans. D. Flach (Darmstadt, 1996–2002), vols. I–III.

Virgil, *P. Vergili Maronis opera*, ed. F. A. Hirtzel (Oxford, 1900).

Vitas Sanctorum Patrum Emeretensium, ed. M. Sánchez (Corpus Christianorum Series Latina 116, Turnhout, 1992).

Ps.-Zachariah Rhetor, *The Syriac Chronicle Known as that of Zachariah of Mitylene*, trans. F. J. Hamilton and E. W. Brooks (London, 1899).

Zosimus, *Histoire nouvelle* ed. F. Paschoud (Paris, 1971–89).

SECONDARY SOURCES

Adams, W. Y., 'The Vintage of Nubia', *Kush* 14 (1966), 262–83.

Akhruf, C., 'Syria', in International Association of Agricultural Economists (ed.), *World Atlas of Agriculture*, II (Novara, 1973), 475–90.

Algaze, G., Breuninger, R., and Knudstad, J., 'The Tigris – Euphrates Archaeological Reconnaissance Project: Final Report of the Birecik and Carchemish Dam Survey Areas', *Anatolica* 20 (1994), 1–96.

Algaze, G., et al., 'The Tigris–Euphrates Archaeological Reconnaissance Project: A Preliminary Report of the 1989–1990 Seasons', *Anatolica* 17 (1991), 175–240.

Allen, H. W., *A History of Wine: Great Vintage Wines from the Homeric Age to the Present Day* (London, 1961).

Amiram, D., 'The Madaba Mosaic Map as a Climate Indicator for the Sixth Century', *Israel Exploration Journal* 47 (1997), 97–9.

Amouretti, M.-C., 'Des agronomes latins aux agronomes provençaux: les moulins à huile', *Provence Historique* 24 (1981), 83–100.

——*Le pain et l'huile dans la Grèce antique: de l'araire au moulin* (Annales littéraires de l'Université de Besançon 328; Paris, 1986).

André, J., *L'alimentation et la cuisine à Rome* (2nd edn, Collection d'études anciennes; Paris, 1981).

——*Les noms de plantes dans la Rome antique* (Paris, 1985).

Arar, A., 'The Role of Rainfed Agriculture in the Near East Region: Summary of the Present Situation, Potential and Constraints', in FAO (ed.), *Rainfed Agriculture in the Near East and North Africa* (Rome, 1980), 10–21.

Arberry, A. J., *The Seven Odes: The First Chapter in Arabic Literature* (London and New York, 1957).

Argoud, G., 'Un huilerie à Salamine', *Salamine de Chypre IV* (Paris, 1973).

Arjava, 'Zum Gebrauch der griechischen Rangprädikate des Senatorenstandes in den Papyri und Inschriften', *Tyche* 6 (1991), 17–35.

Ashburner, W., 'The Farmer's Law', *Journal of Hellenic Studies* 30 (1910), 85–108.

——'The Farmer's Law Continued', *Journal of Hellenic Studies* 31 (1912), 38–95.

Aupert, P., *Guide d'Amathonte* (Sites et monuments 15; Paris, 1996).

Aviam, M., 'Horvat Hesheq: A Church in Upper Galilee', in Yoram Tsafrir (ed.), *Ancient Churches Revealed* (Jerusalem, 1993), 54–65.

Avni, G., *Map of Har Saggi Northeast (225)* (Archaeological Survey of Israel; Jerusalem, 1991).

Ayalon, E., 'The Jar Installation of Khirbet Sabiya', *Israel Exploration Journal* 29 (1979), 175–81.

——'Typology and Chronology of Water-Wheel (saqiya) Pottery Pots from Israel (Methodological Research on Roman Period Hydraulic Techniques)', *Israel Exploration Journal* 50 3/4 (2000), 216–26.

Bagnall, R. S., 'Landholding in Late Roman Egypt: The Distribution of Wealth', *Journal of Roman Studies* 82 (1992), 128–49.

——*Egypt in Late Antiquity* (Princeton, 1993).

Baillie, M. G. L., 'Dendrochronology Raises Questions About the Nature of the AD 536 Dust-veil Event', *The Holocene* 4/2 (1994), 212–17.

——'Patrick, Comets and Christianity', *Emania* 13 (1995), 69–78.

Balty, J., *La Mosaïque de Sarrîn* (Osrhoène) (Paris, 1990).

Banaji, J., *Agrarian Change in Late Antiquity: Gold, Labour, and Aristocratic Dominance* (Oxford, 2001).

Banning, E. B., 'Peasants, Pastoralists and Pax-Romana: Mutualism in the Southern Highlands of Jordan', *Bulletin of the American Schools of Oriental Research* 261 (1986), 25–50.

Bar, D., 'Population, Settlement and Economy in Late Roman and Byzantine Palestine', *Bulletin of the School of Oriental and African Studies* 67/3 (2004), 307–20.

——'Frontier and Periphery in Late Antique Palestine', *Greek, Roman and Byzantine Studies* 44/4 (2004), 69–92.

Bar-Yosef, O. and Meadow, R. H., 'The Origins of Agriculture in the East', in T. Douglas Price and Anne Birgitte Gebauer (eds.), *Last Hunters—First Farmers: New Perspectives on the Prehistoric Transition to Agriculture* (Santa Fe, 1995), 39–94.

Barker, G. W., 'The Archaeology of Samnite Settlement in Molise', *Antiquity* 51 (1977), 20–4.

Barker, G. W., Lloyd, J., and Webley, D., 'A Classical Landscape in Molise', *Papers of the British School at Rome* 46 (1978), 35–51.

Bartl, K., 'The Balīḫ Valley, Northern Syria, During the Islamic Period, Remarks Concerning the Historical Topography', *Berytus* 41 (1993–4), 29–38.

Bartl, K. and Hauser, S. R., *Continuity and Change in Northern Mesopotamia from the Hellenistic to the Early Islamic Period: Proceedings of a Colloquium Held at the Seminar für Vorderasiatische Altertumskunde, Freie Universität Berlin, 6th–9th April, 1994* (Berliner Beiträge zum Vorderen Orient 17; Berlin, 1996).

Bass, G. F. and Van Doorninck, F. H., *Yassi Ada* (1st edn, The Nautical Archaeology Series 1; College Station, 1982).

Baumgarten, Y. A., *Mapat Shivṭah (166)* (Archaeological Survey of Israel; Jerusalem, 2004).

Baumstark, A., *Lucubrationes syro-graecae* (Leipzig, 1894).

Beckh, H., 'De Geoponicorum codicibus manuscriptis', thesis (Erlangae, 1886).

Bedal, L.-A., 'A Paradeisos at Petra: New Light on the "Lower Market"', *Annual of the Department of Antiquities of Jordan* 43 (1999), 227–39.

Beit-Arieh, I., *Map of Tel Malhatah (144)* (Archaeological Survey of Israel; Jerusalem, 2003).

Berman, A., *Map of Nizzanim: West (87); Map of Nizzanim: East (88)* (Archaeological Survey of Israel; Jerusalem, 2005).

Berman, A., Barda, L., and Stark, H., *Map of Ashdod (84)* (Archaeological Survey of Israel; Jerusalem, 2005).

Berman, A., Stark, H., and Barda, L., *Map of Ziqim (91)* (Archaeological Survey of Israel; Jerusalem, 2004).

Berthier, S., *Peuplement rural et aménagements hydroagricoles dans la moyenne vallée de l'Euphrate, fin VIIe–XIXe siècle: Région de Deir ez Zōr-Abu Kemāl, Syrie: Mission Mésopotamie syrienne, archéologie islamique*, 1986–9 (Damas, 2001).

Billiard, R., *La vigne dans l'antiquité* (Marseille, 1997).

Biscop, J.-L., Orssaud, D., and Mango, M. M., *Deir Déhès, Monastère d'Antiochène: étude architecturale* (Bibliothèque archéologique et historique 148; Beirut, 1997).

Bisheh, G., 'Excavations at Qasr al-Hallabat, 1979', *Annual of the Department of Antiquities of Jordan* 24 (1980), 69–77.

——'The Second Season of Excavation at Hallabat, 1980', *Annual of the Department of Antiquities of Jordan* 26 (1982), 133–43.

——'Qasr al-Hallabat: An Umayyad Desert Retreat or Farm Land?', *SHAJ* 2 (1985), 263–5.

——'Qasr al-Hallabat: A Summary of the 1984 and 1985 Excavations', *Archiv für Orientforschung* 38 (1986), 158–62.

Blanton, R. E., *Hellenistic, Roman and Byzantine Settlement Patterns of the Coast Lands of the Western Rough Cilicia* (BAR International Series 879; Oxford, 2000).

Bonsor, J. E. and Penney, C. L., *The Archaeological Expedition Along the Guadalquivir: 1889–1901* (New York, 1931).

Boserup, E., *The Conditions of Agricultural Growth* (Chicago, 1965).

Bovill, E. W., *Caravans of the Old Sahara: An Introduction to the History of the Western Sudan* (London, 1933).

——*The Golden Trade of the Moors* (London and New York, 1958).

Bowersock, G. W., 'Social and Economic History of Syria Under the Roman Empire', in Jean-Marie Dentzer and Winifred Orthmann (eds.), *Archéologie et histoire de la Syrie* (Saarbrücken, 1992), 63–80.

Bowman, A., *Egypt After the Pharoahs: 332 BC–AD 642 from Alexander to the Arab Conquest* (London, 1986).

Bowman, A. K., 'Landholding in the Hermopolite Nome in the Fourth Century A.D.', *Journal of Roman Studies* 75 (1985), 137–63.

Braun, T., 'Barley Cakes and Emmer Bread', in John Wilkins, David Harvey, and Mike Dobson (eds.), *Food in Antiquity* (Exeter, 1995), 25–37.

Bréhier, L., 'Les colonies d'Orientaux en Occident au commencement du moyen-âge', *Byzantinische Zeitschrift* 12 (1903), 1–39.

Brengle, K. G., *Principles and Practices of Dryland Farming* (Boulder, Colo., 1982).

Breton, J. F., *Les inscriptions forestières d'Hadrien dans le Mont Liban* (Bibliothèque archéologique et historique 104; Paris, 1980).

Brock, S., 'The Life of Simeon of the Olives', *Ostkirchliche Studien* 29 (1979), 174–9.

Brooks, R., *Peah* (The Talmud of the Land of Israel: A Preliminary Translation and Explanation 2; Chicago, 1990).

Broshi, M., 'The Population of Western Palestine in the Roman–Byzantine Period', *Bulletin of the American Schools of Oriental Research* 236 (1976), 1–10.

Brothwell, D. and Brothwell, P., *Food in Antiquity: A Survey of the Diets of Ancient Peoples* (Baltimore, 1969).

Broughton, T. R. S., 'Roman Asia Minor', in Tenney Frank (ed.), *An Economic Survey of Ancient Rome* (Baltimore, 1938), 503–916.

Brown, A. D. F., 'Black Wine', *Classical Review* 12 (1962), 192–5.

Brown, M. R., 'A Large Residence (House XVIII)', in Bert De Vries (ed.), *Umm el-Jimal: A Frontier Town and its Landscape in Northern Jordan* (Journal of Roman Archaeology. Supplementary Series 26; Portsmouth, RI, 1998), 195–204.

Browning, R., 'Minor Figures', in E. J. Kenney (ed.), *Cambridge History of Classical Literature*, II (Cambridge, 1982), 770–3.

Bruins, H. J., 'Comparative Chronology of Climate and Human History in the Southern Levant from the Late Chalcolithic to the Early Arab Period', in O. Bar-Yosef and R. S. Kra (eds.), *Late Quaternary Chronology and Paleoclimates of the Eastern Mediterranean* (Tucson, Ariz., 1994), 301–14.

Brun, J.-P., *L'oléiculture antique en Provence: les huileries du département du Var* (Revue archéologique de Narbonnaise. Supplément 15; Paris, 1986).

——*Archéologie du vin et de l'huile dans l'Empire romain* (Collection des Hespérides; Paris, 2004).

——*Archéologie du vin et de l'huile de la préhistoire á l'époque hellénistique* (Collection des Hespérides; Paris, 2004).

——*Archéologie du vin et de l'huile en Gaule romaine* (Collection des Hespérides; Paris, 2005).

Brunt, P. A., 'Review of Roman Farming', *Journal of Roman Studies* 62 (1972), 153–8.

Bryer, A., 'The Estates of the Empire of Trebizond', *Archeion Pontou* 35 (1979), 370–477.

——'Byzantine Agricultural Implements: The Evidence of Medieval Illustrations of Hesiod's Works and Days', *The Annual of the British School at Athens* 81 (1986), 45–80.

——'The Means of Agricultural Production: Muscle and Tools', in A. E. Laiou and C. Bouras (eds.), *The Economic History of Byzantium* (Washington, DC, 2002), 101–14.

Bryson, R. U. and Bryson, R. A., 'Application of a Global Volcanicity Time-series on High-resolution Paleoclimatic Modeling of the Eastern Mediterranean', in A. Issar and N. Brown (eds.), *Water, Environment and Society in Times of Climatic Change: Contributions from an International Workshop Within the Framework of International Hydrological Program (IHP) UNESCO, held at Ben-Gurion University, Sede Boker, Israel from 7–12 July 1996* (Dordrecht, 1998), 1–20.

Bulliet, R. W., *The Camel and the Wheel* (2nd edn; New York, 1990).

Butler, H. C., *Architecture and Other Arts* (New York, 1904).

Butler, H. C., et al., *Syria: Publications of the Princeton University Archaeological Expedition to Syria in 1904–1905 and 1909*, 4 vols. (Leiden, 1907–49).

Butzer, K. W., et al., 'Irrigation Agrosystems in Eastern Spain: Roman or Islamic Origins?', *Annals of the Association of American Geographers* 75 (1985), 479–509.

Cadell, H., 'Le vocabulaire de l'agriculture d'après les papyrus grecs d'Égypte: problèmes et voies de recherche', in Deborah H. Samuel (ed.), *Proceedings of the Twelfth International Congress of Papyrology* (Ann Arbor, 1970), 69–83.

Callot, O., *Huileries antiques de Syrie du Nord* (Paris, 1984).

Calvet, Y. and Geyer, B., *Barrages antiques de Syrie* (1992).

Campbell, W. A. and Stillwell, R., *The Excavations, 1933–1936* (Antioch-on-the-Orontes 2; Princeton, NJ, 1938).

Cappel, A. J., 'Emphyteusis', in A. P. Kazhdan et al. (eds.), *ODB* (Oxford, 1991), 693–4.

Cappers, R. T. J., 'Archaeobotanical Remains', in S. E. Sidebotham and W. Wendrich (eds.), *Berenike 1995: Preliminary Report of the Excavations at Berenike (Egyptian Red Sea Coast) and the Survey of the Eastern Desert* (Leiden, 1996), 319–35.

——'Archaeobotanical Remains', in S. E. Sidebotham, W. Wendrich, and F. Aldsworth (eds.), *Berenike 1996: Report of the 1996 Excavations at Berenike (Egyptian Red Sea Coast) and the Survey of the Eastern Desert* (Leiden, 1998), 289–330.

Carrara, A. A., 'Geoponica and Nabatean Agriculture: A New Approach into their Sources and Authorship', *Arabic Sciences and Philosophy* 16 (2006), 103–32.

Carruthers, D., 'The Great Desert Caravan Route: Aleppo to Basra', *Geographical Journal* 52/3 (1918), 157–84.

Casana, J., 'The Archaeological Landscape of Late Roman Antioch', in I. Sandwell and J. Huskinson (eds.), *Culture and Society in Later Roman Antioch: Papers from a Colloquium, London, 15th December 2001* (Oxford, 2004), 102–25.

Catling, H., 'An Early Byzantine Pottery Factory at Dhiorios in Cyprus', *Levant* 4 (1972), 1–85.

Catling, H. W., 'The Ancient Topography of the Yialias Valley', *RDAC* (1982), 227–36.

Cerati, A., *Caractère annonaire et assiette de l'impôt foncier au Bas-Empire* (Paris, 1975).

Chang-Ho, C. J. and Lee, J. K., 'Archaeological Survey and Settlement Patterns in the Region of 'Iraq al-'Amir: A Preliminary Report', *Annual of the Department of Antiquities of Jordan* 42 (1998), 587–606.

——'The 1998 Season of Archaeological Survey in the Regions of 'Iraq al-'Amir and Wadi al-Kafraya: A Preliminary Report', *Annual of the Department of Antiquities of Jordan* 43 (1999), 521–39.

Chastagnol, A. and Tantillo, I., *Aspects de l'antiquité tardive* (Saggi di storia antica 6; Rome, 1994).

Chatterton, B. A. and Chatterton, L., 'Medicago—its Possible Role in Romano-Libyan Dryfarming and its Positive Role in Modern Dryfarming', *Libyan Studies* 15 (1984), 157–60.

——'A Hypothetical Answer to the Decline of the Granary of Rome?', *Libyan Studies* 16 (1985), 95–9.

Chatterton, L., *Sustainable Dryland Farming: Combining Farmer Innovation and Medic Pasture in a Mediterranean Climate* (Cambridge, 1996).

Chéhab, M. H., *Mosaiques du Liban* (Bulletin du Musée de Beyrouth 14; Paris, 1957).

Cheynet, J.-C., 'Sceaux byzantines des Musées d'Antioche et de Tarse', *Travaux et Mémoire* 12 (1994), 391–478.

Chopra, K. R., *Technical Guideline for Sorghum and Millet Seed Production* (Rome, 1982).

Christensen, P., *The Decline of Iranshahr: Irrigation and Environments in the History of the Middle East, 500 B.C. to A.D. 1500* (Copenhagen, 1993).

Christmann, E., 'Gargilius 4', in H. Cancik and H. Schneider (eds.), *Brill's New Pauly Handbuch der Lateinischen Literatur der Antike*, V (Leiden, 2004), 700.

Clark-Maxwell, W. C., 'The Roman Towns in the Valley of the Baetis Between Córdoba and Sevilla', *Archaeological Journal* 56 (1899), 245–305.

Coello, T., *Unit Sizes in the Late Roman Army* (Oxford, 1996).

Cohen, R., *Map of Sede-Boqer—East (168)* (Archaeological Survey of Israel; Jerusalem, 1981).

——*Map of Sede-Boqer—West (167)* (Archaeological Survey of Israel; Jerusalem, 1985).

Corbo, V., *Gli scavi di Kh. Siyar el-Ghanam (Campo dei pastori) e i monasteri di dintorni* (Jerusalem, 1955).

Cotte, J. and Cotte, C., *Etude sur les blés de l'antiquité classique* (Paris, 1912).

Crawford, D. J., 'Imperial Estates', in M. I. Finley (ed.), *Studies in Roman Property* (Cambridge, 1976), 35–70.

Crawford, J. S., *The Byzantine Shops at Sardis* (Cambridge, Mass., 1990).

Crone, P., *Meccan Trade and the Rise of Islam* (Oxford, 1987).

Cumont, F., 'The Population of Syria', *Journal of Roman Studies* 24 (1934),187–90.

Dagan, Y., *Map of Lakhish (98)* (Archaeological Survey of Israel; Jerusalem, 1992).

Dagron, G. and Feissel, D., 'Inscriptions inédites du Musée d'Antioche', *Travaux et Mémoires* 9 (1985), 421–61.

——*Inscriptions de Cilicie* (Travaux et Mèmoires du Cèntre de Recherche d'histoire et civilisation; Paris, 1987).

Dahari, U., 'Remote Monasteries in Southern Sinai and their Economic Base', in Yoram Tsafrir (ed.), *Ancient Churches Revealed* (Jerusalem, 1993), 341–50.

Dalman, G., *Arbeit und sitte in Palästina* (Hildesheim, 1964).

Daniel, R. W., 'P. Petra Inv. 10 and its Arabic', in Isabella Andorlini (ed.), *Atti del XXII Congresso internazionale di papirologia, Firenze, 23–29 agosto 1998* (Florence, 2001), 331–41.

Dar, S., 'The History of the Hermon Settlements', *Palestine Exploration Quarterly* 120 (1988), 26–44.

——*Settlements and Cult Sites on Mount Hermon, Israel* (Oxford, 1993).

——*Sumaqa: A Roman and Byzantine Jewish village on Mount Carmel, Israel* (BAR International Series 815; Oxford, 1999).

——*Ruqit: Marinus' Estate on the Carmel, Israel* (Oxford, 2004).

Dar, S. and Applebaum, S., *Landscape and Pattern: An Archaeological Survey of Samaria 800 B.C.E.–636 C.E.* (BAR International Series 308; Oxford, 1986).

Dark, K. R., 'Early Byzantine Mercantile Communities in the West', in Chris Entwistle (ed.), *Through a Glass Brightly: Studies in Byzantine and Medieval Art and Archaeology Presented to David Buckton* (Oxford, 2003), 75–81.

Dauphin, C., 'A Seventh Century Measuring Rod from the Ecclesiastical Farm at Shelomi in Western Galilee [Israel]', *Jarhbuch der Österreichischen Byzantinistik* 32/3 (1982), 513–22.

——'A Byzantine Ecclesiastical Farm at Shelomi', in Yoram Tsafrir (ed.), *Ancient Churches Revealed* (Jerusalem, 1993), 43–8.

——*La Palestine byzantine: peuplement et populations* (BAR International Series 726; Oxford, 1998).

Dauphin, C. and Kingsley, S., 'Ceramic Evidence for the Rise and Fall of a Late Antique Ecclesiastical Estate at Shelomi in Phoenicia Maritima', in G. C. Bottini, L. Di Segni, and L. D. Chrupcala (eds.), *One Land—Many Cultures: Archaeological Studies in Honour of S. Loffreda* (Jerusalm, 2003), 61–74.

Dauphin, C. and Schonfield, J. J., 'Settlements of the Roman and Byzantine Periods in the Golan Heights: Preliminary Report of Three Seasons of Survey (1979–1981)', *Israel Exploration Journal* 33/3–4 (1983), 189–206.

Davis, R., *The Book of Pontiffs (Liber pontificalis): The Ancient Biographies of the First Ninety Roman Bishops to AD 715* (Translated Texts for Historians. Latin Series 5; Liverpool, 1989).

De Lagarde, P. A., *Geoponicon in sermonem Syriacum versonum quæ supersunt*, ed P. A. de Largarde (Leipzig and London, 1860).

De Ligt, L., *Fairs and Markets in the Roman Empire: Economic and Social Aspects of Periodic Trade in a Pre-industrial Society* (Amsterdam, 1993).

De'maffei, F., 'Il Palazzo di Qasr ibn-Wardan: dopo gli scavi e i restauri', in Antonio Iacobini and Enrico Zanini (eds.), *Arte profana e arte sacra a Bisanzio* (Rome, 1995), 105–87.

De Ruyt, C., *Macellum: marché alimentaire des Romains* (Louvain-la-Neuve, 1983).

De Vogüé, C. J. M., *Syrie centrale: architecture civile et religieuse du Ier au VIIe siècle* (Paris, 1865–77).

De Vries, B., 'The Umm el-Jimal Project 1972–1977', *Annual of the Department of Antiquities of Jordan* 26 (1982), 97–116.

——'Location, Plan and Design of the Late-antique Town', in B. De Vries and J. Wilson Betlyon (eds.), *Umm el-Jimal: A Frontier Town and its Landscape in Northern Jordan* (Journal of Roman Archaeology Supplementary Series 26; Portsmouth, RI, 1998), 91–127.

——'Research Goals and Implementation', in B. De Vries and J. Wilson Betlyon (eds.), *Umm el-Jimal: A Frontier Town and its Landscape in Northern Jordan* (Journal of Roman Archaeology Supplementary Series 26; Portsmouth, RI, 1998), 15–26.

De Vries, B. and Betlyon, J. W., *Umm el-Jimal: A Frontier Town and its Landscape in Northern Jordan* (Journal of Roman Archaeology Supplementary Series 26; Portsmouth, RI, 1998).

Decker, M. 'The Authorship and Context of Early Byzantine Farming Manuals', *Byzantion* 77 (2007), 106–15.

——'Agricultural Production and Trade in Oriens, 4th–7th Centuries AD', D.Phil. thesis (Oxford, 2001).

——'The Wine Trade of Cilicia in Late Antiquity', *Aram* 15 (2005), 51–9.

——'Towers, Refuges, and Fortified Farms in the Late Roman East', *Liber Annuus* 56 (2006), 499–520.

——'Water into Wine: Trade and Technology in Late Antiquity', in Luke Lavan, Enrico Zanini, and Alexander Sarantis (eds.), *Technology in Transition A.D. 300–650* (Late Antique Archaeology 4; Leiden, 2007), 65–91.

——'Frontier Settlement and Economy in the Byzantine East', *Dumbarton Oaks Papers* 61 (2007), 117–67.

Déléage, A., *La capitation du Bas-Empire* (Mâcon, 1945).

Delpech, A., *Les norias de l'Oronte: analyse technologique d'un élément du patrimoine syrien* (Damascus, 1997).

Demandt, A., *Die Spätantike: römische Geschichte von Diocletian bis Justinian, 284–565 n. Chr* (Handbuch der Altertumswissenschaft 3.6; Munich, 1989).

Dendy, D. A. V., 'Sorghum and the Millets: Production and Importance', in David A. V. Dendy (ed.), *Sorghum and Millets: Chemistry and Technology* (St Paul, 1995), 11–25.

Denis, L., 'The Present Economic Importance of Olive Production', in FAO (ed.), *Modern Olive Production* (Rome, 1977), 1–7.

Dentzer, J.-M., 'Khans ou casernes à Palmyre? À propos de structures visibles sur des photographies aériennes anciennes', *Syria* 71 (1994), 45–112.

——'Les recherches archéologiques françaises dans la perspective de l'exploration de la Syrie du Sud basaltique', *Annales archéologiques arabes syriennes* 42 (1997), 297–318.

——(ed.), *Hauran I: Recherches archéologiques sur la Syrie du sud à l'époque héllenistique et romaine* (Paris, 1985).

——(ed.), *Hauran II: Recherches archéologuqes sur la Syrie du sud à l'époque héllenistique et romaine* (Paris, 1986).

Dressel, H., 'Ricerche sul monte testaccio', *Annali dell'instituto di corrispondenza archeologica* (1878), 118–92.

Dulière, C., *Mosaïques des portiques de la grande colonnade: section VII, 16–17* (Fouilles d'Apamée de Syrie. Miscellanea, fasc. 3; Brussels, 1974).

Duncan-Jones, R., *The Economy of the Roman Empire: Quantitative Studies* (London, 1974).

——'The Size of the Modius Castrensis', *Zeitschrift für Papyrologie und Epigraphik* 21 (1976), 53–62.

——'Some Configurations of Landholding in the Roman Empire', in M. I. Finley (ed.), *Studies in Roman Property* (Cambridge, 1976), 7–32.

——*Structure and Scale in the Roman Economy* (Cambridge, 1990).

Durliat, J., *De la ville antique à la ville byzantine. Le problème des subsistances* (Rome, 1990).

Durliat, J. and Guillou, A., 'Le tarif d'Abydos', *Bulletin de Correspondence Hellénique* 108 (1984), 581–91.

Dussaud, R. and Macler, F., *Voyage archéologique au Safâ et dans Djebel ed Drûz avec 1 itinéraire* (Paris, 1901).

Duval, P. R., *La littérature syriaque* (Paris, 1899).

El Faïz, M., 'Salinité et histoire de la Mésopotamie pré-islamique', *Journal of the Economic and Social History of the Orient* 33/1 (1990), 105–16.

——*L'agronomie de la Mésopotamie antique: analyse du 'Livre de l'agriculture nabatéenne' de Qûtâmä* (Leiden, 1995).

El Fakhry, A. K., 'Studies in Crop Rotations and Tillage Practices on Wheat Production Under Rainfed Conditions in Northern Iraq', in FAO (ed.), *Rainfed Agriculture in the Near East and North Africa* (Rome, 1980), 59–65.

Empereur, J. Y. and Picon, M., 'Les régions de production d'amphores impériales en Méditerranée orientale', in M. Lenoir, D. Manacorda, and C. Panella (eds.), *Amphores romaines et histoire économique: dix ans de recherche: actes du colloque* Collection de l'Ecole française de Rome, 114 (Rome, 1989), 223–48.

Evenari, M., Shanan, L., and Tadmor, N., 'The Ancient Desert Agriculture in the Negev IV: Chain Well Systems in the Wadi Arava', *Ktavim* 9 (1958), 223–40.

Evenari, M., Shanan, L., and Tadmor, N., *The Negev: The Challenge of a Desert* (2nd edn; Cambridge, Mass., and London, 1982).

Eyice, S., 'Un site byzantine de la Cilicie: Öküzlü et ses basiliques', in Lydie Hadermann-Misguich, Georges Raepsaet, and Guy Cambier (eds.), *Rayonnement Grec: Hommages à Charles Delvoye* (Brussels, 1982), 355–67.

Fahd, T., *Matériaux pour l'histoire de l'agriculture en Irak: Al-filâha n-nabatiyya* (Handbüch der Orientalistik; Leiden and Cologne, 1977).

——'Traductions en Arabes d'écrits geoponiques', Ciencias de la naturaleza en al-Andalus: textos y estudios (Granada, 1990), 11–21.

——'Botany and Agriculture', in Roshdi Rashed (ed.), *Encyclopedia of the History of Arab Science* (London, 1996), 813–52.

FAO, *Improving Productivity of Dryland Areas* (Rome, 1987).

Farquharson, P., 'Byzantium, Planet Earth and the Solar System', in Pauline Allen and Elizabeth Jeffreys (eds.), *The Sixth Century: End or Beginning?* (Byzantina Australiensia 10; Brisbane, 1996), 263–9.

Fehrle, E., 'Richtlinien zur textgestaltung der griechischen geoponica', *Sitzungsberichte der Heidelberger Akademie der Wissenschaften, Philosophisch-Historische Klasse* 11 (1920), 15.

——*Studien zu dem griechischen geoponikern* (Leipzig, 1920).

——'Studien zu dem Griechischen Geoponikern', *ΣΤΟΙΚΕÎΑ* 3 (1920).

Feissel, D., 'Magnus, Mégas et les curateurs des "maisons divines" de Justin II à Maurice', *Travaux et Mémoires* 9 (1985), 465–76.

Fiema, Z. T., 'Economics, Administration and Demography of Late Roman and Byzantine Southern Transjordan', Ph.D. diss. (Salt Lake City, 1991).

——'Byzantine Petra—A Reassessment', in Thomas S. Burns and John W. Eadie (eds.), *Urban Centers and Rural Contexts in Late Antiquity* (East Lansing, Mich., 2001), 111–31.

Fiensy, D. A., *The Social History of Palestine in the Herodian Period: The Land is Mine* (Lewiston, Queenstown, and Lampeter, 1991).

Finkelstein, I. and Magen, Y., *Archaeological Survey of the Hill Country of Benjamin* (Jerusalem, 1993).

Finley, M. I., *The Ancient Economy* (2nd edn; London, 1985).

Fisher, M., Isaac, B. H., and Roll, I., *Roman Roads in Judaea*, II: *The Jaffa–Jerusalem Roads* (BAR International Series 628; Oxford, 1996).

Fitzgerald, G. M., *Beth-Shan Excavations 1921–1923: The Arab and Byzantine Levels* (Philadelphia, 1931).

Foote, R. M., 'Umayyad Markets and Manufacturing: Evidence for a Commercialized and Industrializing Economy in Early Islamic Bilad al-Sham', Ph.D. diss. (Cambridge, Mass., 1999).

Forbes, H., 'The Ethnoarchaeological Approach to Ancient Greek Agriculture: Olive Cultivation as a Case Study', in B. Wells (ed.), *Agriculture in Ancient Greece: Proceedings of the Seventh International Symposium at the Swedish Institute at Athens, 16–17 May 1990* (Stockholm, 1992), 87–101.

Foss, C., *Ephesus after Antiquity: A Late Antique, Byzantine and Turkish City* (Cambridge, 1979).

——'The Lycian Coast', *Dumbarton Oaks Papers* 48 (1994), 1–52.

——'Syria in Transition, A.D. 550–750: An Archaeological Approach', *Dumbarton Oaks Papers* 51 (1997), 189–269.

Fowden, E. K., *The Barbarian Plain: Saint Sergius Between Rome and Iran* (Transformation of the Classical Heritage 28; Berkeley, 1999).

Foxhall, L. and Forbes, H. A., '*ΣΙΤΟΜΕΤΡΙΑ*: The Role of Grain as a Staple Food in Classical Antiquity', *Chiron* 12 (1982), 41–90.

Frankel, R., *Wine and Oil Production in Antiquity in Israel and Other Mediterranean Countries* (Sheffield, 1999).

Frankel, R. and Getzov, N., *Map of Akhziv, Map of Hanitah* (Archaeological Survey of Israel; Jerusalem, 1997).

Frankel, R., Avitsur, S., and Ayalon, E., *History and Technology of Olive Oil in the Holy Land* (Arlington, Va., 1994).

Frankel, Y., 'Roads and Stations in Southern Bilad ash-Sham in the 7th–8th Centuries', *Aram* 8 (1996), 177–88.

Frayn, J. M., *Subsistence Farming in Italy* (Fontwell, 1979).

Frézouls, E., 'La vie rurale au Bas-Empire d'après l'oeuvre de Palladius', *Ktèma*, 5 (1980), 193–210.

Frumkin, A., et al., 'Middle Holocene Environmental Change Determined from the Salt Caves of Mount Sedom, Israel', in O. Bar-Yosef and R. S. Kra (eds.), *Late Quaternary Chronology and Paleoclimates of the Eastern Mediterranean* (Tucson, Ariz., 1994), 315–32.

Fuller, M. J., 'Abila of the Decapolis: A Roman–Byzantine City in Transjordan', Ph.D. diss. (St Louis, 1989).

Gagos, T. and Frösén, J., 'Petra Papyri', *Annual of the Department of Antiquities of Jordan* 42 (1998), 473–81.

Gal, Z., *Map of Gazit (19–22)* (Archaeological Survey of Israel; Jerusalem, 1991).

——*Map of Har Tavor (41) and Map of En Dor (45)* (Archaeological Survey of Israel; Jerusalem, 1998).

Gallant, T. W., *Risk and Survival in Ancient Greece: Reconstructing the Rural Domestic Economy* (Cambridge, 1991).

García Moreno, L. A., 'Colonias de commerciantes orientales en la península ibérica S. V–VII', *Habis* 3 (1972), 127–54.

García Sánchez, E., 'Agriculture in Muslim Spain', in S. K. Jayyusi (ed.), *The Legacy of Muslim Spain* (New York, 1992), 987–99.

Garnsey, P., *Famine and Food Supply in the Graeco-Roman World: Responses to Risk and Crisis* (Cambridge, 1988).

——*Food and Society in Classical Antiquity* (Cambridge, 1999).

Garnsey, P. and Whittaker, C. R., 'Trade, Industry and the Urban Economy', in Averil Cameron and Peter Garnsey (eds.), *The Cambridge Ancient History*, XIII: *The Late Empire, A.D. 337–425* (Cambridge, 1998), 312–37.

Gascou, J. and Maccoull, L., 'Le cadastre d'Aphroditô', *Travaux et Memoires* 10 (1987), 103–58.

Gawlikowski, M., 'Palmyra as a Trading Center', *Iraq* 56 (1994), 27–33.

——'Palmyra and its Caravan Trade', *Annales archéologiques arabes syriennes* (1997), 42.

Gazit, D., *Map of Urim (125)* (Archaeological Survey of Israel; Jerusalem, 1996).

Gebreab, F., et al., 'Donkey Utilisation and Management in Ethiopia', in Paul Starkey and Denis Fielding (eds.), *Donkeys, People and Development: A Resource Book of the Animal Traction Network for Eastern and Southern Africa (ATNESA)* (Wageningen, 2004), 46–52.

Gemoll, F. W. K., *Untersuchungen über die Quellen, den Verfasser und die Abfassungszeit der Geoponica* (Berlin, 1883).

Genequand, D., 'Some Thoughts on Qasr al-Hayr al-Gharbi, its Dam, its Monastery, and the Ghassanids', *Levant* 38 (2006), 45–62.

Geyer, B., 'Aux origines de l'archéologie aérienne', in Lévon Nordiguian and Jean-François Salles (eds.), *Aux origines de l'archéologie aérienne* (Beirut, 2000), 109–22.

Geyer, B. and Monchambert, J. Y., 'Prospection dans la basse vallée de l'Euphrate Syrien', *Annales archéologiques arabes syriennes* 33 (1983), 261–5.

——*La basse vallée de l'Euphrate syrien du Néolithique à l'avènement de l'islam: géographie, archéologie et histoire*, 2 vols. (Beirut, 2003).

Geyer, B. and Rousset, M.-O., 'Les steppes arides de la Syrie du Nord à l'époque byzantine ou la rouée vers l'est', in B. Geyer (ed.), *Conquête de la steppe et appropriation des terres sur les marges arides du Croissant fertile* (Lyon, 2001), 111–21.

Gibson, S., Ibbs, B., and Kloner, A., 'The Sataf Project of Landscape Archaeology and the Judean Hills: A Preliminary Report on Four Seasons of Survey and Excavation (1987–89)', *Levant* 23 (1991), 23–54.

Gilbertson, D. D., 'Explanations: Environment as Agency', in Graeme Barker (ed.), *Farming the Desert: The UNESCO Libyan Valleys Archaeological Survey* (Paris, 1996), 291–317.

Giorgi, J., 'Plant Remains from the Sediments of the Inner Harbour (Area I14)', in K. G. Holum, Avner Raban, and J. Patrich (eds.), *Caesarea Papers*, II: *Herod's Temple, the Provincial Governor's Praetorium and Granaries, the Later Harbor, a Gold Coin Hoard, and Other Studies* (Providence, RI, 1999), 352–6.

Given, M., Knapp, A. B., and Coleman, D., *The Sydney Cyprus Survey Project: Social Approaches to Regional Archaeological Survey* (Los Angeles, 2003).

Glick, T. F., *Irrigation and Society in Medieval Valencia* (Cambridge, Mass., 1970).

Glucker, C. A. M., *The City of Gaza in the Roman and Byzantine Periods* (BAR International Series 325; Oxford, 1987).

Glueck, N., *Rivers in the Desert* (London, 1959).

González-Andrés, F. and María-Ortiz, J., 'Potential of Cytisus and Allied Genera (Genisteae: Fabaceae) as Forage Shrubs: 1. Seed Germination and Agronomy', *New Zealand Journal of Agricultural Research* 39 (1996), 195–204.

Gophna, R., *Map of Herziliyya (69)* (Archaeological Survey of Israel; Jerusalem, 1998).

Gough, M., 'Anazarbus', *Anatolian Studies* 2 (1954), 85–150.

Govrin, Y., *Map of Nahal Yattir (139)* (Archaeological Survey of Israel; Jerusalem, 1991).

Grabar, O., *City in the Desert: Qasr al-Hayr East* (Harvard Middle Eastern Monographs 23–4; Cambridge, Mass, 1978).

Graf, D., 'Roman Roads East of the Jordan', in Michele Piccirillo and Eugenio Alliata (eds.), *The Madaba Map Centenary, 1897–1997: Travelling through the Byzantine Umayyad Period: Proceedings of the International Conference Held in Amman, 7–9 April 1997* (Jerusalem, 1999), 227–9.

Grant, M., 'Oribasios and Medical Text', in John Wilkins, David Harvey, and Mike Dobson (eds.), *Food in Antiquity* (Exeter, 1995), 371–80.

——*Dieting for an Emperor: A Translation of Books 1 and 4 of Oribasius' Medical Compilations with an Introduction and Commentary* (Studies in Ancient Medicine 15; Leiden, 1997).

Greene, J. A. and Kehoe, D. P., 'Mago the Carthaginian', in M. Hassine and M. Ghaki (eds.), *Actes du IIIe congrès international des études phéniciennes et puniques: Tunis, 11–16 novembre 1991* (Tunis, 1995), 110–17.

Grossmann, P., 'Zu den Bogen und Gewölben in dem Wüstenpalast von Qaṣr ibn Wardān', *Damaszener Mitteilungen* 12 (2000), 291–302.

Grove, A. T. and Rackham, O., *The Nature of Mediterranean Europe: An Ecological History* (New Haven, 2001).

Guichand, P., 'Review of T. F. Glick, *Irrigation and Hydraulic Technology: Medieval Spain and Its Legacy*', *Technology and Culture* 39/3 (1998), 564–6.

Gunn, J. D., *The Years Without Summer: Tracing A.D. 536 and its Aftermath* (Oxford, 2000).

Gutwein, K. C., *Third Palestine: A Regional Study in Byzantine Urbanization* (Washington, DC, 1981).

Haase, C.-P., 'Madīnat al-Fār: the Regional Late Antique Tradition of an Early Islamic Foundation', in K. Bartl and S. Hauser (eds.), *Continuity and Change in Northern Mesopotamia from the Hellenistic to the Early Islamic Period* (Berlin, 1996), 349–61.

Hadjisavvas, S., 'The Economy of the Olive', in V. Karageorghis and D. Michaelides (eds.), *The Development of the Cypriot Economy* (Nicosia, 1996), 127–38.

——'Olive Oil Production in Ancient Cyprus', *Report of the Department of Antiquities Cyprus* (1988), 111–19.

——'Windlass vs. Screw: A Case-Study for the Reconstruction of an Olive Press', *Report of the Department of Antiquities Cyprus* (1990), 181–5.

——'Perforated Monoliths: Myths and Reality', in M.-C. Amouretti, J.-P. Brun, and D. Eitam (eds.), *Oil and Wine Production in the Mediterranean Area from the Bronze Age to the End of the XVIth Century* (Paris, 1991), 69–80.

——*Olive Oil Production in Cyprus from the Bronze Age to the Byzantine Period* (Nicosia, 1992).

Haiman, M., *Map of Har Hamran—Southwest (198)* (Archaeological Survey of Israel; Jerusalem, 1986).

——*Map of Mizpé Ramon—Southwest (200)* (Archaeological Survey of Israel; Jerusalem, 1991).

——*Map of Har Hamran—Southeast (199)* (Archaeological Survey of Israel; Jerusalem, 1993).

——'Agriculture and Nomad–State Relations in the Negev Desert in the Byzantine and Early Islamic Periods', *Bulletin of the American Schools of Oriental Research* 297 (1995), 29–54.

——*Map of Har Ramon* (203) (Archaeological Survey of Israel; Jerusalem, 1999).

Hall, L. J., *Roman Berytus: Beirut in Late Antiquity* (London, 2004).

Harlan, J. R., 'The Early History of Wheat: Earliest Traces to the Sack of Rome', in L. T. Evans and W. J. Peacock (eds.), *Wheat Science—Today and Tomorrow* (Cambridge, 1981), 1–20.

Harper, R. P., 'Neocaesareia—Qasrin—Dibsi Faraj', in J. Cl. Margueron (ed.), *Le Moyen Euphrate: Actes du Colloque de Strasbourg 10–12 mars 1977* (Strasbourg, 1980), 327–48.

Harper, R. P. and Wilkinson, T. J., 'Excavations at Dibsi Faraj, Northern Syria 1972–1974: A Preliminary Note on the Site and its Monuments', *Dumbarton Oaks Papers* 29 (1975), 319–38.

Harris, A., *Byzantium, Britain and the West: The Archaeology of Cultural Identity AD 400–800* (Stroud, 2003).

Harris, C. P., *The Syrian Desert: Caravans, Travel and Exploration* (London, 1937).

Hayes, J. W., *Excavations at Saraçhane in Istanbul*, II (Princeton and Washington, DC, 1992).

——'The Pottery', in S. E. Sidebotham and W. Wendrich (eds.), *Berenike 1994: Preliminary Report of the 1994 Excavations at Berenike (Egyptian Red Sea Coast) and the Survey of the Eastern Desert* (Leiden, 1995), 147–73.

Heather, P. J., 'New Men for New Constantines? Creating an Imperial Elite in the Eastern Mediterranean', in P. Magdalino (ed.), *New Constantines: The Rhythm of Imperial Renewal in Byzantium, 4th–13th Centuries: Papers from the Twenty-sixth Spring Symposium of Byzantine Studies, St Andrews, March 1992* (Aldershot, and Brookfield, Vt., 1994), 11–34.

Heichelheim, F. M., 'Roman Syria', in Tenney Frank and T. R. S. Broughton (eds.), *Economic Survey of Ancient Rome* (Baltimore, 1933–40), III.123–257.

Heim, C., Nowaczyk, N., and Negendank, J. F. W., 'Near East Desertification: Evidence from the Dead Sea', *Naturwissenschaften* 84 (1997), 398–401.

Heine, P., *Weinstudien: Untersuchungen zu Anbau, Produktion und Konsum des Weins im arabisch-islamischen Mittelalter* (Wiesbaden, 1982).

Hellenkemper, H. and Hild, F., *Neue Forschungen in Kilikien* (Tabula Imperii Byzantini 4; Vienna, 1986).

Hendy, M. F., *Studies in the Byzantine Monetary Economy, c. 300–1450* (Cambridge, 1985).

——*The Economy, Fiscal Administration and Coinage of Byzantium* (Northampton, 1989).

Hennessy, B., McNicoll, A., and Smith, R. H., *Pella in Jordan* (Canberra, 1982).

Herr, L. G., *Madaba Plains Project: The 1989 Season at Tell el-Umeiri and Vicinity and Subsequent Studies* (Berrien Springs, Mich., 1997).

Hild, F. and Hellenkemper, H., *Kilikien und Isaurien* (Tabula Imperii Byzantini 5; Vienna, 1990).

Hill, D. R., 'Engineering', in Roshdi Rashed (ed.), *Encyclopedia of the History of Arab Science*, III (London, 1996), 751–95.

Hillman, G. C., 'Traditional Husbandry and Processing of Archaic Cereals in Modern Times: Part 1, the Glume-Wheats', *Bulletin on Sumerian Agriculture* 1 (1984), 114–52.

Hirschfeld, Y., *Map of Herodium (108/2)* (Archaeological Survey of Israel; Jerusalem, 1985).

——*The Judean Desert Monasteries in the Byzantine Period* (New Haven, 1992).

——'Ramat Ha-Nadiv', in E. Stern (ed.), *The New Encyclopedia of Archaeological Excavations in the Holy Land* (Jerusalem, New York, and London, 1993), 1257–60.

——*The Palestinian Dwelling in the Roman–Byzantine Period* (Collectio Minor Studium Biblicum Franciscanum 34; Jerusalem, 1995).

——'Farms and Villages in Byzantine Palestine', *Dumbarton Oaks Papers* 57 (1997), 33–71.

——*Ramat Hanadiv Excavations: Final Report of the 1984–1998 Seasons* (Jerusalem, 2000).

——'The Water Supply of the Monastery of Chariton', in D. Amit, J. Patrich, and Y. Hirschfeld (eds.), *The Aqueducts of Israel* (Portsmouth, RI, 2002), 425–37.

——'A Climatic Change in the Early Byzantine Period? Some Archaeological Evidence', *Palestine Exploration Quarterly* 136/2 (2004), 133–49.

——'The Crisis of the Sixth Century: Climatic Change, Natural Disasters, and the Plague', *Mediterranean Archaeology and Archaeometry* 6/1 (2006), 19–32.

Hirschfeld, Y. and Birger, R., 'Khirbet-Ed-Deir (Judean-Desert) 1981–1984', *Revue Biblique* 93/2 (1986), 276–84.

Hirschfeld, Y. and Birgercalderon, R., 'Early Roman and Byzantine Estates Near Caesarea', *Israel Exploration Journal* 41/1–3 (1991), 81–111.

Hitchner, R. B. and Mattingly, D. J., 'Ancient Agriculture', *National Geographic Research & Exploration* 7/1 (1991), 36–55.

Holum, K. G., *King Herod's Dream: Caesarea on the Sea* (New York and London, 1988).

Honigmann, E., *Historische Topographie von Nordsyrien im Alterum* (Leipzig, 1923–4).

——'Studien zur Notitia Antiochena', *Byzantinische Zeitschrift* 25 (1925), 60–88.

Hopkins, D. C., 'Agriculture', in Eric M. Meyers (ed.), *The Oxford Encyclopedia of Archaeology in the Near East* (Oxford, 1996), 23–30.

Hopkins, K., 'Taxes and Trade in the Roman Empire (200 B.C.–A.D. 400)', *Journal of Roman Studies* 70 (1980), 101–25.

Horden, P. and Purcell, N., *The Corrupting Sea: A Study of Mediterranean History* (Oxford, 2000).

Horsley, R. A., *Galilee: History, Politics, People* (Valley Forge, Pa., 1995).

Hoyland, R., *Arabia and the Arabs: From the Bronze Age to the Coming of Islam* (London, 2001).

Husson, G., *Oikia: le vocabulaire de la maison privée en Egypte d'après les papyrus grecs* (Paris, 1983).

Ibach, R. D., *Archaeological Survey of the Hesban Region: Catalogue of Sites and Characterization of Periods* (Hesban 5; Berrien Springs, Mich., 1987).

Ionides, M. G., *The Régime of the Rivers Euphrates and Tigris* (London, 1937).

Isaac, B. H., 'Bandits in Judaea and Arabia', *Harvard Studies in Classical Philology* 88 (1984), 171–205.

——*The Limits of Empire: the Roman Army in the East* (rev. edn; Oxford, 1992).

——and Roll, I., *Roman Roads in Judaea*, I: *The Legio-Scythopolis Road* (BAR International Series 141; Oxford, 1982).

Israel, Y. (1993), 'Ashqelon', *Excavations and Surveys in Israel* 13 (1993), 100–5.

Issar, A., *Water Shall Flow from the Rock: Hydrogeology and Climate in the Lands of the Bible* (Berlin, 1990).

Issar, A. and Yakir, D., 'Isotopes from Wood Buried in the Roman Siege Ramp of Masada: The Roman Period's Colder Climate', *Biblical Archaeologist* 60/2 (1997), 101–6.

Jacobini, N., Battaglini, M., and Preziosi, P., 'Propagation of the Olive Tree', in *Modern Olive Growing* (Rome, 1977), 27–59.

Jasny, N., 'Competition Among Grains in Classical Antiquity', *American Historical Review* 47/4 (1942), 747–64.

——*The Wheats of Classical Antiquity* (Baltimore, 1944).

Jaubert, R. et al., *Land Use and Vegetation Cover: Semi-arid Areas of Aleppo and Hama Provinces (Syria)* (Lyon, 1999).

Johnson, A. C. and West, L. C., *Byzantine Egypt: Economic Studies* (Amsterdam, 1967).

Jones, A. H. M., 'Census Records of the Later Roman Empire', *Journal of Roman Studies* 49 (1953), 49–64.

——*The Later Roman Empire, 284–602: A Social, Economic and Administrative Survey* (Oxford, 1964).

——*The Cities of the Eastern Roman Provinces* (Oxford, 1971).

Jones, A. H. M. and Brunt, P. A., *The Roman Economy: Studies in Ancient Economic and Administrative History* (Oxford, 1974).

Jones, A. H. M., Martindale, J. R., and Morris, J. (eds.), *The Prosopography of the Later Roman Empire*, vol. I: *A.D. 260–395* (Cambridge, 1971).

Kaplan, M., *Les propriétés de la couronne et de l'Église dans l'empire byzantin, Ve–VIe siècles: documents* (Paris, 1976).

——'Les grands propriétaires de Cappadoce (VIe–XIe siècles)', in C. Fonseca (ed.), *Le aree omogenee della civiltà rupestre nell'ambito dell'impero bizantino: la Cappadocia* (Lecce, 1981), 125–58.

——'Nouvelle de Tibère II sur les «maisons divines»', *Travaux et Memoire* 8 (1981), 237–45.

——'L'exploitation paysanne byzantine entre l'antiquité et le moyen-âge (VIème–VIIIème siècles). Affirmation d'une structure économique et sociale', in V. Vavrínek (ed.), *The Byzantinological Symposium in the 16th International Eirene Conference* (Prague, 1985), 101–6.

——*Les hommes et la terre à Byzance du VI au XI siècle: propriété et exploitation du sol* (Paris, 1992).

Karageorghis, V., 'Chronique de fouilles à Chypre', *Bulletin de Correspondence Hellénique* 113 (1989), 789–853.

Karagiorgou, O., 'LR2: A Container for the Military Annona on the Danubian Border?', in S. A. Kingsley and M. Decker (eds.), *Economy and Exchange in the East Mediterranean during Late Antiquity* (Oxford, 2001), 129–66.

Karnapp, W., 'Der Khan in der syrischen Ruinenstadt Resafa', *Archäologischer Anzeiger* 93/2 (1978), 136–50.

Kehoe, D. P., *The Economics of Agriculture on Roman Imperial Estates in North Africa* (Hypomnemata 89; Göttingen, 1988).

——'Allocation of Risk and Investment on the Estates of Pliny the Younger', *Chiron* 18 (1988), 15–42.

——'The Sale of the Vintage in Pliny, Ep. 8.2', *Annals of the American Philological Association* 51 (1988), 310.

——'Approaches to Economic Problems in the Letters of Pliny the Younger: The Question of Risk in Agriculture', in H. Temporini (ed.), *Aufstieg und Niedergang der römischen Welt* II/33.1 (Berlin and New York, 1989), 555–90.

Keil, J. and Wilhelm, A., *Denkmäler aus dem Rauhen Kilikien* (Monumenta Asiae Minoris antiqua 3; Manchester, 1931).

Kennedy, D. L., *The Roman Army in Jordan* (London, 2000).

Khamis, E., 'Two Wall Mosaic Inscriptions from the Market Place in Bet Shean/Baysān', *Bulletin of the School of Oriental and African Studies* 64/2 (2001), 159–76.

Kiliç, S., Senol, S., and Evrendilek, F., 'Evaluation of Land Use Potential and Suitability of Ecosystems in Antakya for Reforestation, Recreation, Arable Farming, and Residence', *Türk tarim ve ormancilik dergisi* 27 (2003), 15–22.

Killebrew, A., 'Qasrin: the Village', in Ephraim Stern (ed.), *The New Encyclopedia of Archaeological Excavations in the Holy Land* (Jerusalem, New York, and London, 1993), 1222–4.

Killebrew, A. and Fine, S., 'Qatzrin: Reconstructing Village Life in Talmudic Times', *Biblical Archaeology Review* 17/3 (1991), 44–56.

King, R., Proudfoot, L. J., and Smith, B. J., *The Mediterranean: Environment and Society* (London, 1997).

Kingsley, S., 'The Sumaqa Pottery Assemblage: Classification and Quantification', in S. Dar (ed.), *Sumaqa: A Roman and Byzantine Jewish Village on Mount Carmel, Israel* (Oxford, 1999), 263–330.

——'Specialized Production and Long-Distance Trade in Byzantine Palestine', D.Phil. thesis (Oxford, 1999).

——'The Economic Impact of the Palestinian Wine Trade in Late Antiquity', in S. A. Kingsley and M. Decker (eds.), *Economy and Exchange in the East Mediterranean during Late Antiquity* (Oxford, 2001), 44–68.

——'The Dor D Shipwreck and Holy Land Wine Trade', *International Journal of Nautical Archaeology* 32/1 (2003), 85–90.

Kingsley, S. A., Johnson, A., and Williams, D., *A Sixth-century AD Shipwreck off the Carmel Coast, Israel: Dor D and Holy Land Wine Trade* (BAR International Series 1065; Oxford, 2002).

Kislev, M., 'Triticum paricoccum sp. nov., the Oldest Naked Wheat', *Israel Journal of Botany* 28 (1979–80), 95–107.

——'Contenu d'un silo a blé de l'époque du fer ancien', in J. Friend and J.-B. Humbert (eds.), *Tell Keisan 1971–1976: une cité phénicienne en Galilée* (Paris, 1980), 361–80.

Klaimi, Y. Y., 'Importance of Winter Cereals in Rainfed Agriculture of the Near East and North African Regions—Present Situation, Potential and Constraints', in FAO (ed.), *Rainfed Agriculture in the Near East and North Africa* (Rome, 1980), 87–92.

Kochavi, M. and Beit-Arieh, I., *Map of Rosh ha-Ayin (78)* (Archaeological Survey of Israel; Jerusalem, 1994).

Koder, J., 'Climatic Change in the Fifth and Sixth Centuries?', in Pauline Allen and Elizabeth Jeffreys (eds.), *The Sixth Century End or Beginning?* (Byzantina Australiensia 10; Brisbane, 1996), 270–86.

Kosso, C., 'Public Policy and Agricultural Practice: An Archaeological and Literary Study of Late Roman Greece', Ph.D. diss. (Chicago, 1993).

——*The Archaeology of Public Policy in Late Roman Greece* (BAR International Series 1126; Oxford, 2003).

Kraemer, C., *Excavations at Nessana*, III: *The Non-literary Papyri* (Princeton, 1958).

Kron, G., 'Roman Ley-Farming', *Journal of Roman Archaeology* 13 (2000), 277–87.

Lambert-Gócs, M., *The Wines of Greece* (London, 1990).

Lassus, J., *Inventaire archéologique de la région au nord-est de Hama* (Damascus, 1935).

Lauffray, J., *Halabiyya-Zenobia: Place forte du limes oriental et la haute-Mésopotamie au VIe siècle*, 2 vols. (Paris, 1983–91).

Lauffray, J. and Van Liere, W., 'Nouvelle prospection archéologique dans la Haute-Jézireh syrienne', *Annales archéologiques arabes syriennes* 4–5 (1954–5), 129–48.

Lavee, H. L., Poesent, J., and Yair, A., 'Evidence of High Efficiency Water-Harvesting by Ancient Farmers in the Negev Desert, Israel', *Journal of Arid Environments* 35 (1997), 341–8.

Le Clerc, J. A., Bailey, L. H., and Wessling, H. L., 'Milling and Baking Tests of Einkorn, Emmer, Spelt and Polish Wheat', *Journal of the American Society of Agronomy* 10 (1918), 215–18.

Le Strange, G., *Palestine Under the Moslems: A description of Syria and the Holy Land from A.D. 650 to 1500* (London, 1890).

Lemerle, P., *The Agrarian History of Byzantium: From the Origins to the Twelfth Century, the Sources and Problems* (Galway, 1979).

——*Byzantine Humanism* (Sydney, 1987).

Lender, Y. Y., *Map of Har Nafhah (196)* (Archaeological Survey of Israel; Jerusalem, 1990).

Levy, E., *West Roman Vulgar Law: The Law of Property* (Memoirs of the American Philosophical Society 29; Philadelphia, 1951).

Lewis, M. J. T., *Millstone and Hammer: The Origins of Water Power* (Hull, 1997).

Lewit, T., *Agricultural Production in the Roman Economy A.D. 200–400* (BAR International Series 568; Oxford, 1991).

Liebeschuetz, J. H. W. G., *Antioch: City and Imperial Administration in the Later Roman Empire* (Oxford, 1972).

Lightfoot, D. R., 'Syrian Qanat Romani: History, Ecology, Abandoment', *Journal of Arid Environments* 33 (1996), 321–36.

Lipschitz, N. and Waisel, Y., 'The Dendroarchaeological Remains at En Boqeq', in M. Gichon (ed.), *En Boqeq: Ausgraben in einer Oase am Toten Meer, Band I: Geographie und Geschichte der Oase das Spätrömisch-Byzantinische Kastell* (Mainz, 1993), 389–96.

Lipshits, E. E., *Geoponiki: vizantiiskaia sel'skokhoziaistvennaia entsiklopediia X veka* (Moscow, 1960).

Little, L. K., *Plague and the End of Antiquity: The Pandemic of 541–750* (New York and Cambridge, 2007).

Lombard, M., *The Golden Age of Islam* (Amsterdam, 1975).

Loulloupis, M., 'Excavations at Amathus', *Annual Report of the Department of Antiquities of Cyprus* (1975), 29–39.

Lutz, H. F., *Viticulture and Brewing in the Ancient Orient* (Leipzig and New York, 1922).

Lyall, C. J., *Translations of Ancient Arabian Poetry: Chiefly Pre-Islamic, with an Introduction and Notes* (New York, 1930).

Ma'oz, Z. U., 'Golan', in Ephraim Stern (ed.), *The New Encyclopedia of Archaeological Excavations in the Holy Land* (Jerusalem, New York, and London, 1993), 525–46.

Macadam, H. I., 'Epigraphy and Village Life in Southern Syria During the Roman and Byzantine Periods', *Berytus* 31 (1983), 103–15.

——'Some Aspects of Land Tenure and Social Development in the Roman Near East: Arabia, Phoenicia and Syria', in T. Khalidi (ed.), *Land Tenure and Social Transformation in the Middle East* (Beirut, 1984), 45–62.

——'Cities, Villages and Veteran Settlements: Roman Administration of the Syrian Hawran', in D. Panzac (ed.), *Histoire économique et sociale de l'Empire ottoman et de la turquie (1326–1960): Actes du sixième congrès international, Aix-en-Provence, 1–4 juillet 1992* (Paris, 1995), 641–52.

Macdonald, B., *The Wadi el Hasa Archaeological Survey, 1979–1983, West-Central Jordan* (Waterloo, Ont., 1988).

Magness, J., *The Archaeology of the Early Islamic Settlement in Palestine* (Winona Lake, Ind., 2003).

Mango, C. A., 'A Late Roman Inn in Eastern Turkey', *Oxford Journal of Archaeology* 5 (1986), 223–31.

Mango, M., 'The Commercial Map of Constantinople', *Dumbarton Oaks Papers* 54 (2000), 189–207.

Mango, M. M., 'Artistic Patronage in the Roman Diocese of Oriens, 313–641 AD', D.Phil. thesis (Oxford, 1984).

——*Silver from Early Byzantium: The Kaper Koraon and Related Treasures* (Walters Art Gallery Publication in the History of Art; Baltimore, Md., 1986).

——'The Origins of the Syrian Ecclesiastical Silver Treasures of the Sixth–Seventh Centuries', in N. Duval et al. (eds.), *Argenterie romaine et byzantine: actes de la table ronde, Paris 11–13 octobre 1983* (Paris, 1988), 163–84.

——'Excavations and Survey at Androna, Syria: The Oxford Team 1999', *Dumbarton Oaks Papers* 56 (2002), 307–15.

——'Excavations and Survey at Androna, Syria: The Oxford Team 2000', *Dumbarton Oaks Papers* 57 (2003), 293–7.

Mango, M. M. et al., 'Oxford Excavations at Andarin (Androna): September 1998', *Annales Archéologiques Arabes Syriennes* (1999).

——'Oxford Excavations at Andarin (Androna): September 1999', *Annales Archéologiques Arabes Syriennes* (2000).

Manning, S., et al., 'Late Roman Type 1a Amphora Production at the Late Roman Site of Zygi-Petrini, Cyprus', *Report of the Department of Antiquities Cyprus* (2000), 233–58.

Marcone, A., 'Late Roman Social Relations', in Averil Cameron and Peter Garnsey (eds.), *The Cambridge Ancient History*, XIII: *The Late Empire, A.D. 337–425* (Cambridge, 1998), 338–70.

Marfoe, L., 'Between Qadesh and Kumidi: A History of Frontier Settlement and Land Use in the Biqa', Lebanon', Ph.D. diss. (Chicago, 1978).

Matthews, E., Neidinger, W., and Ayalon, E., 'Preliminary Report on the Excavations at Zur Natan' (unpublished excavation report: Houston, Publication of the Texas Foundation for Archaeological and Historical Research, 1990), 1–21.

Matthews, J., 'The Tax Law of Palmyra: Evidence for Economic History in a City of the Roman East', *Journal of Roman Studies* 74 (1984), 157–80.

Mattingly, D. J., 'Libyans and the "Limes": Culture and Society in Roman Tripolitania', *Antiquités Africaines* 23 (1987), 71–94.

——'Olea Mediterranea?', *Journal of Roman Archaeology* 1 (1988), 153–61.

——'Oil for export? A Comparison of Libyan, Spanish and Tunisian Olive Oil Production in the Roman Empire', *Journal of Roman Archaeology* 1 (1988), 33–56.

——'Megalithic Madness and Measurement or How Many Olives Could an Olive Press Press?', *Oxford Journal of Archaeology* 7 (1988), 177–95.

——'Farmers and Frontiers: Exploiting and Defending the Countryside of Roman Tripolitania', *Libyan Studies* 20 (1989), 135–53.

Mattingly, D. J., 'Maximum Figures and Maximizing Strategies of Oil Production?', in M.-C. Amouretti and J.-P. Brun (eds.), *La production du vin et de l'huile en Méditerranée* (Athens, 1993), 463–98.

——*Tripolitania* (London, 1995).

Mattingly, D. J. and Hayes, J. W., 'Nador and Fortified Farms in North Africa', *Journal of Roman Archaeology* 5 (1992), 408–18.

Mattingly, D. J. and Salmon, J., *Economies Beyond Agriculture in the Classical World* (Leicester-Nottingham Studies in Ancient Society 9; London, 2001).

Mayer, W. and Allen, P., *John Chrysostom* (London, 2000).

Mayerson, P., *The Ancient Agricultural Regime of Nessana and the Central Negeb* [sic] (London, 1960).

——'Wheat in the Roman World: An Addendum', *Classical Quarterly* 34/1 (1984), 243–5.

——'The Wine and Vineyards of Gaza in the Byzantine Period', *Bulletin of the American Schools of Oriental Research* 257 (1985), 75–108.

——'The Gaza "Wine" Jar (Gazition) and the "Lost" Ashkelon Jar (Askalonion)', *Israel Exploration Journal* 42 (1992), 76–80.

——'The Use of Ascalon Wine in the Medical Writers of the Fourth to the Seventh Centuries', *Israel Exploration Journal* 43 (1993), 169–73.

McC. Adams, R., *Land Behind Baghdad: A History of Settlement on the Diyala Plains* (Chicago, 1965).

——'Historic Patterns of Mesopotamian Irrigation Agriculture', in R. McC. Adams, T. E. Downing, and M. Gibson (eds.), *Irrigation's Impact on Society* (Tuscon, Ariz., 1974), 1–6.

——*Heartland of Cities: Surveys of Ancient Settlement and Land Use on the Central Floodplain of the Euphrates* (Chicago, 1981).

McCabe, A. E., A *Byzantine Encyclopaedia of Horse Medicine: The Sources, Compilation, and Transmission of the Hippiatrica* (Oxford Studies in Byzantium; Oxford, 2007).

McCormick, M., *Origins of the European Economy: Communications and Commerce, A.D. 300–900* (Cambridge, 2001).

——'Rats, Communications, and Plague: Toward an Ecological History', *Journal of Interdisciplinary History* 34/1 (2003), 1–25.

McGovern, P. E., et al., 'The Beginnings of Winemaking and Viniculture in the Ancient Near East', *Expedition* 39/1 (1997), 3–21.

Melena, J. L., 'Olive Oil and Other Sorts of Oil in the Mycenaean Tablets', *Minos* n.s. 18 (1983), 89–123.

Merrill, S., *East of the Jordan: A Record of Travel and Observation in the Countries of Moab, Gilead, and Bashan* (London, 1881).

Meyer, E. H. F. and Verdoorn, F., *Geschichte der Botanik* (Amsterdam, 1965).

Michaelides, D., 'The Economy of Cyprus During the Hellenistic and Roman Periods', in V. Karagheorghis and D. Michaelides (eds.), *The Development of the Cypriot Economy* (Nicosia, 1996), 139–52.

Millar, F., 'Caravan Cities: The Roman Near East and Long Distance Trade by Land', in M. Austin, J. Harries and C. Smith (eds.), *Modus Operandi: Essays in Honour of Geoffrey Rickman* (London: Institute of Classical Studies, School of Advanced Study, University of London), 119–38.

Miller, D., 'From the Archives of Architecture: Agrarian Settlement and Society in the Roman Hawran', in D. Panzac (ed.), *Histoire économique et sociale de l'Empire ottoman et de la turquie (1326–1960): actes du sixième congrès international tenu à Aix-en-Provence du 1er au 4 juillett 1992* (Paris, 1995), 653–60.

Miller, J. M., *Archaeological Survey of the Kerak Plateau: Conducted During 1978–1982 Under the Direction of J. Maxwell Miller and Jack M. Pinkerton* (American Schools of Oriental Research Archaeological Reports 1; Atlanta, Ga., 1991).

Miller, N. F. and Gleason, K. L., *The Archaeology of Garden and Field* (Philadelphia, 1994).

Moore, F. G., 'Three Canal Projects, Roman and Byzantine', *American Journal of Archaeology* 54/2 (1950), 97–111.

Moors, S. M., 'De Decapolis: steden en dorpen in de Romeinse provincies Syria en Arabia', Ph.D. diss. (Leiden, 1992).

Morony, M., 'Land Use and Settlement Patterns in Late Sassanian and Early Islamic Iraq', in G. R. D. King and A. Cameron (eds.), *The Byzantine and Early Islamic Near East*, II: *Land Use and Settlement Patterns* (Princeton, 1994), 221–9.

——'Michael the Syrian as a Source for Economic History', *Hugoye: Journal of Syriac Studies* [http://syrcom.cua.edu/syrcom/Hugoye], 3/2 (2000).

Morrisson, C., 'Le trésor byzantin de Nikertai', *Revue belge de numismatique et de sigillographie* 118 (1972), 29–91.

——*Monnaie et finances à Byzance: analyses, techniques* (Collected studies 461; Aldershot, 1994).

Mouterde, R. and Poidebard, A., *Le Limes de Chalcis: organisation de la steppe en haute Syrie romaine* (Paris, 1945).

Musil, A., *Arabia Petraea* (Vienna, 1907).

——*The Middle Euphrates* (New York, 1927).

——*Palmyrena, a Topographical Itinerary* (New York, 1928).

Ne'eman, Y., *Map of Ma'anit (54)* (Archaeological Survey of Israel; Jerusalem, 1990).

Ne'eman, Y., Sender, S., and Oren, E., *Map of Mikhmoret (52); Map of Hadera (53)* (Archaeological Survey of Israel; Jerusalem, 2000).

Negev, A., *The Architecture of Mampsis Final Report*, I: *The Middle and Late Nabatean Periods* (Qedem 26.1; Jerusalem, 1988).

——*The Architecture of Oboda* (Qedem 36; Jerusalem, 1997).

Negev, N., 'Be'er Sheva', *Excavations and Surveys in Israel* 12 (1992), 95–6.

Nesbitt, J. W., 'The Life of St. Philaretos and its Significance for Byzantine Agriculture', *Greek Orthodox Theological Review* 14 (1969), 150–8.

Netzer, E., 'Population Growth and Decline in the Northern Part of Eretz-Israel during the Historical Period as Related to Climatic Changes', in A. Issar and Neville Brown (eds.), *Water, Environment and Society in Times of Climatic Change: Contributions from an International Workshop Within the Framework of International*

Hydrological Program (IHP) UNESCO, Held at Ben-Gurion University, Sede Boker, Israel from 7–12 July 1996 (Dordrecht, 1998), 129–45.
Neubauer, A., *La Géographie du Talmud* (Paris, 1868).
Neusner, J., *Introduction to Rabbinic Literature* (New York and London, 1994).
Nevo, Y. D., *Pagans and Herders: A Re-examination of the Negev Runoff Cultivation Systems in the Byzantine and Early Arab Periods* (Negev, 1991).
Newman, J., *The Agricultural Life of the Jews in Babylonia Between 200 C.E. and 500 C.E.* (London, 1932).
Nöldeke, T., 'Noch Einiges über die "nabataïsche Landwirtschaft"', *Zeitschrift der deutschen morgenländischen Gesellschaft* 29 (1875), 445–55.
Northedge, A., Bennett, C. M., and Bowsher, J., *Studies on Roman and Islamic Amman: The Excavations of Mrs C-M Bennett and Other Investigations* (British Academy Monographs in Archaeology 3; Oxford, 1992).
O'Dea, W., *The Social History of Lighting* (London, 1958).
Oakman, D. E., *Jesus and the Economic Questions of His Day* (Lewiston and Queenstown, 1986).
Oder, E., 'Beiträge zur Geschichte der Landwirtschaft bei den Griechen I', *Rheinisches Museum für Philologie* 45 (1890), 58–99.
——'Beiträge zur Geschichte der Landwirtschaft bei den Griechen II', *Rheinisches Museum für Philologie* 45 (1890), 212–22.
——'Beiträge zur Geschichte der Landwirtschaft bei den Griechen III', *Rheinisches Museum für Philologie* 48 (1893), 1–40.
——*Ein angebliches Bruchstück Democritus über die Entdeckung unterirdischer Quellen* (Leipzig, 1898).
Olami, Y. A., *Map of Daliya (31)* (Archaeological Survey of Israel; Jerusalem, 1981).
Olami, Y. A. and Gal, Z., *Map of Shefar'am (24)* (Jerusalem, 2003).
Olami, Y. A., Ronen, A., and Romano, A., *Map of Haifa—West (22)* (Archaeological Survey of Israel; Jerusalem, 2003).
Olami, Y. A., Sender, S., and Oren, E., *Map of Yagur (27)* (Archaeological Survey of Israel; Jerusalem, 2004).
Oleson, J. P., *Greek and Roman Mechanical Water-Lifting Devices: The History of a Technology* (Toronto, 1984).
Orssaud, D., 'Notice sur la céramqiue', in J. Lauffray (ed.), *Halabiyya-Zenobia II* (Paris, 1991), 260–75.
Ovadiah, A., 'Excavations in the Area of the Ancient Synagogue of Gaza (preliminary report)', *Israel Exploration Journal* 19 (1969), 193–8.
Ovadiah, A. and Mucznik, S., 'The Mosaic Pavement of Kissufim, Israel', in Henri Stern (ed.), *Mosaïque: recueil d'hommages à Henri Stern* (Paris, 1983), 273–80.
Pacetti, F., 'Appunti su alcuni tipi di anfore orientali della prima età bizantine centri di produzione, contenuti, cronologia e distribuzione', in Lorenzo Quilici and Stefania Quilici Gigli (eds.), *Agricoltura e commerci nell'Italia antica* (Atlante tematico di topografia antica. Supplemento 1; Rome, 1995), 273–94.
Palmer, A., *Monk and Mason on the Tigris Frontier: The Early History of Tur Abdin* (Cambridge, 1990).

Palumbo, G., et al., 'The Wadi Az-Zarqa/Wadi Ad-Dulayl Excavations and Survey Project: Report on the October–November 1993 Fieldwork Season', *Annual of the Department of Antiquities of Jordan* 40 (1996), 375–427.

Papacostas, T., 'The Economy of Late Antique Cyprus', in S. A. Kingsley and M. Decker (eds.), *Economy and Exchange in the East Mediterranean During Late Antiquity* (Oxford, 2001), 107–28.

Papageorghiou, A., 'Chronique de fouilles et découvertes archéologiques à Chypre', *Bulletin de Correspondence Hellénique*, 114 (1990), 941–85.

Papathomas, A., 'Das erste antike Zeugnis für die veterinärmedizinische Exzerptensammlung des Anatolios von Berytos', *Wiener Studien* 113 (2000), 135–51.

Paret, R., 'Les villes de Syrie du Sud et les routes commerciales d'Arabie à la fin du VIe siècle', *Berichte zum XI. Internationalen Byzantinisten-Kongress, München, 1958* (Munich, 1960), 438–44.

Parker, A. J., *Ancient Shipwrecks of the Mediterranean and the Roman Provinces* (BAR International Series 580; Oxford, 1992).

Parker, S. T., 'Peasants, Pastoralists, and Pax-Romana: A Different View', *Bulletin of the American Schools of Oriental Research* 265 (1987), 35–51.

——*The Roman Frontier in Central Jordan: Interim Report on the Limes Arabicus Project, 1980–1985* (BAR International Series 340 (i–ii); Oxford, 1987).

——'The Limes Arabicus Project: The 1987 Campaign', *ADAJ* 32 (1988), 171–88.

——'The Limes and Settlement Patterns in Central Jordan in the Roman and Byzantine Periods', *SHAJ* 4 (1992), 321–5.

Parker, S. T. and Betlyon, J. W., *The Roman Frontier in Central Jordan: Final Report on the Limes Arabicus Project, 1980–1989* (Dumbarton Oaks Studies; Washington, DC, 2006).

Paterson, J., 'Trade and Traders in the Roman World: Scale, Structure, and Organisation', in H. Parkins and C. J. Smith (eds.), *Trade, Traders and the Ancient City* (London, 1998), 149–67.

Patlagean, E., *Pauvreté économique et pauvreté sociale à Byzance 4e–7e siècles* (Mouton, 1977).

Patrich, J., *Map of Deir Mar Saba (109/7)* (Archaeological Survey of Israel; Jerusalem, 1995).

Patrich, J. and Tsafrir, Y., 'A Byzantine Church Complex at Horvat Beit Loya', in Y. Tsafrir (ed.), *Ancient Churches Revealed* (Jerusalem, 1993), 265–76.

Peacock, D. P. S. and Williams, D. F., *Amphorae and the Roman Economy: An Introductory Guide* (London, 1986).

Peña, I., Castellana, P., and Fernandez, R., *Les reclus syriens: recherches sur les anciennes formes de vie solitaire en Syrie* (Collectio Minor: Studium Biblicum Franciscanum 23; Jerusalem, 1980).

——*Inventaire du Jebel El-A'la: recherches archéologiques dans la région des villes mortes de la Syrie du nord* (Collectio Minor: Studium Biblicum Franciscanum 31; Milan, 1990).

Pentz, P., *The Invisible Conquest: The Ontogenesis of Sixth and Seventh Century Syria* (Copenhagen, 1992).

Pentz, P., *The Medieval Citadel and its Architecture* (Hama: Fouilles et recherches de la Fondation Carlsberg 1931–1938 4.1; Aarhus, 1998).

Percival, J., *The Wheat Plant: A Monograph* (London, 1921).

Piccirillo, M., 'Rural Settlements in Byzantine Jordan', in A. Hadidi (ed.), *Studies in the History and Archaeology of Jordan*, II (Amman, 1985), 257–61.

Piganiol, A., *L'impôt de capitation sous le Bas-Empire romain* (Chambéry, 1916).

Poidebard, A., *La trace de Rome dans le désert de Syrie, Le limes de Trajan à la conquête arabe* (Paris, 1934).

Pollard, N., *Soldiers, Cities, and Civilians in Roman Syria* (Ann Arbor, 2000).

Ponsich, M., *Implantation rurale antique sur le Bas-Guadalquivir* (Madrid, 1974).

Raban, A., *Map of Nahalal (28)* (Archaeological Survey of Israel; Jerusalem, 1982).

——*Map of Mishmar Ha-'Emeq (32)* (Archaeological Survey of Israel; Jerusalem, 1999).

Rachie, K. O., *The Millets: Importance, Utilization and Outlook* (Begumpet, 1975).

Rathbone, D., *Economic Rationalism and Rural Society in Third-Century A.D. Egypt: The Heroninos Archive and the Appianus Estate* (Cambridge, 1991).

Rautman, M., 'The Busy Countryside of Late Roman Cyprus', *Report of the Department of Antiquities Cyprus* (2000), 317–31.

Rautman, M. L., McClellan, M. C., and Benson, L., *A Cypriot Village of Late Antiquity: Kalavasos-Kopetra in the Vasilikos Valley* (Journal of Roman Archaeology Supplementary Series 52; Portsmouth, RI, 2003).

Reich, S. S., *Études sur les villages araméens de l'Anti-Liban* (Damascus, 1937).

Reynolds, P., *Trade in the Western Mediterranean, AD 400–700: The Ceramic Evidence* (BAR international Series 604; Oxford, 1995).

Reynolds, P. J., *Iron Age Farm: The Butser Experiment* (London, 1979).

Riley, J. A., 'The Coarse Pottery from Benghazi', in J. A. Lloyd (ed.), *Excavations at Sidi Khrebish Benghazi (Berenice)*, II (Tripoli, 1979), 91–497.

Robert, L., *Noms indigènes dans l'Asie-Mineure gréco-romaine* (Paris, 1963).

Rodgers, R. H., 'Hail, Frost, and Pests in the Vineyard: Anatolius of Berytus as a Source for the Nabataean Agriculture', *Journal of the American Oriental Society* 80/1 (1980), 1–11.

——'*Κεποποιΐα*', in A. Littlewood, H. Maguire, and J. Wolschke-Bulmahn (eds.), *Byzantine Garden Culture* (Washington, DC, 2002), 150–75.

Roll, I., 'The Roads in Roman–Byzantine Palaestina and Arabia', in M. Piccirillo and E. Alliata (eds.), *The Madaba Map Centenary, 1897–1997: Travelling Through the Byzantine Umayyad Period: Proceedings of the International Conference Held in Amman, 7–9 April 1997* (Jerusalem, 1999), 109–13.

Roll, I. and Ayalon, E., 'Two Large Wine Presses in the Red Soil Regions of Israel', *Palestine Exploration Quarterly* 130 (1981), 111–25.

Ronen, A. and Olami, Y. A., *Map of Atlit (1)* (Archaeological Survey of Israel; Jerusalem, 1978).

Ronen, A. and Olamy, Y. A., *Map of Haifa—East (23)* (Archaeological Survey of Israel; Jerusalem, 1983).

Rose, J. C. and Burker, D. E. (eds.), *Sa'ad: A Late Roman/Byzantine Site in North Jordan* (Irbid, Jordan, 2004).

Rose, V. and Plommer, W. H., *Vitruvius and Later Roman Building Manuals* (Cambridge, 1973).

Rosen, B. and Finklestein, I., 'Subsistence Patterns, Carrying Capacity and Settlement Oscillations in the Negev Highlands', *Palestine Exploration Quarterly* 124 (1992), 42–58.

Rosen, S. A., *Map of Makhtesh Ramon (204)* (Archaeological Survey of Israel; Jerusalem, 1994).

Rosenfeld, B.-Z. and Menirav, J., 'Methods of Pricing and Price Regulation in Roman Palestine in the Third and Fourth Centuries', *Journal of the American Oriental Society* 121/3 (2001), 351–69.

Rossiter, J. J., 'Roman Villas of the Greek East and the Villa of Gregory of Nyssa Ep. 20', *Journal of Roman Archaeology* 2 (1989), 101–10.

Rossiter, J. J. and Freed, J., 'Canadian–Turkish Excavations at Domuztepe, Cilicia (1989)', *Echoes du monde classique/Classical Views* n.s. 10 (1991), 145–74.

Rostovtzeff, M. I. and Fraser, P. M., *The Social and Economic History of the Roman Empire* (2nd edn; Oxford, 1957).

Rougé, J., 'La navigation intérieure dans le Proche Orient antique', in P. Louis (ed.), *L'homme et l'eau en Méditerranée et au Proche Orient* (Lyon, 1986).

Rowlandson, J., *Landowners and Tenants in Roman Egypt: The Social Relations of Agriculture in the Oxyrhynchite Nome* (Oxford, 1996).

Rowley-Conwy, P., 'Sorghum from Qasr Ibrim, Egyptian Nubia, c. 800 BC – AD 1811: A Preliminary Study', in J. Renfrew (ed.), *New Light on Early Farming* (Edinburgh, 1991), 191–212.

Rowley-Conwy, P., Deakin, O. J., and Shaw, C. H., 'Ancient DNA from Archaeological Sorghum (Sorghum bicolor) from Qasr Ibrim Nubia: Implications for Domestication and Evolution', *Sahara* 9 (2000), 23–34.

Rubin, R., 'Urbanization, Settlement and Agriculture in the Negev Desert—The Impact of the Roman–Byzantine Empire on the Frontier', *Zeitschrift des deutschen Palästina-Vereins* 112 (1996), 49–60.

——'The Romanization of the Negev, Israel: Geographical and Cultural Changes in the Desert Frontier in Late Antiquity', *Journal of Historical Geography* 23/3 (1997), 267–83.

Ruggini, L., *Ebrei e orientali nell'Italia settentrionale fra il IV e il VI secolo* (Rome, 1959).

Ruska, J., 'Cassianus Bassus Scholasticus und die arabischen Versionen der griechischen Landwirtschaft', *Der Islam* 5 (1914), 174–9.

Russel, J. C., 'Ancient and Medieval Population', *Transactions of the American Philosophical Society* 48 (1958), 63–70.

Russell, J., 'Byzantine Instrumenta Domestica from Anemurium: The Significance of Context', in R. Hohlfelder (ed.), *City, Town and Countryside in the Early Byzantine Era* (Boulder, Colo., 1982), 133–63.

Safrai, Z., *The Economy of Roman Palestine* (London, 1994).

Salame-Sarkis, H., 'Syria grammata kai algamata', *Syria* 61 (1989), 322–5.

Saliby, N., 'Les installations hydrauliques à Qasr el-Heir el-Gharbi', in B. Geyer (ed.), *Techniques et pratiques hydro-agricoles traditionnelles en domaine irrigué: approche pluridisciplinaire des modes de culture avant le motorisation en Syrie: actes du Colloque de Damas 27 juin–1er juillet 1987* (Damascus, 1990).

Saller, R. P., 'Framing the Debate Over Growth in the Ancient Economy', in Walter Scheidel (ed.), *The Ancient Economy* (Edinburgh, 2002), 251–69.

Saller, S. J. and Schneider, H. R., *The Memorial of Moses on Mount Nebo* (Publications of the Studium Biblicum Franciscanum 1; Jerusalem, 1941).

Samuel, D., 'Plant Remains from the Northwest Tell at Busra', *Berytus* 34 (1986), 83–96.

Sarre, F. P. T., Herzfeld, E., and Berchem, M. V., *Archäologische Reise im Euphrat- und Tigris-Gebiet* (Forschungen zur islamischen Kunst 1; Berlin, 1911).

Sarris, P., 'Rehabilitating the Great Estate: Aristocratic Property and Economic Growth in the Late Antique East', in W. Bowden, L. Lavan, and C. Machado (eds.), *Recent Research on the Late Antique Countryside* (Leiden and Boston, 2004), 55–71.

——'The Origins of the Manorial Economy: New Insights from Late Antiquity', *English Historical Review* 129/2 (2004), 279–311.

——*Economy and Society in the Age of Justinian* (Cambridge, 2006).

Scaramuzzi, F., 'Intensive Olive-Growing', in FAO (ed.), *Modern Olive-Growing* (Rome, 1977), 61–84.

Scheidel, W., *Death on the Nile: Disease and the Demography of Roman Egypt* (Leiden and Boston, 2001).

Schick, R., *The Christian Communities of Palestine from Byzantine to Islamic Rule: A Historical and Archaeological Study* (Princeton, 1995).

Schiere, H. and Kater, L., *Mixed Crop–Livestock Farming: A Review of Traditional Technologies Based on Literature and Field Experiences* (Rome, 2001).

Schilbach, E., *Byzantinische Metrologie* (Handbuch der Altertumswissenschaft Byzantinisches Handbuch 12.4; Munich, 1970).

Schilbach, E. and Kazhdan, A. P., 'Modios', in A. P. Kazhdan (ed.), *The Oxford Dictionary of Byzantium* (Oxford, 1991), 1388.

Schlumberger, D., 'Les fouilles de Qasr el-Heir el-Gharbi (1936–1938), rapport préliminaire', *Syria* 20 (1939), 195–238.

Schlumberger, D., et al., *Qasr el-Heir el Gharbi: textes et planches* (Bibliothèque archéologique et historique 120; Paris, 1986).

Schuler, C., *Ländliche Siedlungen und Gemeinden im hellenistischen und römischen Kleinasien* (Munich, 1998).

Scranton, R. L., *Mediaeval Architecture in the Central Area of Corinth* (Corinth: Results of Excavations Conducted by the American School of Classical Studies at Athens; Princeton, 1957).

Segal, A., *The Byzantine City of Shivta (Esbeita), Negev Desert, Israel* (Oxford, 1983).

Segal, J. B., 'Mesopotamian Communities from Julian to the Rise of Islam', *Proceedings of the British Academy* 41 (1955), 109–40.

——*Edessa: 'The Blessed City'* (Oxford, 1970).

Segrè, A., 'Essays on Byzantine Economic History, I: The Annona Civica and the Annona Militaris', *Byzantion* 16 (1942–3), 393–444.

Seyrig, H., 'Antiquités syriennes: postes romains sur la route de Médine', *Syria* 24 (1941), 218–23.

Sezgin, F., *Geschichte des arabischen Schrifttums* (Leiden, 1967).

Shahîd, I., 'Byzantino-Arabica: The Conference of Ramla, A.D. 524', *Journal of Near Eastern Studies* 23/2 (1964), 115–31.

Shaw, B. D., 'Lamasba: An Ancient Irrigation Community', *Antiquités africaines* 18 (1982), 61–103.

——'Water and Society in the Ancient Mahgrib: Technology, Property and Development', *Antiquités africaines* 20 (1984), 121–73.

Shehadeh, N., 'The Climate of Jordan in the Past and Present', in Adnan Hadidi (ed.), *Studies in the History and Archaeology of Jordan*, II (Amman, 1985), 25–38.

Shereshevski, J., *Byzantine Urban Settlements in the Negev Desert* (Beer-Sheva, Israel, 1991).

Sidebotham, S. E., Wendrich, W., and Aldsworth, F., *Berenike 1995: Preliminary Report of the 1995 Excavations at Berenike (Egyptian Red Sea Coast) and the Survey of the Eastern Desert* (CNWS Publications Special Series 2; Leiden, 1996).

Sly, D., *Philo's alexandria* (London, 1996).

Smith, N., 'Roman Canals', *Transactions of the Newcomen Society* 49 (1977–8), 75–86.

Smith, R. R. R. and Ratté, C., 'Archaeological Research at Aphrodisias in Caria, 1996', *American Journal of Archaeology* 102 (1998), 225–50.

Sodini, J.-P., 'L'artisanat urbain à l'Epoque Paléochrétienne (IV–VIIe S.)', *Ktèma* 4 (1979), 71–119.

——'Les églises de la Syria du Nord', in J. M. Dentzer and W. Orthmann (eds.), *Archéologie et histoire de la Syrie*, II: *La Syrie de l'époque achéménide à l'avènement de l'Islam* (Saarbrücken, 1989), 347–72.

Sodini, J.-P., et al., 'Déhès (Syrie du Nord) Campagnes I–III (1976–1978)', *Syria* 57 (1980), 1–300.

Sørensen, L. W., Rupp, D. W., and Bilde, P. G., *The Land of the Paphian Aphrodite* (Studies in Mediterranean Archaeology 104; Göteborg, 1993).

Sperber, D., *Roman Palestine, 200–400: Money and Prices* (Ramat-Gan, Israel, 1974).

——'Objects of Trade Between Palestine and Egypt in Roman Times', *Journal of the Economic and Social History of the Orient* 19 (1976), 113–47.

Sprenger, G., 'Darlegung der Grundsätze nach denen die syrische Übertragung der griechischen Geoponika gearbeitet worden ist', doctoral thesis (Göttingen, 1889).

Spurr, M. S., 'The Cultivation of Millet in Roman Italy', *Papers of the British School at Rome* 51 (1983), 1–15.

——*Arable Cultivation in Roman Italy: c.200 B.C.–c.A.D. 100* (Journal of Roman Studies Monographs 3; London, 1986).

——'Palladius (1)', in S. Hornblower and A. Spawforth (eds.), *The Oxford Classical Dictionary* (3rd edn; Oxford, 1995), 1101.

Stathakopoulos, D., *Famine and Pestilence in the Late Roman and Early Byzantine Empire: A Systematic Survey of Subsistence Crises and Epidemics* (Birmingham Byzantine and Ottoman Monographs; Aldershot, 2004).

Stevens, C. E., 'Agricultural and Rural Life in the Later Roman Empire', in J. H. Clapham and E. Power (eds.), *The Cambridge Economic History of Europe: The Agrarian Life of the Middle Ages*, I (Cambridge, 1949), 89–117.

Stieglitz, R. R., 'A Late Byzantine Reservoir and Piscina at Tell Tanninim', *Israel Exploration Journal* 48/1–2 (1998), 54–65.

Strack, H. L., Stemberger, G., and Bockmuehl, M. N. A., *Introduction to the Talmud and Midrash* (Edinburgh, 1991).

Svennung, J., 'De auctoribus Palladii', *Eranos* 25 (1927), 230–48.

——*Untersuchungen zu Palladius und zur lateinischen Fach- und Volkssprache* (Uppsala, 1935).

Tacoma, L. E., 'Replacement Parts for an Irrigation Machine of the Divine House at Oxyrhynchus P. Columbia Inv. 83, October 12 AD 549 (?)', *Zeitschrift für Papyrologie und Epigraphik* 120 (1998), 123–30.

Tate, G., 'Mutabilité des économies antiques: l'exemple de la Syrie du nord (IVe–VIe siècle)', in P.-L. Gatier, B. Helly, and J.-P. Rey-Coquais (eds.), *Géographie historique au Proche-Orient (Syrie, Phénicie, Arabie, grecques, romaines, byzantines): Actes de la Table Ronde de Vallebonne 16–18 septembre 1985* (Paris, 1988), 249–55.

——'Les campagnes de la Syrie du nord à l'époque proto-byzantine', in C. Abadie-Reynal (ed.), *Hommes et richesses dans l'Empire byzantin* (Paris, 1989), 63–77.

——*Les campagnes de la Syrie du Nord du IIe au VIIe siècle* (Bibliothèque archéologique et historique 133; Paris, 1992).

——'Le Latifundium en Syrie: mythe ou realite?', in *Du Latifundium au Latifondo* (Paris, 1995), 243–54.

——'The Syrian Countryside during the Roman Era', in S. E. Alcock (ed.), *The Early Roman Empire in the East* (Oxford, 1997), 55–71.

Tchalenko, G., *Villages antiques de la Syrie du Nord: le massif du Bélus à l'époque romaine*, 3 vols. (Paris, 1953).

Tchernia, A., *Le vin de l'Italie romaine: essai d'histoire économique d'après les amphores* (Paris, 1986).

Teall, J., 'The Grain Supply of the Byzantine Empire', Ph.D. diss. (New Haven, 1956).

——'The Byzantine Agricultural Tradition', *Dumbarton Oaks Papers* 25 (1971), 35–59.

Teja, R., *Organización económica y social de Capadocia en el siglo IV, según los padres capadocios* (Acta Salmanticensia. Filosofía y letras 78; Salamanca, 1974).

Thoumin, R. L., *La maison syrienne dans la plaine hauranaise, le bassin du Barada et sur les plateaux du Qalamun* (Paris, 1932).

——*Géographie humaine de la Syrie centrale* (Paris, 1936).

Tomber, R., 'Pottery from the Sediments of the Inner Harbour', in K. G. Holum, A. Raban, and J. Patrich (eds.), *Caesarea Papers*, II: *Herod's Temple, the Provincial Governor's Praetorium and Granaries, the Later Harbor, a Gold Coin Hoard, and Other Studies* (JRA Supplementary Series 35; Providence, RI, 1999), 295–322.

Tomber, R. S., 'The Pottery', in Steven E. Sidebotham and Willemina Wendrich (eds.), *Berenike 1996: Report of the 1996 Excavations at Berenike (Egyptian Red Sea Coast) and the Survey of the Eastern Desert*, (CNWS Publications Special Series 3; Leiden, 1998), 163–80.

Treadgold, W. T., *Byzantium and its Army, 284–1081* (Stanford, 1995).

Trombley, F., 'Korykus in Cilicia Trachis: The Economy of a Small Coastal City in Late Antiquity (saec. V–VI)—A précis', *Ancient History Bulletin* 1/1 (1987), 16–23.

——'War and Society in Rural Syria c. 502–613 A.D.', *Byzantine and Modern Greek Studies* 21 (1997), 154–209.

Tsafrir, Y. and Foerster, G., 'The Dating of the "Earthquake of the Sabbatical Year" of 749 C.E. in Palestine', *Bulletin of the School of Oriental and African Studies*, 55 (1992), 231–5.

——'Urbanism at Scythopolis-Bet Shean in the Fourth to Seventh Centuries', *Dumbarton Oaks Papers* 51 (1997), 85–146.

Tsafrir, Y., Di Segni, L., and Green, J., *Tabula Imperii Romani: Iudea Palaestina* (Jerusalem, 1994).

Tyree, E. L. and Stefanoubaki, E., 'The Olive Pit and Roman Oil Making', *Biblical Archaeologist* 59/3 (1996), 171–8.

Ullmann, M., *Die Natur- und Geheimwissenschaften im Islam* (Handbuch der Orientalistik, 1.2: Nahe und der Mittlere Osten 6; Leiden, 1972).

Unwin, T., *Wine and the Vine* (London, 1996).

Urman, D., *The Golan: A Profile of a Region During the Roman and Byzantine Periods* (Oxford, 1985).

Uscatescu, A., *La cerámica del Macellum de Gerasa* (*Ŷaraš, Jordania*) (Madrid, 1996).

Van Alfen, P. G., 'New Light on the 7th-c. Yassi Ada Shipwreck: Capacities and Standard Sizes of LRA1 Amphoras', *Journal of Roman Archaeology* 9 (1996), 189–213.

Van Der Veen, M. and Hamilton-Dyer, S., 'A Life of Luxury in the Desert? The Food and Fodder Supply to Mons Claudianus', *Journal of Roman Archaeology* 11 (1998), 101–16.

Van Zeist, W., 'Evidence for Agricultural Change in the Balikh Basin, Northern Syria', in C. Gosden and J. Hather (eds.), *The Prehistory of Food: Appetites for Change* (London, 1999), 350–74.

Vasiliev, A. A., *History of the Byzantine Empire, 324–1453* (2nd English edn; Madison, 1958).

Vázquez De Benito, M. C. (ed.), *El manuscrito no. XXX de la Colección Gayangos* (*fols. 1–98*) (Madrid, 1974).

Villeneuve, E. and Watson, P. (eds.), *La céramique byzantine et proto-islamique en Syrie-Jordanie: IVe–VIIIe siècles apr. J.-C.: actes du colloque tenu à Amman les 3, 4 et 5 décembre 1994* (Beirut, 2001).

Villeneuve, F., 'Les villages antiques du Hauran', *Le Monde de la Bible* 28 (1983), 123–8.

——'Prospection archéologique et géographiquie historique: la région d'Iraq al-Amir (Jordanie)', in P.-L. Gatier, B. Helly, and J.-P. Rey-Coquais (eds.), *Géographie historique au Proche-Orient* (*Syrie, Phénicie, Arabie, grecques, romaines, byzantines*)*: Actes de la Table Ronde de Vallebonne 16–18 septembre 1985* (Paris, 1988), 257–88.

Vita-Finzi, C., 'Roman Dams in Tripolitania', *Antiquity* 35 (1961), 14–20.

——*The Mediterranean Valleys: Geological Changes in Historical Times* (London, 1969).

Wallace, S. L., *Taxation in Egypt from Augustus to Diocletian* (Princeton University Studies in Papyrology 2; Princeton and London, 1938).

Walmsley, A., *Early Islamic Syria: An Archaeological Assessment* (Duckworth Debates in Archaeology; London, 2007).

Ward-Perkins, B., 'The Cities', in Averil Cameron and Peter Garnsey (eds.), *The Cambridge Ancient History*, XIII: *The Late Empire, A.D. 337–425* (Cambridge, 1998), 371–410.

——*The Fall of Rome and the End of Civilization* (Oxford, 2005).

Ware, E. F., *Roman Water Law* (Littleton, Colo., 1985).

Watson, A. M., 'The Arab Agricultural Revolution and its Diffusion', *Journal of Economic History* 34 (1974), 8–35.

——*Agricultural Innovation in the Early Islamic World* (Cambridge, 1983).

Watson, P., 'Change in Foreign and Regional Economic Links with Pella in the Seventh Century A.D: The Ceramic Evidence', in P. Canivet and J.-P. Rey-Coquais (eds.), *La Syrie de Byzance à l'Islam: VIIe–VIIIe siècles: actes du colloque international Lyon—Maison de l'Orient méditerranéen, Paris—Institut du monde arabe, 11–15 Septembre 1990* (Damascus, 1992), 233–47.

Weinberg, G. D., *Excavations at Jalame: Site of a Glass Factory in Late Roman Palestine* (Columbia, Mo., 1988).

Weiss, D., Zissu, B., and Solimany, G., *Mapat Nes Harim (104)* (Archaeological Survey of Israel; Jerusalem, 2004).

Wellmann, M., 'Didymus 7', in *Paulys Real-Encyclopädie der classischen Altertumswissenschaft*, IV (Stuttgart, 1895), 445.

——'Die Georgika des Demokritos', *Abhandlungen der Preussischen Akademie der Wissenschaften, Philosophisch-historische Klasse* 4 (1921).

Wensinck, A., 'Khamr', in E. van Donzel, B. Lewis, and Ch. Pellat (eds.), *Encyclopaedia of Islam* (2nd edn; Leiden, 1960), IV.994–7.

Weulersse, J., 'Antioche, essai de géographie urbaine', *Bulletin d'études orientales* 4 (1934), 27–79.

——*L'Oronte: étude de fleuve* (Tours, 1940).

——*Le pays des Alaouites* (Tours, 1940).

——*Paysans de Syrie et du Proche-Orient* (Le paysan et la terrre; Paris, 1944).

White, K. D., *Agricultural Implements of the Roman World* (Cambridge, 1967).

——*Roman Farming* (Ithaca, 1970).

——*Farm Equipment of the Roman World* (Cambridge, 1975).

——*Greek and Roman Technology* (London, 1984).

——'Cereals, Bread and Milling in the Roman World', in J. Wilkins, D. Harvey, and M. Dobson (eds.), *Food in Antiquity* (Exeter, 1995), 38–43.

Whittow, M., 'Ruling the Late Roman and Early Byzantine City: A Continuous History', *Past and Present* 129/4 (1990), 3–29.

Wickham, C., *Framing the Early Middle Ages: Europe and the Mediterranean 400–800* (Oxford, 2005).

Wikander, C., 'Canals', in Örjan Wikander (ed.), *Handbook of Ancient Water Technology* (Leiden, 2000), 321–31.

Wilfong, T. G., 'Agriculture among the Christian Population of Early Islamic Egypt', in A. K. Bowman and E. L. Rogan (eds.), *Agriculture in Egypt from Pharaonic to Modern Times* (Oxford, 1999), 217–66.

Wilkinson, J., *Jerusalem Pilgrims Before the Crusades* (Warminster, 1977).

Wilkinson, T. J., *Town and Country in Southeastern Anatolia*, 2 vols. (Oriental Institute Publications; Chicago, 1990).

——'The Structure and Dynamics of Dry-farming States in Upper Mesopotamia', *Current Anthropology* 35/4 (1994), 483–520.

——'The Archaeological Landscape of the Balikh Valley, Syria: 1994–95 Annual Report' (1999), http://www-oi.uchicago.edu/OI/AR/94–95/94–95_Jazira.html (updated 14 November 1997), accessed 21 May 1999.

——'Water and Human Settlement in the Balikh Valley, Syria: Investigations from 1992–1995', *Journal of Field Archaeology* 25 (1998), 63–87.

Will, E., 'Antioch sur l'Oronte: métropole de l'Asie', *Syria* 74 (1997), 99–114.

Will, E., et al., *Iraq al Amir: le château du Tobiade Hyrcan* (Bibliothèque archéologique et historique 132; Paris, 1991).

Wilson, A., 'Land Drainage', in Ö. Wikander (ed.), *Handbook of Ancient Water Technology* (Leiden, 2000), 303–17.

Wilson, A. D., 'A Greek Treatise on Agriculture', *British Museum Quarterly* 13 (1939), 10–11.

Wilson, A. I., 'Classical Water Technology in the Islamic World', in C. Brunn and A. Saastamoinen (eds.), *Technology, Ideology, Water: From Frontinus to the Renaissance and Beyond* (Rome, 2001), 115–41.

Winkler, A. J., *General Viticulture* (Berkeley, 1962).

Youtai, Y., 'Millet and Sorghum—The Ancients of Crops', in Sylvan Wittwer et al. (eds.), *Feeding a Billion: Frontiers of Chinese Agriculture* (East Lansing, 1987), 201–8.

Zacos, G., Veglery, A., and Nesbitt, J. W., *Byzantine Lead Seals* (Basel, 1972).

Zakariyya, A. W., *Jawla athariyya fi ba'd al-balad al-shamiyya* (Damascus, 1934).

Zenebe, S. and Fekade, T., 'The Role of Donkey Pack-Transport in the Major Grain Market of Addis Ababa', in Paul Starkey and Denis Fielding (eds.), *Donkeys, People and Development: A Resource Book of the Animal Traction Network for Eastern and Southern Africa (ATNESA)* (Wageningen, The Netherlands, 2004), 71–8.

Zissu, B., 'The Dovecote at Horvat 'Eleq', in Y. Hirschfeld (ed.), *Ramat Hanadiv Excavations: Final Report of the 1984–1998 Seasons* (Jerusalem, 2000), 617–27.

Zohary, D. and Hopf, M., *Domestication of Plants in the Old World: The Origin and Spread of Cultivated Plants in West Asia, Europe, and the Nile Valley* (2nd edn; Oxford Science Publications; Oxford, 1993).

Index

CPSIA information can be obtained
at www.ICGtesting.com
Printed in the USA
BVOW06*0948301016
465996BV00018B/33/P

9 780199 565283